Organs-on-chips

Organs-on-chips

Special Issue Editors

Yu-suke Torisawa
Yi-Chung Tung

MDPI • Basel • Beijing • Wuhan • Barcelona • Belgrade • Manchester • Tokyo • Cluj • Tianjin

Special Issue Editors

Yu-suke Torisawa
Hakubi Center for Advanced
Research and Department of
Micro Engineering,
Kyoto University
Japan

Yi-Chung Tung
Research Center for Applied
Sciences, Academia Sinica
Taiwan

Editorial Office
MDPI
St. Alban-Anlage 66
4052 Basel, Switzerland

This is a reprint of articles from the Special Issue published online in the open access journal *Micromachines* (ISSN 2072-666X) (available at: https://www.mdpi.com/journal/micromachines/special_issues/Organs_On_Chips).

For citation purposes, cite each article independently as indicated on the article page online and as indicated below:

LastName, A.A.; LastName, B.B.; LastName, C.C. Article Title. *Journal Name* **Year**, *Article Number*, Page Range.

ISBN 978-3-03928-917-2 (Hbk)
ISBN 978-3-03928-918-9 (PDF)

Contents

About the Special Issue Editors

Yu-suke Torisawa is an Associate Professor at the Hakubi Center for Advanced Research and the Department of Micro Engineering at Kyoto University. He received his PhD in Chemical Engineering from Tohoku University in 2006. He then conducted postdoctoral research at the University of Michigan where he worked on developing microfluidic cell culture devices to control cellular microenvironment. In 2009, he joined Donald Inger's lab at the Wyss Institute at Harvard University where he developed a bone marrow-on-a-chip platform that permits culture of living bone marrow with a functional hematopoietic niche in vitro. His current research interests are Microfluidics, Biomaterials, Tissue engineering, Mechanobiology, and Cellular microenvironments and niches. His research focuses primarily on developing organs-on-chips by combining with iPS cell technology to model bone marrow, blood vessels, immune responses, lung diseases, and cancer as well as to generate hematopoietic and blood cells.

Yi-Chung Tung is a Research Fellow/Professor in the Research Center for Applied Sciences (RCAS) at Academia Sinica, Taipei, Taiwan. Before joining Academia Sinica, he was a Postdoctoral Research Fellow in the Department of Biomedical Engineering at the University of Michigan, Ann Arbor for three years. He received his Ph.D. degree in Mechanical Engineering from the University of Michigan, Ann Arbor in 2005, and his B.S. and M.S. degrees in Mechanical Engineering from National Taiwan University, Taipei, Taiwan in 1996 and 1998, respectively. His research interests include developing microfluidic devices to mimic physiological microenvironments for in vitro cell culture, microfluidic devices for cell analysis, development of biomedical instruments and biomedical microelectromechanical systems (Bio-MEMS).

 *micromachines*

Editorial

Editorial for the Special Issue on Organs-on-Chips

Yu-suke Torisawa [1,2,*] **and Yi-Chung Tung** [3,4,*]

1. Hakubi Center for Advanced Research, Kyoto University, Kyoto 615-8540, Japan
2. Department of Micro Engineering, Kyoto University, Kyoto 615-8540, Japan
3. Research Center for Applied Sciences, Academia Sinica, Taipei 11529, Taiwan
4. College of Engineering, Chang Gung University, Taoyuan 33302, Taiwan
* Correspondence: torisawa.yusuke.6z@kyoto-u.ac.jp (Y.-s.T.); tungy@gate.sinica.edu.tw (Y.-C.T.)

Received: 25 March 2020; Accepted: 27 March 2020; Published: 1 April 2020

Recent advances in microsystems technology and cell culture techniques have led to the development of organ-on-chip microdevices to model functional units of organs. By recapitulating natural tissue architecture and microenvironmental chemical and mechanical cues, organs-on-chips reconstitute tissue-level functionalities in vitro, which is infeasible with conventional culture methods. Organs-on-chips are typically microfluidic culture devices made of optically transparent materials that permit high-resolution imaging and real-time monitoring of cellar responses. These microdevices allow to mimic the biomechanical forces observed inside the body, playing important roles in controlling cell fates and functions. Recently, stem cell technologies including induced pluripotent stem (iPS) cells have been leveraged to develop organs-on-chips, which enable various types of models of organs and diseases not possible with primary cells and cell lines. Organs-on-chips replicate tissue-level responses with human cells, enabling more accurate prediction of human responses to drugs and diseases. Since the cost of drug discovery is constantly increasing due to the limited predictability of conventional monolayer culture methods and animal models, this technology has great potential to promote drug discovery and development as well as to model human physiology and disease.

This Special Issue is themed to provide insight and advancements in organ-on-chip microdevices. There are fifteen papers including three review papers, covering a novel material to fabricate microfluidic organs-on-chips [1], methods to deliver mechanical stimuli [2,3], methods to measure mechanical forces [4,5], methods to evaluate cellular functions in 3D cultures [6–8], and specific organ models; lung chips [3,9], liver chips [10,11], blood vessel chips [12–15] including models of the outer blood-retina barrier [14] and ischemia-reperfusion injury [15].

Inside the body, cells are exposed to biomechanical forces, including fluidic shear stress and mechanical strain, which regulate cell function and contribute to disease. Kaarj et al. reviewed methods to produce mechanical stimuli focusing on the technical details of devices [2]. This paper shows organ-on-chip systems that incorporate various types of mechanical stimuli and their potential applications to develop physiologically relevant models and to study mechanobiology. Lin et al. developed a simple yet powerful microfluidic device that can generate hydrostatic pressure and cyclic strain to mimic the lung physiological microenvironment [3]. This device paves the way to better understand the cellular behaviors under various lung physiological conditions for future translational studies. It is also important to quantitatively measure mechanical forces generated by cells and tissues. A strain gauge has been developed to measure mechanical strain in situ [4]. This strain gauge can be used to measure mechanical strain on a membrane where cells are cultured. This method enables the integration of the strain gauges in monolithically fabricated organ-on-chip devices. Another approach using a micro vacuum chuck has been developed to measure biomechanical properties of a three dimensional (3D) tissue [5]. This tool can position a cardiac tissue by vacuum pressure, which enables the measurement of beating force of a cardiac tissue assembled from human iPSC-derived cardiomyocytes. These methods to measure mechanical forces can be useful for heart-on-chip applications.

By combining 3D cell culture and microdevice technologies, 3D culture devices have been developed to replicate in vivo-like microenvironment as well as to develop high-throughput systems. Bastiaens et al. developed a 3D neuronal culture device by combining nanogrooved substrates with a 3D hydrogel culture [6]. This method permits the formation of an aligned 3D neural cell network. The use of nanogrooves enhances the structural complexity of 3D neuronal cell cultures, providing a way to develop a brain-on-a-chip model. Chen et al. developed an imaging method for 3D cultures [7]. By using lattice light-sheet microscopy, individual neuronal cells in a 3D hippocampal neuron can be monitored. This method makes quantification of voltage responses and calcium dynamics in individual neurons in 3D culture feasible. Choi et al. developed a chip to form an array of 3D cell spheroids for drug testing [8]. A 12-by-36 array of alginate hydrogels containing cancer spheroids was formed using micropillars and microwells. This system can test seventy drugs at six replicates on a chip providing a useful platform for drug screening.

Various types of organs-on-chips have been reported in this issue. Frost et al. reported a microfluidic lung-on-a-chip by recreating the epithelial-endothelial interface of the lung to evaluate drug permeability [9]. This microfluidic device allows to evaluate the effect of fluid shear stress on tissue permeability. Deng et al. reviewed strategies to build liver-on-chip models [10]. Liver chips consisting of human cells could potentially correlate clinical testing. These chips enable to predict hepatotoxicity and metabolism of drugs in humans and can be connected to other organ chips to recapitulate physiological interactions between multiple organs. A biomimetic method has also been developed to mimic drug metabolism in the liver [11]. Catalysts immobilized onto magnetic nanoparticles could efficiently produce drug metabolites in very small volumes.

Blood vessel chips have been developed to study angiogenesis and to develop disease models. Wang et al. reviewed current strategies to engineer microvessels on-chip focusing on the generation of 3D microvascular networks [12]. Akbari et al. reported the role of the flow conditions that occur due to vessel bifurcations on endothelial sprouting using a microfluidic 3D culture device [13]. This model revealed the importance of local flow dynamics due to branched vessel geometry in determining the location of sprouting angiogenesis. A microfluidic co-culture model has been developed to recapitulate the outer blood-retina barrier [14]. The device consists of an upper microchannel and multiple lower microchannels to form co-culture with 3D blood vessels. By integrating platinum electrodes into the device, this system allows to measure trans-epithelial electrical resistance (TEER) in real time, enabling to assess the epithelial barrier integrity on-chip. Nemcovsky et al. developed a novel microfluidic system to model ischemia-reperfusion injury [15]. This system consists of a vascular compartment lined with human endothelial cells that can be obstructed with a human blood clot and then re-perfused by thrombolytic treatment. Restoration of blood supply is essential to salvage ischemic tissue; however, reperfusion paradoxically causes further damage, even in remote tissues. The microfluidic mode of ischemia-reperfusion injury permits to recapitulate key features following restoration of flow upon removal of vascular embolic occlusion and thus this system can potentially serve as a powerful platform to study new therapeutic approaches for treatment of ischemia-reperfusion injury.

These organs-on-chips are mostly made of polydimethylsiloxane (PDMS) because it is easy to use, biocompatible, highly gas permeable, optically clear, and flexible. Although PDMS is very useful, one serious drawback is that small hydrophobic molecules are strongly absorbed into PDMS. This limitation is critical for drug testing because PDMS soaks up small hydrophobic drugs. Sano et al. reported a novel method to fabricate microfluidic devices using a fluoroelastomer which is resistant to absorption of small hydrophobic drugs comparable with standard culture plates [1]. This method could be a useful platform to construct organs-on-chips for drug discovery and development. Since organs-on-chips have now attracted great attention from the pharmaceutical industry, it is very important to identify suitable materials to develop commercialization-ready organs-on-chips.

We thank all the authors who submitted their papers to this Special Issue. We would also like to acknowledge all the reviewers whose careful and timely reviews ensured the quality of this Special Issue.

Conflicts of Interest: The authors declare no conflict of interest.

References

1. Sano, E.; Mori, C.; Matsuoka, N.; Ozaki, Y.; Yagi, K.; Wada, A.; Tashima, K.; Yamasaki, S.; Tanabe, K.; Yano, K.; et al. Tetrafluoroethylene-Propylene Elastomer for Fabrication of Microfluidic Organs-on-Chips Resistant to Drug Absorption. *Micromachines* **2019**, *10*, 793. [CrossRef] [PubMed]
2. Kaarj, K.; Yoon, J.Y. Methods of Delivering Mechanical Stimuli to Organ-on-a-Chip. *Micromachines* **2019**, *10*, 700. [CrossRef] [PubMed]
3. Lin, T.R.; Yeh, S.L.; Peng, C.C.; Liao, W.H.; Tung, Y.C. Study Effects of Drug Treatment and Physiological Physical Stimulation on Surfactant Protein Expression of Lung Epithelial Cells Using a Biomimetic Microfluidic Cell Culture Device. *Micromachines* **2019**, *10*, 400. [CrossRef] [PubMed]
4. Quirós-Solano, W.F.; Gaio, N.; Silvestri, C.; Pandraud, G.; Dekker, R.; Sarro, P.M. Metal and Polymeric Strain Gauges for Si-Based, Monolithically Fabricated Organs-on-Chips. *Micromachines* **2019**, *10*, 536. [CrossRef] [PubMed]
5. Uesugi, K.; Shima, F.; Fukumoto, K.; Hiura, A.; Tsukamoto, Y.; Miyagawa, S.; Sawa, Y.; Akagi, T.; Akashi, M.; Morishima, K. Micro Vacuum Chuck and Tensile Test System for Bio-Mechanical Evaluation of 3D Tissue Constructed of Human Induced Pluripotent Stem Cell-Derived Cardiomyocytes (hiPS-CM). *Micromachines* **2019**, *10*, 487. [CrossRef] [PubMed]
6. Bastiaens, A.; Xie, S.; Luttge, R. Nanogroove-Enhanced Hydrogel Scaffolds for 3D Neuronal Cell Culture: An Easy Access Brain-on-Chip Model. *Micromachines* **2019**, *10*, 638. [CrossRef] [PubMed]
7. Chen, C.Y.; Liu, Y.T.; Lu, C.H.; Lee, P.Y.; Tsai, Y.C.; Wu, J.S.; Chen, P.; Chen, B.C. The Applications of Lattice Light-sheet Microscopy for Functional Volumetric Imaging of Hippocampal Neurons in a Three-Dimensional Culture System. *Micromachines* **2019**, *10*, 599. [CrossRef] [PubMed]
8. Choi, J.W.; Lee, S.Y.; Lee, D.W. A Cancer Spheroid Array Chip for Selecting Effective Drug. *Micromachines* **2019**, *10*, 688. [CrossRef] [PubMed]
9. Frost, T.S.; Jiang, L.; Lynch, R.M.; Zohar, Y. Permeability of Epithelial/Endothelial Barriers in Transwells and Microfluidic Bilayer Devices. *Micromachines* **2019**, *10*, 533. [CrossRef] [PubMed]
10. Deng, J.; Wei, W.; Chen, Z.; Lin, B.; Zhao, W.; Luo, Y.; Zhang, X. Engineered Liver-on-a-Chip Platform to Mimic Liver Functions and Its Biomedical Applications: A Review. *Micromachines* **2019**, *10*, 676. [CrossRef] [PubMed]
11. Decsi, B.; Krammer, R.; Hegedűs, K.; Ender, F.; Gyarmati, B.; Szilágyi, A.; Tőtős, R.; Katona, G.; Paizs, C.; Balogh, G.T.; et al. Liver-on-a-Chip-Magnetic Nanoparticle Bound Synthetic Metalloporphyrin-Catalyzed Biomimetic Oxidation of a Drug in a Magnechip Reactor. *Micromachines* **2019**, *10*, 668. [CrossRef] [PubMed]
12. Wang, X.; Sun, Q.; Pei, J. Microfluidic-Based 3D Engineered Microvascular Networks and Their Applications in Vascularized Microtumor Models. *Micromachines* **2018**, *9*, 493. [CrossRef] [PubMed]
13. Akbari, E.; Spychalski, G.B.; Rangharajan, K.K.; Prakash, S.; Song, J.W. Competing Fluid Forces Control Endothelial Sprouting in a 3-D Microfluidic Vessel Bifurcation Model. *Micromachines* **2019**, *10*, 451. [CrossRef] [PubMed]
14. Chen, L.J.; Raut, B.; Nagai, N.; Abe, T.; Kaji, H. Prototyping a Versatile Two-Layer Multi-Channel Microfluidic Device for Direct-Contact Cell-Vessel Co-Culture. *Micromachines* **2020**, *11*, 79. [CrossRef] [PubMed]
15. Nemcovsky Amar, D.; Epshtein, M.; Korin, N. Endothelial Cell Activation in an Embolic Ischemia-Reperfusion Injury Microfluidic Model. *Micromachines* **2019**, *10*, 857. [CrossRef] [PubMed]

Review

Methods of Delivering Mechanical Stimuli to Organ-on-a-Chip

Kattika Kaarj and Jeong-Yeol Yoon *

Department of Biosystems Engineering, The University of Arizona, Tucson, AZ 85721, USA;
kkaarj@email.arizona.edu
* Correspondence: jyyoon@email.arizona.edu

Received: 7 September 2019; Accepted: 10 October 2019; Published: 14 October 2019

Abstract: Recent advances in integrating microengineering and tissue engineering have enabled the creation of promising microengineered physiological models, known as organ-on-a-chip (OOC), for experimental medicine and pharmaceutical research. OOCs have been used to recapitulate the physiologically critical features of specific human tissues and organs and their interactions. Application of chemical and mechanical stimuli is critical for tissue development and behavior, and they were also applied to OOC systems. Mechanical stimuli applied to tissues and organs are quite complex in vivo, which have not adequately recapitulated in OOCs. Due to the recent advancement of microengineering, more complicated and physiologically relevant mechanical stimuli are being introduced to OOC systems, and this is the right time to assess the published literature on this topic, especially focusing on the technical details of device design and equipment used. We first discuss the different types of mechanical stimuli applied to OOC systems: shear flow, compression, and stretch/strain. This is followed by the examples of mechanical stimuli-incorporated OOC systems. Finally, we discuss the potential OOC systems where various types of mechanical stimuli can be applied to a single OOC device, as a better, physiologically relevant recapitulation model, towards studying and evaluating experimental medicine, human disease modeling, drug development, and toxicology.

Keywords: organ-on-a-chip (OOC); microfluidic device; mechanical cue; shear flow; compression; stretch; strain; syringe pump; integrated pump; passive delivery

---

1. Introduction

Organ-on-a-chip (OOC) has enabled new opportunities in cell biology research through reproducing key aspects of an in vivo cellular microenvironment. One of these parameters is mechanical force, which imparts strain on cells and tissues. Such mechanical force and subsequent strain are integral parts of the cellular microenvironment that modulates the proliferation, migration, phenotype, and/or differentiation of cells. There have been extensive studies describing the cellular mechanisms where mechanical forces are transduced into biochemical signals (called cellular mechanotransduction) [1–3]. These responses affect the function of multicellular systems, which is critical in tissue- and organ-level health and disease. Mechanical forces provide cues for morphogenesis during organ development [4], tissue homeostasis [5], and wound healing [6], to name a few. Disease processes of fibrosis and cancer metastasis are also linked to the abnormal mechanical properties applied to tissues [7,8]. Recent studies on cellular mechanotransduction suggest that mechanical modulation of the cellular microenvironment can be used to improve the wound healing response [9], either by promoting faster wound closure or by decreasing fibrosis [10]. Generally, it is crucial to characterize the influence of mechanical forces in order to clearly understand in vivo physiology.

Mechanotransduction operates through numerous pathways that are often complex and unpredictable. OOCs can be used to recreate the in vivo biomechanical environment for studying and evaluating such mechanotransduction. In this review, we describe the latest approaches and methods used to generate mechanical stimuli on OOC systems and how they can be used towards diverse applications. Specifically, technical details of OOC system designs and equipment used are summarized and classified into several different categories. We also discuss on the potential advancement in combining different types of mechanical stimuli that can be delivered to a single OOC system.

2. Types of Mechanical Stimuli Utilized in Current Organ-on-a-Chip (OOC)

In vivo cells and tissues experience various mechanical stimuli, which have been recognized as key determinants of differentiated cellular functions. For example, fluid shear stress presented in the kidney induces the morphological polarization and allows epithelium to perform the transportation function, including the transport of water and sodium in response to hormonal stimulation. Mechanical stimuli commonly applied in OOC systems can be classified into three categories: shear flow, compression, and stretch/strain (Figure 1).

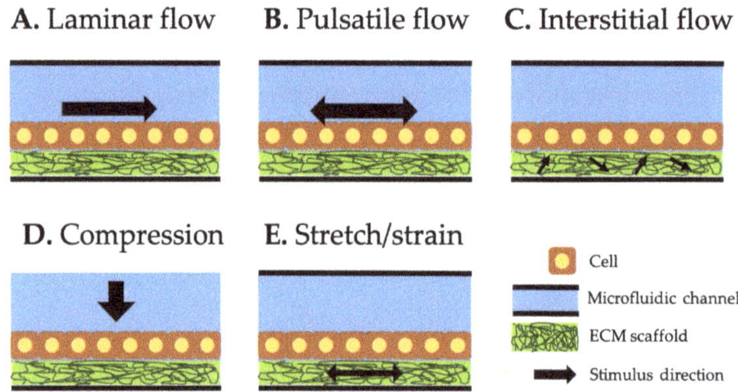

Figure 1. Types of mechanical stimuli integrated in organ-on-a-chip (OOC) systems. Black arrow represents the direction of each stimulus. Laminar (**A**) and pulsatile (**B**) flows along the microfluidic channel and generates laminar and pulsatile shear stresses on the cell surfaces. Interstitial flow (**C**) through the extracellular matrix (ECM) scaffold generates the shear stress to the cells attached to it. Compression (**D**) generates compressive force on top of the cells. Stretch/strain applied to the ECM scaffold (**E**) generates the force to the cells attached to it.

Liquid flow within microchannels of OOC systems usually induce the shear stress on the cells or tissues cultured in the system, thus called shear flow. Flow-induced shear stress has initially been used to study the effects of this stress on cell adhesion, mechanics, morphology, and growth in early microfluidic cell culture systems. Recent studies are focused on reproducing physiologically relevant shear stresses to understand their effects in the specific tissues and organs. Shear forces can be generated by laminar, pulsatile, or interstitial flow.

2.1. Laminar Flow

Laminar flow is dominant in tissues and organs and subsequently OOC systems, as the dimensions are in micrometer scale (Figure 1A). This leads to a very small Reynolds number (= $Dv\varrho/\mu$, where, D is tube diameter, v is flow velocity, ϱ is fluid density, and μ is fluid viscosity) [11]. Such flow is characterized as laminar flow, which is a series of parallel straight paths of flow lines with parabolic profile and no mixing among them. As such, laminar flow has commonly been used in various OOC systems (Figure 2A). For example, one kidney-on-chip device reproduced luminal fluid shear stress of

0.2–20 dyn/cm^2 (which is ~10% of the shear stress experienced by endothelial cells) in the collecting duct system of the kidney, to study the role of fluid shear stress of 1 dyn/cm^2 in the reorganization of actin cytoskeleton [12] and translocation of water transport proteins (aquaporin-2) of inner medullary collecting duct (IMCD) cells of the kidney [13]. The kidney-on-chip device developed by Jang et al. was composed of two polydimethylsiloxane (PDMS) layers separated by a thin porous polyester membrane. The top layer was covered by IMCD cells to mimic the inside tubular system and the bottom layer was left empty for fluid flow to mimic the outside tubular system. Various environmental factors and their combinations, including 1–12 h of exposure time, 0.2–5 dyn/cm^2 of shear stress, different concentrations of hormones, and transepithelial osmotic gradient conditions across a porous membrane, were investigated to generate various dynamic conditions for kidney tubular cells. Fluid shear stress, together with the presence of hormone and osmotic gradient, triggered the F-actin polymerization and depolymerization in both apical and basal regions of the cells, and the process is reversible. These kidney-on-chip models can be used towards studying renal physiology and pathophysiology.

Laminar flow was also used for an OOC system to study how fluid forces modulate angiogenesis (formation of new blood vessel) [14,15]. The OOC system for an angiogenesis study invented by Vickerman et al. was comprised of two parallel microfluidic channels, lined with human umbilical vein endothelial cells (HUVECs), and a central microchannel of a three-dimensional (3D) collagen extracellular matrix (ECM) that separated the two parallel microfluidic channels, into which HUVECs could migrate [14]. Similarly, Zheng et al. fabricated a microvessel network of HUVECs within collagen matrix and the 0.1 dyn/cm^2 shear stress was delivered to the cells [15]. These OOC systems recreated the physiological environment of vascular endothelial cells for assessing angiogenesis. Laminar flow used in this system generated the tangential shear stress on the HUVECs and induced cell sprouting into the central channel.

Another example was a liver-on-chip device, where the hepatocytes were 3D-cultured within a microfluidic channel to study drug toxicity in vitro [16]. The device was composed of highly fluidic-resistant endothelial barrier channels, which was connected to the flow channel and the cell loading channel. Media was pumped as laminar flow at the steady rate of 10–20 nL/min to mimic the mass transport properties of the hepatic microcirculation. The endothelial barrier was introduced to minimize the damage to hepatocytes caused by high shear stress. The 3D morphology of hepatocytes and liver metabolically specific functions were maintained under this continuous, laminar flow. Hepatotoxicity of the anti-inflammatory drug, diclofenac, was also investigated using this liver-on-chip model to assess the short-term (<4 h) and long-term (24 h) exposures to the liver cells and compared with the reported drug-induced injury to liver.

2.2. Pulsatile Flow

When the heart beats, blood is pumped to vessels throughout the body due to cycles of contraction and relaxation of the ventricles and atria, which resulted in a pulsatile flow within the transportation part of cardiovascular system (Figure 1B) [17]. Pulsatile flow has commonly been utilized in blood vessel-on-chip systems to simulate the actual pulsatile blood flow in human circulation [18]. For example, a hemodynamic microfluidic system with endothelial cells was developed mimicking the physiological pulsatile nature of the vascular system, to investigate the effect of biological factors in the blood from diabetic patients [18]. This microfluidic chip consisted of three rows for triplicate experiments to be performed at the same time. Each row has three different microchannel designs that mimic different blood vessels shapes: 600 µm-wide channel to mimic typical human arteries, 600 µm-wide at the beginning and gradually narrowing down to 300 µm, and a half-circle block with a radius of 300 µm in the middle of the 600-µm channel to mimic the blood vessel with an atherosclerotic plaque. All three channels were 150 µm deep. Pulsatile shear stress was applied at the average flow rate of 1.41 µL/s (corresponding to the average shear stress of 15 dyn/cm^2 in 600 µm-wide channels) and at a pulse rate of 70 beats per min (bpm), both of which corresponded to the physiological conditions of typical arteries in a healthy human body. The other profile is a pulsatile flow at the elevated average flow rate of

2.22 µL/s (corresponding to the average shear stress of 23.6 dyn/cm^2 in 600 µm-wide channels) with the elevated pulse rate of 110 bpm, to simulate the blood flow in diabetic patients. The rate of endothelial cell apoptosis increased by a factor of 1.4–2.3 with increasing glucose concentration and shear stress.

Similarly, an OOC system mimicking a microvascular transportation connecting multiple organ-on-chip systems was designed and constructed to study the long-term homeostasis of human blood vasculature and its transport function [17]. Such a multi-organ-on-chip (MOC) device consisted of two separate microvascular circuits and the insert compartments were filled with organ-equivalent culture. Pulsatile flow was delivered to the MOC at the periodically varied rates from 7 to 70 µL/min with the frequency of 144 bpm. The measured average shear stress was 25 dyn/cm^2 in the microvascular circuit, which is within the physiological shear stress of 10–40 dyn/cm^2 [19].

An integrated microfluidic chip cultured with endothelial cells was developed to study the endothelial cell's morphology, cytoskeleton, and barrier formation under the self-contained loop of induced pulsatile shear stress at a physiologically-relevant level (Figure 2B) [20]. The chip is composed of three layers: a bottom layer for microgap, a middle layer for microchannels and microchamber, and a top layer for reshaping the microchannel. The average pulsatile flow rate of 0.34 nL/s was delivered to the systems, which correlated to the instantaneous shear stress from +8.15 to −5.92 dyn/cm^2. Several in vivo physiological structures and processes were similarly represented in this chip, such as endothelium barrier, albumin transport, soluble factors diffusion, and pulsatile and oscillatory shear stress exposed to endothelial cells. These processes were utilized towards comprehensive understanding of the endothelial cell's behavior and functions.

2.3. Interstitial Flow

Interstitial flow is fluid flow through or around a 3D ECM, where interstitial cells such as fibroblasts, tumor cells, tissue immune cells, and adipocytes can be found. It differs from open-channel flow, such as blood flow within vessels, in several ways (Figure 1C). For example, it generally flows at a much slower velocity because of the high flow resistance of the ECM, it moves around the cell-matrix interface in all directions rather than only on the apical side, and it can have important effects on pericellular protein gradients, particularly those that are matrix binding [21]. Microfluidic systems have been considered as a promising platform to investigate the effects of interstitial flow on cell motility. For example, a microfluidic device with two regions for ECM gel was proposed to study the effect of interstitial flow on cancer cells' migration (metastasis) [22]. In this work, breast cancer cells were used as a cancer model and the use of two ECM gels allowed maximum flexibility in adjusting flow rates. Interstitial flow of 10 µm/s increased the number of circulating cancer cells (CTCs) as well as their velocity, but the directionality of such movement varied between cell types. Effects of interstitial flow on cancer cell invasion were also examined in order to provide a better understanding of how cancer cells gain access to the lymphatic system and manipulate their environment [23–25]. Bonvin et al. designed a multichamber radial flow system consisting of nine identical gel-filled chambers sharing inlet and outlet medium reservoirs [23]. Interstitial flow rate of 1–2 µm/s successfully induced the migration of cancer cells into the fibroblast-containing gel region. Similarly, Munson et al. utilized the radial flow chamber to investigate the effect of interstitial flow (average velocity of 0.7 µm/s) on the glioma (brain cancer) invasion as well as flow-induced cellular polarization [24]. A co-culture model with fibroblast and cancer cell was used to study the effect of interstitial flow (~0.5 µm/s) on cancer cell invasion [25]. In this work, interstitial flow enhanced the fibroblast migration through increased transforming growth factor β1 (TGF-β1) activation and collagen degradation, through which fibroblasts eventually reorganized collagen fibers and in turn, enhanced cancer cell invasion.

Interstitial flow also affects angiogenesis (formation of new blood vessels) as well as lymphanogenesis (formation of new lymphatic vessels). These processes are essential to both normal development and pathological processes that are heavily influenced by the dynamic mechanical environment, for example, laminar and interstitial flows experienced by endothelial cells. Effects of interstitial flow on angiogenesis and lymphanogenesis were investigated on OOC models [26,27].

Kim et al. utilized the microfluidic device composed of 6 microchannels: two peripheral media channels, two outermost channels for 3D culture of fibroblasts (secreting angiogenic biochemical cues), one central channel for a fibrin-matrix embedding endothelial cells, and one central channel for an acellular fibrin-matrix for angiogenic sprouting (Figure 2C). In the presence of interstitial flow (0.1–4 µm/s), vasculogenic organization of the microvascular network was significantly facilitated regardless of flow direction, whereas angiogenic sprouting was promoted only when the directions of flow and sprouting were opposite from each other [26]. A similar chip design was used to study the lymphanogenesis in response to interstitial flow (~1 µm/s). The growth of vessel sprouting was significantly augmented in the direction opposite the flow, while sprouting was suppressed along with the flow direction. These results imply that interstitial flow is a key mechanical regulator of tissue vascularization including the rate of morphogenesis and sprout formation [27].

2.4. Compression

Various types of cells experience compression stresses that are applied on top of the cells (Figure 1D). A good example is skin tissue, where the top two layers are epidermis (top, exposed layer) and dermis (just underneath the epidermis). From a physical standpoint, the main function of the dermis is to provide mechanical integrity to the skin. In normal non-injured skin, the majority of compressive and tensile forces imparted onto the skin are borne by the ECM network in the skin dermis, with little force directly applied onto the resident fibroblasts or other structures [28]. The hypodermis, innermost layer of the skin and lining between dermis and skeletal muscle, mainly consists of subcutaneous fat and serves as a shock absorber.

Another good example is cardiac tissue. It displays macroscopic contraction with the entire tissue being compressed at each contraction without collapse of the internal vasculature [29]. Microfluidic devices can be used to mimic the organ functionality through constructing multi-cellular architecture, interfacing the tissues and physiochemical microenvironments along with perfusion of the body. Several different OOCs can fluidically be connected together, mimicking the physiological connections between organs with integrating flow shear stress, mechanical compression, and cyclic strain or other physical forces, which can be used for analyzing drug distribution and organ-specific responses [30]. The complex mechanical microenvironment of organs can be replicated in vitro on OOCs, enhanced pressure can be applied to compress the tissues at a higher level than normal [31]. For example, adipocyte-derived stem cells and bone marrow-derived stem cells were cultured on microfluidic devices and exposed to dynamic hydraulic compression to evaluate the osteogenic abilities [31]. The uniform mechanical compression was provided to the array of cell culture chamber that were located along a concentric circle of the central air inlet on a microfluidic chip (Figure 2D). Dynamic mechanical compression of 1 and 5 psi(= 6.9 and 34 kPa) was evaluated to investigate its effect on the osteogenic differentiation abilities of different types of stem cells. Dynamic compression increased osteogenic ECM formation and integrin levels in both types of stem cells, leading to stimulus-enhanced bone differentiation. Bone marrow-on-chip was also used to study blood cell production under radiation countermeasure drugs [32]. Bone marrow chip exposed to gamma radiation resulted in a reduction of leukocyte production. Treatment of the chips with two potential therapeutics, granulocyte-colony stimulating factor (G-CSF) or bactericidal/permeability-increasing protein (BPI), induced significant increases in the number of hematopoietic stem cells and myeloid cells in the fluidic outflow.

2.5. Stretch/Strain

In addition to flow-induced shear stress, cells and tissues in the human body continuously experience organ-specific tensile and compressive forces during the normal operation of organs (Figure 1E). Compression from outside (skin tissue) and from blood (cardiac tissue) was already described in the previous section. Tensile force applied to the cellular microenvironment leads to stretch or strain to the tissue. One example is a lung-on-a-chip device that uniquely mimics the combined solid and fluid (surface-tension) mechanical stresses induced by the air-flow-induced cellular stretching and the propagation of an air–liquid meniscus in the alveoli of a human lung [33]. Previously reported in vitro models of ventilator-induced lung injury generated either cyclic stretching or air–liquid interface flow over the cells on non-stretching substrates [33–36]. This lung-on-a-chip device created more physiologically relevant mechanical microenvironments for alveolar epithelial cells during ventilation by simulating both fluid and solid mechanical stresses. This study showed that combined solid and fluid mechanical stresses (cyclic stretch and surface tension forces, respectively) significantly increased cell death and detachment compared to solid mechanical stress alone, supporting the clinical observations that cyclic stretch alone is insufficient to induce the level of the cell injury, as seen in ventilator-induced lung injury. Another example is a human-breathing lung-on-a-chip device that reconstituted a mechanically active microenvironment of the alveolar–capillary interface of the human lung [37]. Additionally, the lung-on-a-chip device reproduced the cyclic strain from the breathing movement in a human lung. The microfluidic device consisted of a two-layer central channel separated by a thin membrane for cell cultures and two side channels where a vacuum was applied for suction (Figure 2E). Cyclic stretching was achieved at a frequency of 0.2 Hz (sinusoidal wave form) with 10% strain by applying cyclic suction to the hollow side chambers, thus causing the mechanical stretching of the flexible membrane between the alveolar and capillary compartments. The lung-on-a-chip device was also used to create a human disease model-on-a-chip of pulmonary edema [37]. Such a disease model revealed that mechanical forces associated with physiological breathing motions increased vascular leakage that leads to pulmonary edema.

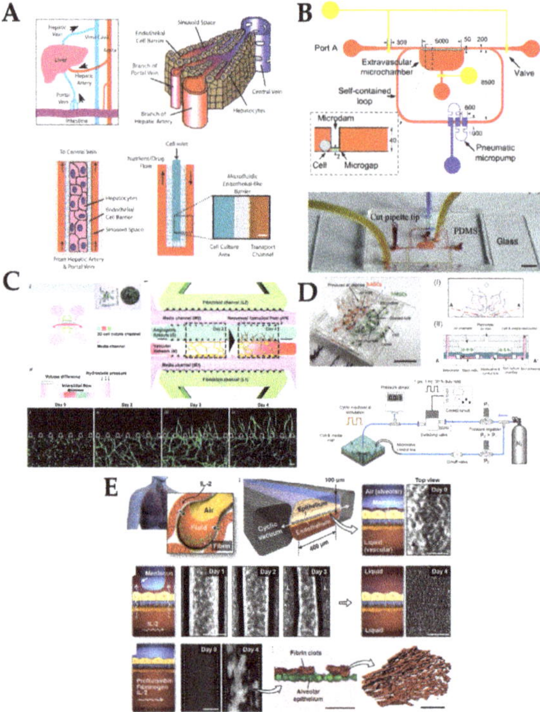

Figure 2. Different types of mechanical stimuli used in OOC systems. (**A**) Laminar flow presented in liver-on-chip, reproduced with permission from Reference [16]. (**B**) Pulsatile flow presented in vessel-on-chip, reproduced with permission from Reference [20]. (**C**) Interstitial flow presented in angiogenesis-on-chip, reproduced with permission from Reference [26]. (**D**) Compression stress presented in bone-on-chip, reproduced with permission from Reference [31]. (**E**) Stretch/strain presented in lung-on-chip, reproduced with permission from Reference [37].

A model mimicking human intestinal microenvironments was developed to evaluate lipid transport across the lymphatic endothelial layer for future applications in screening drug candidates targeting intestinal lymphatics [38]. Incorporation of the presence of peristalsis and the subsequent mechanical stretching on the endothelial layer to this model may be the next critical step for better replicating intestinal lymphatics. As such, there is significant potential for microfluidic techniques to further elucidate the functionality of the lymphatic system, as well as other systems that depend on similar mechanical constraints.

Different types of mechanical stimuli and their application towards different OOC models are summarized in Table 1.

Table 1. Summary of different types of mechanical stimuli and delivering mechanisms towards various organ/tissue models and their applications.

Mechanical Stimuli	Delivery Method	Organ/Tissue Model	Applications	Reference
Laminar flow	Passive delivery (gravity-driven)	Liver	Improvement and maintenance of cell viability under drug exposure	[16]
	Pressure regulator	Liver	Real-time monitoring of metabolic function of liver and drug-induced mitochondrial dysfunction	[39]
	Syringe pump	Kidney	Transportation, absorption, toxicity, and pathophysiology of kidney	[40]
Pulsatile flow	Peristaltic on-chip micropump	Blood vessel	Mimicking blood circulation systems connecting two different organs-on-chip towards long-term homeostasis	[17]
	Syringe pump	Blood vessel (endothelial cells)	Endothelial cell's integrity and apoptosis in the blood of diabetic patients	[18]
	Pneumatic pump	Blood vessel (endothelial cells)	Barrier formation and permeability of endothelial cells	[20]
Interstitial flow	Peristaltic pump	Breast cancer	Cancer cell's invasion in response to interstitial flow	[22]
	Passive delivery (hydrostatic pressure-driven)	Brain cancer (glioblastoma stem cells)	Patient-derived glioblastoma stem cell's invasion in response to interstitial flow	[41]
	Passive delivery (hydrostatic pressure-driven)	Blood vessel (endothelial cells)	Angiogenesis: vasculogenic formation and angiogenic remodeling in response to interstitial flow and vascular morphogens	[26]
Compression	Pressure regulator	Bone (stem cells)	Osteogenic ability of adipose tissue- and human bone marrow-derived stem cells in response to hydraulic compression	[31]
	Compression device	Heart	Formation and functions of cardiac muscle tissue during the stage of compression	[42]
	Pressure controller	Bone (osteoblasts)	Real-time monitoring of single cell response to compressive stimuli	[43]
Stretch/strain	Vacuum and syringe pump	Lung	Visualization and characterization of inflammatory processes in response to bacteria at the alveolar–capillary interface	[37]
	Pneumatic pump	Lung	Conservation of the epithelial barrier property between human lung alveolar cells and primary lung endothelial cells under long-term co-culture and cyclic strain	[44]
	Syringe pump	Connective tissue (fibroblasts)	Effect of static strain on cellular alignment of fibroblasts encapsulated in hydrogel	[45]

3. Current Methods of Delivering Mechanical Stimuli to Organ-on-a-Chips (OOCs)

As described previously, mechanical stimuli found in the human body can be classified into (1) shear flow (laminar, pulsatile, and interstitial), (2) compression, and (3) stretch and strain [45], which have also been demonstrated in OOC systems. The most common stimulus used in OOC systems is shear flow as it can mimic the human body's mechanisms to introduce medium and solution of interests to the body systems. These shear flows have been generated in diverse ways, while they can generally be classified into three categories: external pumping, internal (on-chip) pumping, and passive delivery. Figure 3 graphically illustrates various methods and equipment to deliver mechanical stimuli to OOC systems.

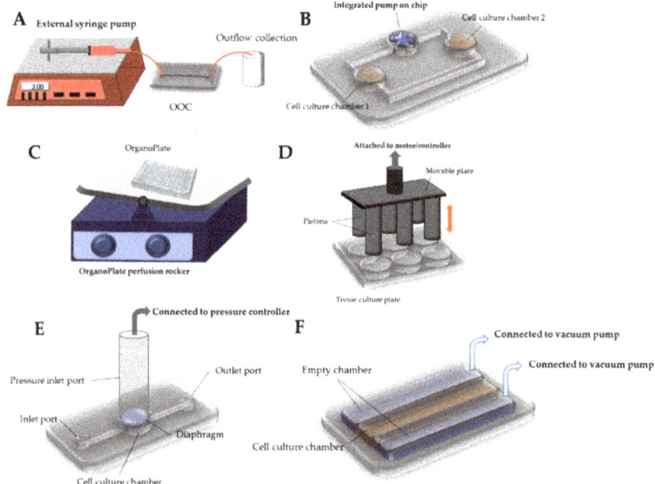

Figure 3. Methods and equipment used to deliver mechanical stimuli to OOC systems. (**A**) External syringe pump is delivering shear flow. (**B**) Micro-pump is integrated within a microfluidic OOC system to drive either laminar or pulsatile flow. (**C**) OrganoPlate perfusion rocker is used to generate the shear flow though the cells cultured within OrganoPlate by tilting the stage. (**D**) Compressive force is delivered to the cells on a tissue culture plate (TCP) by the pistons attached to a moveable plate that moves up and down. (**E**) Compressive force is delivered by the pressure controller to the diaphragm placed on top of the cells. (**F**) Stretch/strain is generated by periodically applying vacuum to the peripheral chambers, allowing the main culture chamber to stretch in a cyclic manner.

3.1. External Pumping

The easiest and most common method for generating shear flow in OOC is the use of an external pump, including a syringe pump or peristaltic pump (Figure 3A). A syringe pump delivers precise, accurate, and small amounts of fluid in a programmable manner, where a stepper motor pushes the plunger flange of a syringe filled with fluid, generating fluid flow to the tube connected from the end of a syringe needle. The syringe pump has popularly been used for the OOCs fabricated with polydimethylsiloxane (PDMS), which is the most frequently used substrate for many microfluidic devices as well as OOCs. For instance, cardiotoxic drug exposure was demonstrated in the PDMS-based microfluidic device cultured with human embryonic stem cell-derived cardiomyocytes with the medium flow generated and manipulated by a syringe pump [46]. Similarly, angiogenesis was demonstrated in the microfluidic device cultured with vascular endothelial cells, recapitulating in vivo blood flow along the PDMS channel [46]. Angiogenesis of vasculature tissue has also been demonstrated on an OOC system to study the angiogenic ability of different cell types, human aortic endothelial cells and HUVECs, and expression level of fibroblast growth factor under the vascular endothelial growth factor (VEGF) stimulation [47]. The syringe pump was also used to perfuse breast cancer cells (MCF7 and MDA, as circulating tumor cells) through different PDMS-based OOC systems, and the metastasis behavior of these cancer cells was observed (Figure 4A) [48]. In this work, co-culture of more than one cell type—lung-liver-bone-muscle co-culture—has been explored to provide better physiological relevance of OOC systems.

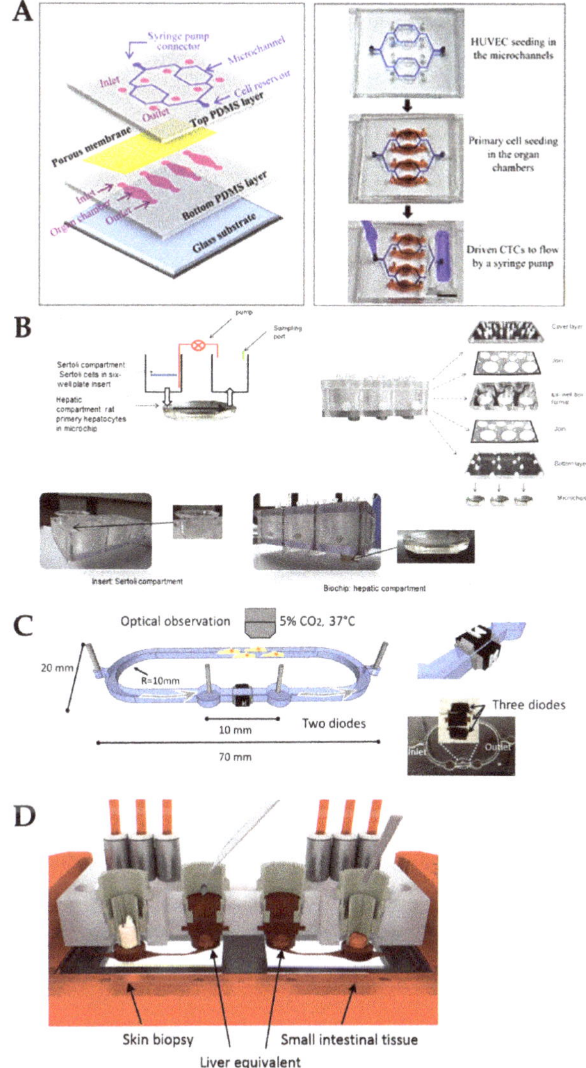

Figure 4. Examples of OOC systems with external and internal pumping. (**A**) External syringe pump for the OOC with endothelial cells. (**B**) External peristaltic pump for the OOC with hepatocytes and Sertoli cells. (**C**) Electro-osmosis (EOS) pumping was integrated onto the OOC with human lung cancer cells. (**D**) A peristaltic valve was integrated onto the OOC with liver-intestine and liver-skin cells. Reproduced with permission from References [49–52].

In a peristaltic pump, fluid is contained within a flexible tube fitted inside the pump casing. Its pumping principle, called peristalsis, is similar to the movements in human throat and intestines: compression and relaxation of the tube is alternated, drawing the new fluid in and propelling the existing fluid away from the pump. The peristaltic pump has also been used, albeit not as frequently as the syringe pump, to generate medium flow in OOC systems. For example, the OOC system with rat primary hepatocyte and primary Sertoli cells co-culture was demonstrated to investigate hepatic metabolism and Sertoli barrier tightness relationship (Figure 4B) [49]. The vacuum pump has occasionally been used as well: for example, in the lung-on-a-chip model, vacuum was applied to the side chambers to generate cyclic stretch (Figure 3F) [50], which is discussed in the later section. All three types of external pumps, commonly utilized for PDMS-based OOC systems, are not suitable to meet the demand of the "miniaturized" nature of OOC, due to their bulky size and requirement of an AC power source.

3.2. Integrated on-Chip Pumping

To overcome the size limitation of external pumping, the pumps were integrated within the microfluidic chips to miniaturize the systems (Figure 3B). For example, an electro-osmosis (EOS, voltage applied to the channel creates the bulk flow of liquid) diode pumping system was used to generate a pulsed, ultra-slow EOS flow (2.0–12.3 nL/s) through lung cancer tissue (Figure 4C) [51]. Briefly, two electro-diodes were embedded along two PDMS side walls. The electric signals were applied to the diodes directly via the electrolyte within the microfluidic channel and required no external electrical connection because the leads of the diodes physically contacted the fluid. The flow rate was adjusted by changing the electric field strength applied to the diodes from 0.5 V_{pp}/cm to 10 V_{pp}/cm. The diode pump-driven fluidics were characterized by intensities and frequencies of electrical inputs, pH of fluids, and fluid type. However, the long-term operation of this electro-osmotic microfluidic system required a change of fluid every hour which is not easy-to-implement for some tissue or organ models.

A miniature peristaltic pump was also integrated on the PDMS chip to generate pulsatile flow along the co-culture of liver-intestine and liver-skin tissue to investigate the efficacy of different drug dosage and administration (Figure 4D) [52]. This microfluidic chip contained the cell culture compartment, peristaltic micropump, and three PDMS membrane valves in a row. These were actuated by applying pressure. The pumping frequency and peristaltic valve pressure were modified to adjust fluid flow rate into a physiological range of a specific organ. The optimized conditions were the pump at frequency of 2 Hz, the median flow rate of 2.6 µL/min, and the median shear stress of 0.7 dyn/cm^2. This system was stable for up to 14 days for the co-cultures of human 3D liver equivalents with human intestinal barriers.

3.3. Passive Delivery

While the internal on-chip pumping allowed for miniaturizing the OOC systems, the integration of such an internal pump onto PDMS chip is extremely complicated for both fabrication and operation. Passive control of the flow has recently emerged in the past three years as an alternative. One example is OrganoPlate, which is a commercialized microtiter plate where the cells are cultured on it and the passive flow is generated by rocking the OrganoPlate (Figure 3C). The OrganoPlate supported up to 96 tissue models on a single platform and each model usually consisted of two layers: a medium layer for passive fluid flow and an ECM gel layer. This device also offered 3D tissue culture where the cells were embedded within the ECM gel layer and also deposited as a monolayer at the inner wall of the medium layer. Continuous perfusion of medium mimicked blood flow and enabled exchange of nutrients and oxygen. This was driven by gravity levelling (difference in fluid level) using its Perfusion Rocker^{TM} and no additional pumps and tubes were necessary. OrganoPlate has been used in many OOC systems. For example, neuron tissues were cultured on OrganoPlate to investigate the stem cell differentiation (Figure 5A) [52] and co-culture of neurons-glia cell for studying neurite outgrowth [53]. While this method is simple, it still requires a moveable stage to tilt the plate for generating a passive flow. An improved method is the use of gravity-driven flow, which is also gaining popularity due to its simplicity. For example, the inlet and outlet of PDMS chip were connected to syringes containing different levels of solution. With gravity, a higher level of the medicine solution in the syringe inlet was driven to flow through liver tumor spheroids, cultured within the PDMS chip for screening the anti-cancer drug toxicity in the liver tissue [54]. Instead of syringes, the PDMS chip was placed on the computer-controlled tilting stage to drive the flow within the chip where the co-culture of liver (HepG2) and cancer (HeLa) cells existed and used to observe the pharmacokinetic and pharmacodynamic of anti-cancer drug activities (Figure 5B) [55]. Similar to the gravity-driven flow, hydrostatic pressure was used to generate interstitial flow along the vasculature tissue culture within the PDMS chip to study the angiogenesis [56]. Nonetheless, these passive flow controls can potentially result in an imprecise rate of flow and may not be practical for long-term experiments.

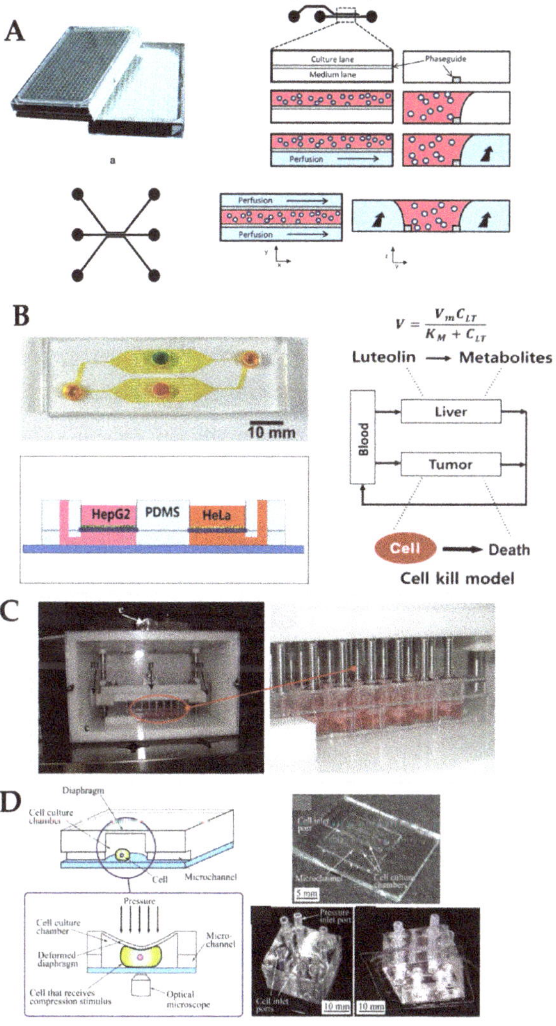

Figure 5. Examples of OOC systems with passive delivery and compression. (**A**) OrganoPlate was rocked to generate a passive flow through network of neurons. (**B**) Gravity flow machine generated flow to the liver and cervical cancer cells on chip. (**C**) Piston chamber generated compressive stress on the cardiac tissue within TCP. (**D**) Diaphragm and pressure controller generated compressive stress on osteoblastic cells on chip. Reproduced with permission from References [26,43,53,56].

3.4. Compression

In addition to shear stress, compression is also an important factor in certain tissues such as heart [26], bone [42], and vessel [57]. For instance, cardiac cells were cultured on alginate scaffold within a 48-well plate. These 3D cell cultures were placed underneath the compression device consisting of 48 pistons to provide the compression and medium perfusion for each culture well to study the cell behavior under mechanical stimuli (Figures 3D and 5C) [42]. Compression stimuli on osteoblasts can be presented in a microfluidic device to study the real-time cellular response, for example, the changes of intracellular calcium concentration, and intracellular strain distribution after exposure to those stimuli. Compression stimuli applied to cells can be controlled by regulating the expansion of the diaphragm via pressure control (Figures 3E and 5D) [58]. Cyclic compression in a sinusoidal waveform with magnitudes of 0.33, 0.5, and 1 MPa were delivered at 1 Hz frequency by a commercially available compression system (FX-5000C^{TM}, Flexcell, Burlington, NC, USA), to study the differentiation of osteoblasts [59]. This commercial system regulated the positive air pressure to compress tissue samples or 3D cell cultures that were placed in between a piston and a stationary platen using their own tissue culture plate (TCP). This device allowed each piston to apply the loads up to 14 lb (=6.4 kg) into an individual well on TCP.

3.5. Stretch/Strain

A vacuum pump was used to generate cyclic stretchin OOCs, specifically lung-on-a-chip, as described in the external pumping section. As PDMS (the most popular material for OOC fabrication) is flexible, periodic changes in the vacuum pressure can result in stretching and/or strains to the substrate (PDMS) and subsequently to the cells attached to it. Stucki et al. fabricated a PDMS-based lung-on-chip device to study the conservation of the epithelial barrier functionality between primary human lung alveolar cells (hAEpCs) and primary lung endothelial cells for long-term co-culture and the effect of physiological cyclic strain on epithelial barrier permeability [53]. In this work, the cyclic strain was delivered to the systems by connecting a chip to an electro-pneumatic setup via access ports which was used to control the opening and closing of the valves by applying negative and positive pressures. The applied pressure deflected the 40 µm thick PDMS membrane (microdiaphragm), resulting in opening and closing the valves. The setup regulated the cyclic motion by applying a negative pressure to the access ports. The movements of the PDMS membrane were transferred to the thin, porous alveolar membrane (Figure 6A). Besides cyclic strain, the effect of gradient static-strain on cellular alignment was studied on microfluidic chip as well [60]. Fibroblasts were cultured on a concentric circular hydrogel pattern and a flexible PDMS membrane. Initially, the outlet of the flow channel on the microfluidic chip was closed with a PDMS plug. Once the liquid was injected in the flow channel, a non-continuous gradient height of hydrogel along the radius was formed. After unplugging the outlet, the PDMS membranes became flat and gradient stress was applied on the cell-encapsulated hydrogels (Figure 6B). Cell behaviors were influenced by a combination of hydrogel shape and a variety of tensile stretch guidance cues.

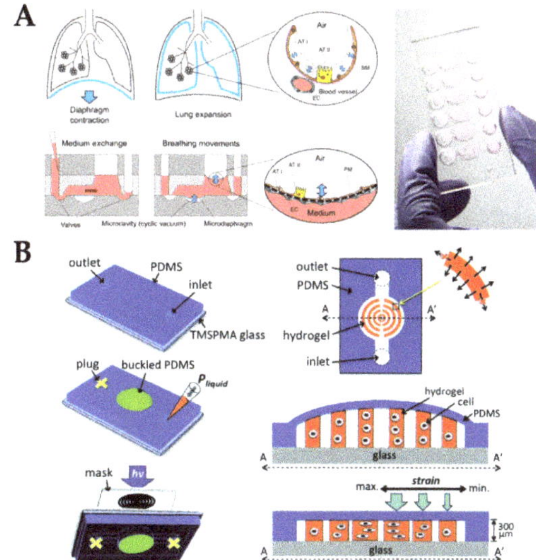

Figure 6. Examples of OOC systems with stretch and strain stimuli. (**A**) Lung-on-chip device was connected to the electro-pneumatic system to provide the negative pressure on the polydimethylsiloxane (PDMS) membrane (microdiaphragm), which subsequently generated the cyclic strain mimicking lung breathing mechanism. Reproduced with permission from Reference [44] (**B**) Gradient strain chip with a simple static strain delivery through the syringe at the inlet and PDMS plug at the outlet. Reproduced with permission from Reference [61].

As mechanobiological studies on OOC systems have been gaining popularity, many commercial tools have been developed that could easily deliver mechanical stimuli to the OOC systems (Figure 7). For example, the fibroblast's ability to align in response to mechanical stimuli was tested through a series of 0% to 10% cyclic uniaxial stretch experiments in 2D and 3D cell culture by using the commercially available MechanoCulture B1 device from CellScale Inc. (Waterloo, ON, Canada) (Figure 7A) [61]. Cyclic stretch generated from commercial devices successfully induced the perpendicular fibroblast alignment in 3D culture. In addition to uniaxial mechanical stretch, biaxial stretch generated by commercialized mechanical tester (BioTester, CellScale Inc.) was used to characterize the heterogeneity and function of the heart valve leaflet (Figure 7B) [62]. The porcine atrioventricular heart valve leaflet was dissected and placed on the device where the various loading ratios and stress-relaxation tests were performed on each region of tissue sample. Briefly, membrane tension applied by the device in the circumferential and radial directions were 100 and 50 N/m for different heart valve leaflet, and varying ratios of membrane tension loading in each direction was applied with eight loading/unloading cycles to investigate all possible physiological tissue deformations. Similarly, proteomic analysis of mitral valve anterior leaflets under 10% cyclic radial strain at a frequency of 1 Hz was investigated by using the commercialized MCT6 uniaxial stretcher (CellScale Inc.), where the tissue was placed between the two clamps of the device [63]. After mechanical stretch exposure, the endothelium of the heart valve was removed to test how endothelium mediates the mitral valve's response to stretch. This work showed a significant finding on new protein groups that might be involved in the mitral valve's response to mechanical stretch.

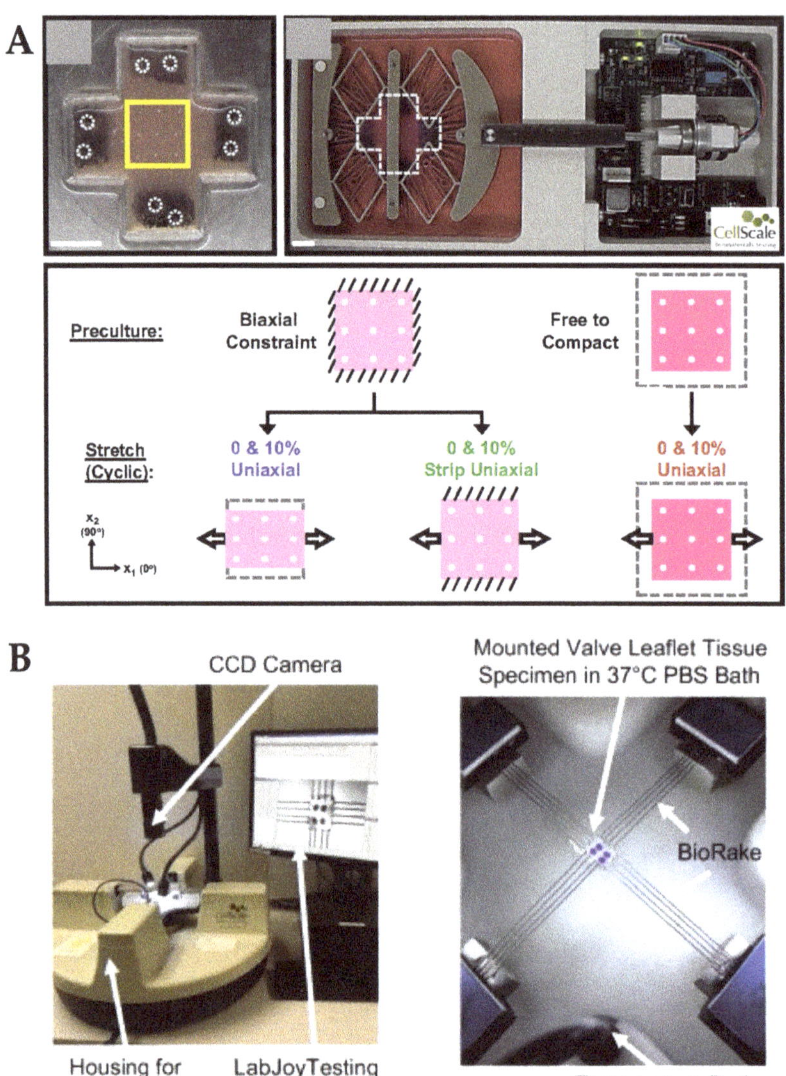

Figure 7. Examples of commercially available OOC systems with stretch and strain stimuli. (**A**) Microfluidic device was inserted to the MechanoCulture B1 system from CellScale Inc. to provide the biaxial mechanical stretch stimulation and evaluate its effect on cell alignment. Reproduced with permission from Reference [61]. (**B**) Porcine heart valves were positioned in the biaxial mechanical tester (BioTester, CellScale Inc.) to characterize heterogeneity of heart valve leaflet. Reproduced with permission from Reference [62].

4. Delivering Multiple Types of Mechanical Stimuli Simultaneously

A specific organ requires a specific microenvironmental stimuli to achieve an organ-level function, for example, lung cells require breathing-derived mechanical stretch [64]. The general approach for fabricating an OOC is to identify key aspects of the geometrical, mechanical and biochemical microenvironment of the tissue. With these key aspects identified, a microengineering approach is used to recapitulate them on a chip, under the researcher's full control. Then, isolated cells are introduced on the chip and subjected to the designed stimuli.

Some tissues experience multiple types of biochemical and/or mechanical stimuli. While multiple biochemical cues can be applied to a single chip relatively easily, it is not easy to do so for mechanical stimuli. For example, lung tissues are exposed to periodic mechanical force within each respiratory cycle: air flow in the air–epithelium interface, blood flow in the endothelium–blood interface, and stretching on the epithelial–endothelial interface. The lung-on-chip model has therefore been developed integrating all of these mechanical stimuli on a single device [64]. The microfabricated lung-on-a-chip consisted of compartmentalized PDMS microchannels to form an alveolar–capillary barrier on a thin, porous, flexible PDMS membrane coated with ECM. The device recreated physiological breathing movements by applying a vacuum at a frequency of 0.2 Hz to the side chambers through a vacuum pump and causing mechanical stretching (5–15% strain) of the PDMS membrane forming the alveolar–capillary barrier. Alveolar epithelial and human pulmonary microvascular endothelial cells were grown on either side of a thin porous membrane and were exerted by 20 µL/h (fluid shear stress of 0.1 dyn/cm^2) media flow via the two channels separated by a membrane (Figure 8A). Culture medium for alveolar epithelial cells was gently aspirated from the upper channel on day five when the cells reached confluency to form an air–liquid interface. After that, culture media for both cell types were delivered to the lower channel to feed the epithelial cells on their basolateral side. This device can serve as a tool for studying lung diseases and related screening of drugs. While the model was still small and relatively simple, complex geometries were utilized that might prevent from easy reproduction and/or long-term operation of the device.

Similarly, Musah et al. developed a glomerulus-on-chip (GOC) that is analogous to the previously described lung-on-chip [65]. This GOC model efficiently facilitated the differentiation of human-induced pluripotent stem (hiPS) cells into functional human podocytes, cells that wrap around capillaries of the glomerulus and play a crucial role in selective permeability in blood filtration. It successfully recapitulated the natural tissue–tissue interface of the glomerulus and the clearance of albumin and inulin when co-cultured with human glomerular endothelial cells. The GOC can be used for drug development and specifically personalized-medicine, for example Adriamycin-induced albuminuria and podocyte injury. The fabrication and design were comparable to the lung-on-chip model developed by Huh et al. [64], where the alveolar–capillary barrier on a thin laminin-coated porous membrane was replaced with the urinary–microvascular interface, which recreated the physiology of human glomerulus in kidney, while the two vacuum channels parallel to the fluidic channel were maintained. The peristaltic pump was used to perfuse the fluidic channels at the volumetric flow rate of 60 µL/h, corresponding to the shear stress of 0.0007 dyn/cm^2 in the 1 mm × 1 mm urinary channel and 0.017 dyn/cm^2 in the 1 mm × 0.2 mm microvascular channel. While the cyclic stretching (10% strain) was applied to the GOC by the vacuum regulator at an amplitude of −85 kPa and frequency of 1 Hz. Adriamycin, a cancer treatment drug, was used as an example of drug toxicity studies by continuously being infused through the microvascular channel for up to five days. The toxicity of this drug was assessed by a series of protein assays.

Another example is the formation of new blood vessels or vasculature, e.g., angiogenesis. Vascular endothelial cells, or simply ECs, are continuously exposed to shear flow (blood and interstitial flows) and thus, experience shear stress tangential to the endothelial surface [66]. Shear stress causes ECs to secrete biomolecular signals [53], leading to vascular structure remodeling [67], cytoskeleton rearrangement [68–70], and transcriptional gene expression. Furthermore, ECs are exposed to interstitial flow, which acts as a transverse force across the vessel wall. Interstitial flow above a certain threshold promotes capillary formation [14,71]. In one OOC model, both tangential shear stress from luminal flow and stress from interstitial flow were applied to the ECs on a chip to study angiogenesis [66]. Briefly, this OOC consisted of two peripheral microfluidic channels for HUVECs cultured on fibronectin coated surface and central channel for collagen/fibronectin gel (Figure 7B). A syringe pump was used to deliver a laminar flow with the shear stress of 3 dyn/cm^2, the physiological level of shear stress in veins or venules, and the interstitial flow within the collagen/fibronectin channel was measured as 0.0045–0.064 dyn/cm^2. The results showed that fluid shear stress enhanced the EC sprouting and interstitial flow promoted morphogenesis and sprout formation.

Finally, osteogenesis, the differentiation of mesenchymal stem cells (MSCs), is another excellent example that can be heavily influenced by multiple types of mechanical stimuli: compression, shear stress, strain, stretch, and hydraulic forces [72]. Virtually all types of mechanical stimuli should be represented in an osteogenesis model, making the OOC design and fabrication extremely challenging. As such, OOC demonstration of osteogenesis is still ongoing.

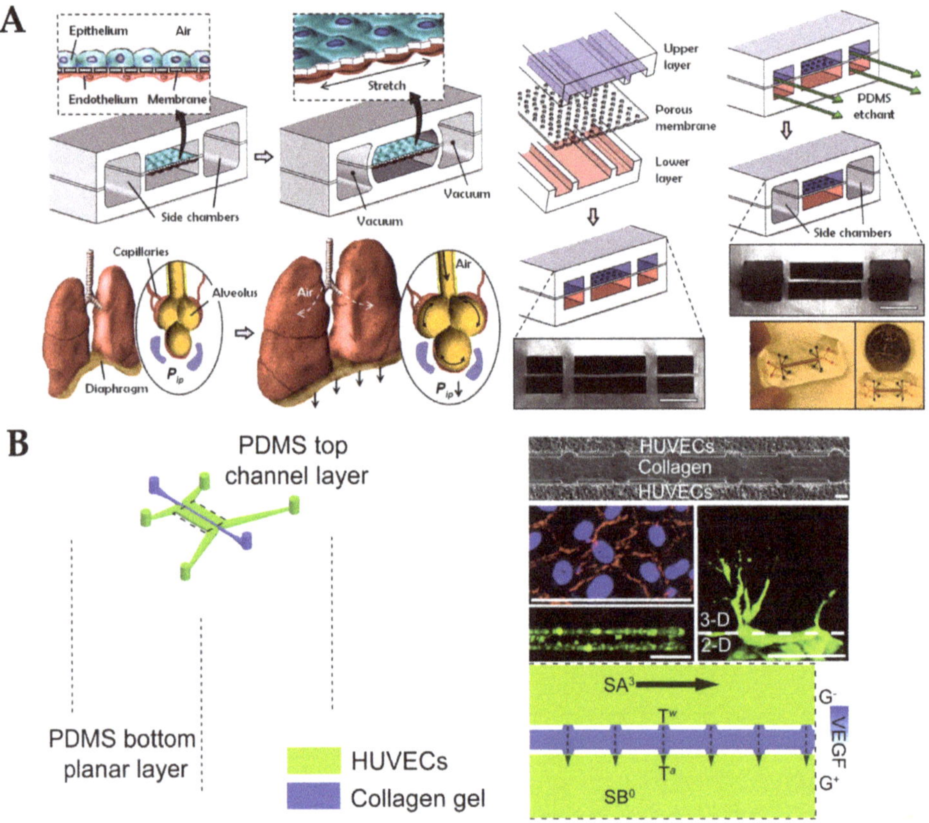

Figure 8. Examples of OOC systems with multiple types of mechanical stimuli. (**A**) Lung-on-chip. Reproduced with permission from Reference [64]. (**B**) Endothelial cells-on-chip. Reproduced with permission from Reference [66].

5. Mechanobiological Studies and Real-time Microscopic Imaging

Mechanical stimuli on OOC systems can also be utilized to improve mechanobiological studies, i.e., how an individual cell responds to the applied mechanical stimuli. To this end, cells and cellular responses must be monitored in real-time without destroying the cells within the OOC systems. For example, a customized miniature incubator was designed and fabricated to allow the real-time microscopic observation of cultured cells in the aforementioned EOS pumping systems, where the microscope lens was positioned above the microfluidic chip and the miniature incubator (Figure 4C). This culturing platform with live-cell imaging was effectively used for assessing cellular biophysical changes under fluid conditions [51].

Cellular response to mechanical stretch can occur as quickly as within a few seconds and sudden disruption of cell cytoskeletons can also occur at any point during such a cyclic stretch. Huang et al. developed a cell stretching device and a real-time imaging system to investigate rapid cellular responses to two phases of cyclic stretch stimulation (Figure 9A) [73]. Three motorized stages were used to deliver the cyclic stretching to a cell culture chamber while the target cells were maintained in the same position. The microscope was positioned underneath the stage system to collect cell images. The cell stretching and imaging system was able to determine the maximum displacement of the target cells in the horizontal and vertical directions and to capture the sudden disruption of cell–cell junctions in response to two phases of cyclic stretch at the frame rate of 30 fps (frames per second). This time resolution was greatly improved from the other cellular mechanobiological studies. Similarly, a commercial microscope-mountable uniaxial stretching system (STB-150W, Strex Inc., San Diego, CA, USA) was used to study the real-time single cell morphological deformation in response to mechanical stretch (Figure 9B) [74]. This commercial device stretches the cell culture chamber from both sides to enable the cells to remain inside the viewing area of a microscope. Nevertheless, the aforementioned systems still required bulky microscope apparatus to image the cells. To resolve this problem, a smartphone-based fluorescence microscope was developed and attached to an OOC device to construct a small and handheld OOC system (Figure 9C) [75]. A microscope attachment was modified with an additional LED light source and bandpass filters and the smartphone captured and processed fluorescent images. This smartphone-based OOC system was used to study nephrotoxicity of drugs with reduced assay time and production cost, while improving assay specificity and ease-of-use.

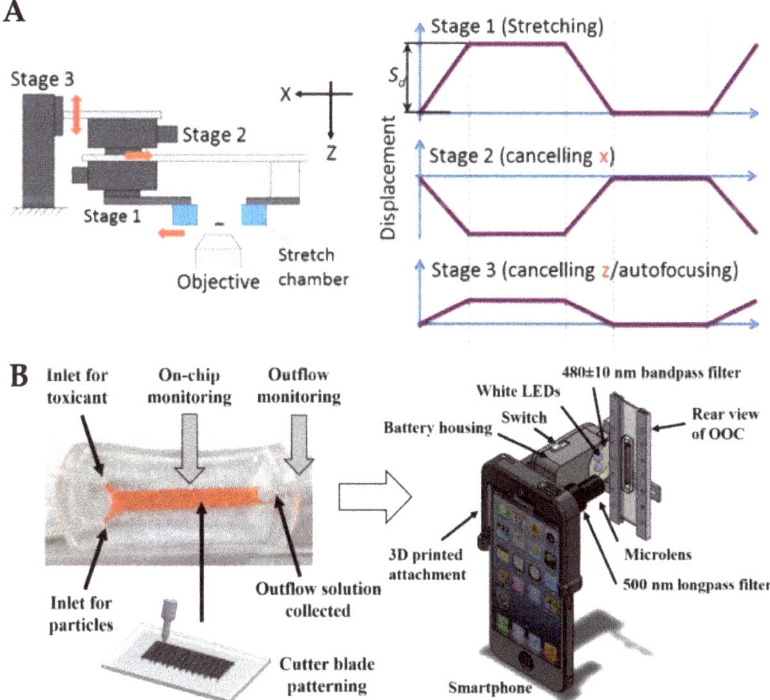

Figure 9. Examples of mechanobiological models and real-time monitoring devices for OOC systems. (**A**) Three motorized stages delivered cyclic stretching and a microscope imaged cellular local displacement. Reproduced with permission from Reference [73]. (**B**) Smartphone-based fluorescence microscope attached to an OOC device for in situ monitoring of nephrotoxicity. Reproduced with permission from Reference [75].

6. Conclusions

The emerging OOC technologies have been widely utilized toward physiologically relevant models for understanding human physiology, studying diseases, drug development, and toxicology. In OOCs, dynamic microenvironments around cells, tissues or organs have been recapitulated within microfluidic chips. Physiologically relevant mechanical stimuli are one of the crucial microenvironments that affect the cellular responses and functions. In order to deliver mechanical stimuli to OOC systems, external syringe pumps have most frequently been used to drive the liquid through the microfluidic channel with high precision and in a programmed manner. Pumps were also integrated within the microfluidic chip to minimize the size. Simpler delivery methods have also been demonstrated, including rocking, passive delivery, or applying hydrostatic pressure. Pistons and pressure controllers on the diaphragm were used to apply the compression stress to OOC systems. Delivery of more than one type of mechanical stimuli has also been suggested and demonstrated to better recapitulate the physiologically relevant microenvironment for tissue and organ, for example, lung-on-a-chip and kidney-on-chip, that is exposed to shear flow as well as cyclic strain. Successful fabrication and operation of OOC can enable us to closely mimic specific physiologically relevant microenvironments, and to provide an accessible, inexpensive, and readily manipulatable platform for rapid and definitive testing of both fundamental and applied biological hypotheses.

Author Contributions: K.K. and J.-Y.Y. jointly conceived the concept, collected and analyzed literature, and wrote the manuscript.

Funding: K.K. acknowledges the scholarship from the Development and Promotion of Science and Technology Talents Project (DPST) of Thailand.

Conflicts of Interest: The authors declare no conflict of interest.

References

1. Liu, M.; Tanswell, A.K.; Post, M. Mechanical force-induced signal transduction in lung cells. *Am. J. Physiol.* **1999**, *277*, L667–L683. [CrossRef] [PubMed]
2. Mammoto, T.; Mammoto, A.; Ingber, D.E. Mechanobiology and developmental control. *Annu. Rev. Cell Dev. Biol.* **2013**, *29*, 27–61. [CrossRef] [PubMed]
3. Waters, C.M.; Roan, E.; Navajas, D. Mechanobiology in lung epithelial cells: Measurements, perturbations, and responses. *Compr. Physiol.* **2012**, *2*, 1–29. [PubMed]
4. Varner, V.D.; Nelson, C.M. Cellular and physical mechanisms of branching morphogenesis. *Development* **2014**, *141*, 2750–2759. [CrossRef] [PubMed]
5. Barnes, J.M.; Przybyla, L.; Weaver, V.M. Tissue mechanics regulate brain development, homeostasis and disease. *J. Cell. Sci.* **2017**, *130*, 71–82. [CrossRef]
6. Rosińczuk, J.; Taradaj, J.; Dymarek, R.; Sopel, M. Mechanoregulation of wound healing and skin homeostasis. *Biomed. Res. Int.* **2016**, *2016*, 3943481. [CrossRef]
7. O'Connor, J.W.; Gomez, E.W. Biomechanics of TGFβ-induced epithelial-mesenchymal transition: Implications for fibrosis and cancer. *Clin. Transl. Med.* **2014**, *3*, 23. [CrossRef]
8. Lampi, M.C.; Reinhart-King, C.A. Targeting extracellular matrix stiffness to attenuate disease: From molecular mechanisms to clinical trials. *Sci. Transl. Med.* **2018**, *10*, eaao0475. [CrossRef]
9. Lancerotto, L.; Orgill, D.P. Mechanoregulation of angiogenesis in wound healing. *Adv. Wound Care* **2014**, *3*, 626–634. [CrossRef]
10. Duscher, D.; Maan, Z.N.; Wong, V.W.; Rennert, R.C.; Januszyk, M.; Rodrigues, M.; Hu, M.; Whitmore, A.J.; Whittam, A.J.; Longaker, M.T.; et al. Mechanotransduction and fibrosis. *J. Biomech.* **2014**, *47*, 1997–2005. [CrossRef]
11. Sosa-Hernández, J.E.; Villalba-Rodríguez, A.M.; Romero-Castillo, K.D.; Aguilar-Aguila-Isaías, M.A.; García-Reyes, I.E.; Hernández-Antonio, A.; Ahmed, I.; Sharma, A.; Parra-Saldívar, R.; Iqbal, H.M.N. Organs-on-a-chip module: A review from the development and applications perspective. *Micromachines* **2018**, *9*, 536. [CrossRef] [PubMed]

12. Jang, K.-J.; Suh, K.-Y. A multi-layer microfluidic device for efficient culture and analysis of renal tubular cells. *Lab Chip* **2010**, *10*, 36–42. [CrossRef] [PubMed]

13. Jang, K.-J.; Cho, H.S.; Kang, D.H.; Bae, W.G.; Kwon, T.-H.; Suh, K.-Y. Fluid-shear-stress-induced translocation of aquaporin-2 and reorganization of actin cytoskeleton in renal tubular epithelial cells. *Integr. Biol.* **2011**, *3*, 134–141. [CrossRef] [PubMed]

14. Vickerman, V.; Kamm, R.D. Mechanism of a flow-gated angiogenesis switch: Early signaling events at cell–matrix and cell–cell junctions. *Intgr. Biol.* **2012**, *4*, 863–874. [CrossRef] [PubMed]

15. Zheng, Y.; Chen, J.; Craven, M.; Choi, N.W.; Totorica, S.; Diaz-Santana, A.; Kermani, P.; Hempstead, B.; Fischbach-Teschl, C.; López, J.A.; et al. In vitro microvessels for the study of angiogenesis and thrombosis. *Proc. Natl. Acad. Sci. USA* **2012**, *109*, 9342–9347. [CrossRef] [PubMed]

16. Lee, P.J.; Hung, P.J.; Lee, L.P. An artificial liver sinusoid with a microfluidic endothelial-like barrier for primary hepatocyte culture. *Biotechnol. Bioeng.* **2007**, *97*, 1340–1346. [CrossRef]

17. Schimek, K.; Busek, M.; Brincker, S.; Groth, B.; Hoffmann, S.; Lauster, R.; Lindner, G.; Lorenz, A.; Menzel, U.; Sonntag, F.; et al. Integrating biological vasculature into a multi-organ-chip microsystem. *Lab Chip* **2013**, *13*, 3588–3598. [CrossRef]

18. Liu, X.F.; Yu, J.Q.; Dalan, R.; Liu, A.Q.; Luo, K.Q. Biological factors in plasma from diabetes mellitus patients enhance hyperglycaemia and pulsatile shear stress-induced endothelial cell apoptosis. *Integr. Biol.* **2014**, *6*, 511–522. [CrossRef]

19. Kamiya, A.; Bukhari, R.; Togawa, T. Adaptive regulation of wall shear stress optimizing vascular tree function. *Bull. Math. Biol.* **1984**, *46*, 127–137. [CrossRef]

20. Shao, J.; Wu, L.; Wu, J.; Zheng, Y.; Zhao, H.; Jin, Q.; Zhao, J. Integrated microfluidic chip for endothelial cells culture and analysis exposed to a pulsatile and oscillatory shear stress. *Lab Chip* **2009**, *9*, 3118–3125. [CrossRef]

21. Rutkowski, J.M.; Swartz, M.A. A driving force for change: Interstitial flow as a morphoregulator. *Trends Cell Biol.* **2007**, *17*, 44–50. [CrossRef] [PubMed]

22. Haessler, U.; Teo, J.C.M.; Foretay, D.; Renaud, P.; Swartz, M.A. Migration dynamics of breast cancer cells in a tunable 3D interstitial flow chamber. *Integr. Biol.* **2012**, *4*, 401–409. [CrossRef] [PubMed]

23. Bonvin, C.; Overney, J.; Shieh, A.C.; Dixon, J.B.; Swartz, M.A. A multichamber fluidic device for 3D cultures under interstitial flow with live imaging: Development, characterization, and applications. *Biotechnol. Bioeng.* **2010**, *105*, 982–991. [CrossRef] [PubMed]

24. Munson, J.M.; Bellamkonda, R.V.; Swartz, M.A. Interstitial flow in a 3D microenvironment Increases Glioma Invasion by a CXCR4-Dependent Mechanism. *Cancer Res.* **2013**, *73*, 1536–1546. [CrossRef]

25. Shieh, A.C.; Rozansky, H.A.; Hinz, B.; Swartz, M.A. Tumor cell invasion is promoted by interstitial flow-induced matrix priming by stromal fibroblasts. *Cancer Res.* **2011**, *71*, 790–800. [CrossRef]

26. Kim, S.; Chung, M.; Ahn, J.; Lee, S.; Jeon, N.L. Interstitial flow regulates the angiogenic response and phenotype of endothelial cells in a 3D culture model. *Lab Chip* **2016**, *16*, 4189–4199. [CrossRef]

27. Kim, S.; Chung, M.; Jeon, N.L. Three-dimensional biomimetic model to reconstitute sprouting lymphangiogenesis in vitro. *Biomaterials* **2016**, *78*, 115–128. [CrossRef]

28. Tomasek, J.J.; Gabbiani, G.; Hinz, B.; Chaponnier, C.; Brown, R.A. Myofibroblasts and mechano-regulation of connective tissue remodelling. *Nat. Rev. Mol. Cell Biol.* **2002**, *3*, 349–363. [CrossRef]

29. Zhang, B.; Korolj, A.; Lai, B.F.L.; Radisic, M. Advances in organ-on-a-chip engineering. *Nat. Rev. Mater.* **2018**, *3*, 257. [CrossRef]

30. Bhatia, S.N.; Ingber, D.E. Microfluidic organs-on-chips. *Nat. Biotechnol.* **2014**, *32*, 760–772. [CrossRef]

31. Park, S.-H.; Sim, W.Y.; Min, B.-H.; Yang, S.S.; Khademhosseini, A.; Kaplan, D.L. Chip-based comparison of the osteogenesis of human bone marrow- and adipose tissue-derived mesenchymal stem cells under mechanical stimulation. *PLoS ONE* **2012**, *7*, e46689. [CrossRef] [PubMed]

32. Torisawa, Y.; Mammoto, T.; Jiang, E.; Jiang, A.; Mammoto, A.; Watters, A.L.; Bahinski, A.; Ingber, D.E. Modeling hematopoiesis and responses to radiation countermeasures in a bone marrow-on-a-chip. *Tissue Eng. C Meth.* **2016**, *22*, 509–515. [CrossRef] [PubMed]

33. Bilek, A.M.; Dee, K.C.; Gaver, D.P. Mechanisms of surface-tension-induced epithelial cell damage in a model of pulmonary airway reopening. *J. Appl. Physiol.* **2003**, *94*, 770–783. [CrossRef] [PubMed]

34. Vlahakis, N.E.; Schroeder, M.A.; Limper, A.H.; Hubmayr, R.D. Stretch induces cytokine release by alveolar epithelial cells in vitro. *Am. J. Physiol.* **1999**, *277*, L167–L173. [CrossRef] [PubMed]

35. Tschumperlin, D.J.; Oswari, J.; Margulies, A.S. Deformation-induced injury of alveolar epithelial cells: Effect of frequency, duration, and amplitude. *Am. J. Respir. Crit. Care Med.* **2000**, *162*, 357–362. [CrossRef] [PubMed]

36. Yalcin, H.C.; Perry, S.F.; Ghadiali, S.N. Influence of airway diameter and cell confluence on epithelial cell injury in an in vitro model of airway reopening. *J. Appl. Physiol.* **2007**, *103*, 1796–1807. [CrossRef] [PubMed]

37. Huh, D.; Leslie, D.C.; Matthews, B.D.; Fraser, J.P.; Jurek, S.; Hamilton, G.A.; Thorneloe, K.S.; McAlexander, M.A.; Ingber, D.E. A human disease model of drug toxicity–induced pulmonary edema in a lung-on-a-chip microdevice. *Sci. Transl. Med.* **2012**, *4*, 159ra147. [CrossRef]

38. Dixon, J.B.; Raghunathan, S.; Swartz, M.A. A tissue-engineered model of the intestinal lacteal for evaluating lipid transport by lymphatics. *Biotechnol. Bioeng.* **2009**, *103*, 1224–1235. [CrossRef]

39. Bavli, D.; Prill, S.; Ezra, E.; Levy, G.; Cohen, M.; Vinken, M.; Vanfleteren, J.; Jaeger, M.; Nahmias, Y. Real-time monitoring of metabolic function in liver-on-chip microdevices tracks the dynamics of mitochondrial dysfunction. *Proc. Natl. Acad. Sci. USA* **2016**, *113*, E2231–E2240. [CrossRef]

40. Jang, K.-J.; Mehr, A.P.; Hamilton, G.A.; McPartlin, L.A.; Chung, S.; Suh, K.-Y.; Ingber, D.E. Human kidney proximal tubule-on-a-chip for drug transport and nephrotoxicity assessment. *Integr. Biol.* **2013**, *5*, 1119–1129. [CrossRef]

41. Kingsmore, K.M.; Logsdon, D.K.; Floyd, D.H.; Peirce, S.M.; Purow, B.W.; Munson, J.M. Interstitial flow differentially increases patient-derived glioblastoma stem cell invasion via CXCR4, CXCL12, and CD44-mediated mechanisms. *Integr. Biol.* **2016**, *8*, 1246–1260. [CrossRef] [PubMed]

42. Shachar, M.; Benishti, N.; Cohen, S. Effects of mechanical stimulation induced by compression and medium perfusion on cardiac tissue engineering. *Biotechnol. Prog.* **2012**, *28*, 1551–1559. [CrossRef] [PubMed]

43. Nakashima, Y.; Yang, Y.; Minami, K. Microfluidic device for monitoring and evaluation of intracellular mechanostress responses. In Proceedings of the 15th International Conference on Biomedical Engineering, Singapore, 4–7 December 2013; Goh, J., Ed.; Springer: Cham, Switzerland, 2014; pp. 868–871.

44. Stucki, J.D.; Hobi, N.; Galimov, A.; Stucki, A.O.; Schneider-Daum, N.; Lehr, C.-M.; Huwer, H.; Frick, M.; Funke-Chambour, M.; Geiser, T.; et al. Medium throughput breathing human primary cell alveolus-on-chip model. *Sci. Rep.* **2018**, *8*, 14359. [CrossRef] [PubMed]

45. Polacheck, W.J.; Li, R.; Uzel, S.G.M.; Kamm, R.D. Microfluidic platforms for mechanobiology. *Lab Chip* **2013**, *13*, 2252–2267. [CrossRef]

46. Shin, S.R.; Zhang, Y.S.; Kim, D.-J.; Manbohi, A.; Avci, H.; Silvestri, A.; Aleman, J.; Hu, N.; Kilic, T.; Keung, W.; et al. Aptamer-based microfluidic electrochemical biosensor for monitoring cell-secreted trace cardiac biomarkers. *Anal. Chem.* **2016**, *88*, 10019–10027. [CrossRef]

47. Tourovskaia, A.; Fauver, M.; Kramer, G.; Simonson, S.; Neumann, T. Tissue-engineered microenvironment systems for modeling human vasculature. *Exp. Biol. Med.* **2014**, *239*, 1264–1271. [CrossRef]

48. Seo, H.-R.; Jeong, H.E.; Joo, H.J.; Choi, S.-C.; Park, C.-Y.; Kim, J.-H.; Choi, J.-H.; Cui, L.-H.; Hong, S.J.; Chung, S.; et al. Intrinsic FGF2 and FGF5 promotes angiogenesis of human aortic endothelial cells in 3D microfluidic angiogenesis system. *Sci. Rep.* **2016**, *6*, 28832. [CrossRef]

49. Kong, J.; Luo, Y.; Jin, D.; An, F.; Zhang, W.; Liu, L.; Li, J.; Fang, S.; Li, X.; Yang, X.; et al. A novel microfluidic model can mimic organ-specific metastasis of circulating tumor cells. *Oncotarget* **2016**, *7*, 78421–78432. [CrossRef]

50. Zeller, P.; Legendre, A.; Jacques, S.; Fleury, M.J.; Gilard, F.; Tcherkez, G.; Leclerc, E. Hepatocytes co-cultured with Sertoli cells in bioreactor favors Sertoli barrier tightness in rat. *J. Appl. Toxicol.* **2017**, *37*, 287–295. [CrossRef]

51. Chang, J.-Y.; Wang, S.; Allen, J.S.; Lee, S.H.; Chang, S.T.; Choi, Y.-K.; Friedrich, C.; Choi, C.K. A novel miniature dynamic microfluidic cell culture platform using electro-osmosis diode pumping. *Biomicrofluidics* **2014**, *8*, 044116.

52. Maschmeyer, I.; Hasenberg, T.; Jaenicke, A.; Lindner, M.; Lorenz, A.K.; Zech, J.; Garbe, L.-A.; Sonntag, F.; Hayden, P.; Ayehunie, S.; et al. Chip-based human liver–intestine and liver–skin co-cultures—A first step toward systemic repeated dose substance testing in vitro. *Eur. J. Pharm. Biopharm.* **2015**, *95*, 77–87. [CrossRef] [PubMed]

53. Moreno, E.L.; Hachi, S.; Hemmer, K.; Trietsch, S.J.; Baumuratov, A.S.; Hankemeier, T.; Vulto, P.; Schwamborn, J.C.; Fleming, R.M.T. Differentiation of neuroepithelial stem cells into functional dopaminergic neurons in 3D microfluidic cell culture. *Lab Chip* **2015**, *15*, 2419–2428. [CrossRef] [PubMed]

54. Wevers, N.R.; van Vught, R.; Wilschut, K.J.; Nicolas, A.; Chiang, C.; Lanz, H.L.; Trietsch, S.J.; Joore, J.; Vulto, P. High-throughput compound evaluation on 3D networks of neurons and glia in a microfluidic platform. *Sci. Rep.* **2016**, *6*, 38856. [CrossRef] [PubMed]

55. Patra, B.; Peng, C.-C.; Liao, W.-H.; Lee, C.-H.; Tung, Y.-C. Drug testing and flow cytometry analysis on a large number of uniform sized tumor spheroids using a microfluidic device. *Sci. Rep.* **2016**, *6*, 21061. [CrossRef]

56. Lee, H.; Kim, D.S.; Ha, S.K.; Choi, I.; Lee, J.M.; Sung, J.H. A pumpless multi-organ-on-a-chip (MOC) combined with a pharmacokinetic–pharmacodynamic (PK–PD) model. *Biotechnol. Bioeng.* **2017**, *114*, 432–443. [CrossRef]

57. Liu, C.; Abedian, R.; Meister, R.; Haasper, C.; Hurschler, C.; Krettek, C.; von Lewinski, G.; Jagodzinski, M. Influence of perfusion and compression on the proliferation and differentiation of bone mesenchymal stromal cells seeded on polyurethane scaffolds. *Biomaterials* **2012**, *33*, 1052–1064. [CrossRef]

58. Chan, E.K.; Sorg, B.; Protsenko, D.; O'Neil, M.; Motamedi, M.; Welch, A.J. Effects of compression on soft tissue optical properties. *IEEE J. Sel. Top. Quantum Electron.* **1996**, *2*, 943–950. [CrossRef]

59. Chen, X.; Guo, J.; Yuan, Y.; Sun, Z.; Chen, B.; Tong, X.; Zhang, L.; Shen, C.; Zou, J. Cyclic compression stimulates osteoblast differentiation via activation of the Wnt/β-catenin signaling pathway. *Mol. Med. Rep.* **2017**, *15*, 2890–2896. [CrossRef]

60. Hsieh, H.-Y.; Camci-Unal, G.; Huang, T.-W.; Liao, R.; Chen, T.-J.; Paul, A.; Tseng, F.-G.; Khademhosseini, A. Gradient static-strain stimulation in a microfluidic chip for 3D cellular alignment. *Lab Chip* **2013**, *14*, 482–493. [CrossRef]

61. Chen, K.; Vigliotti, A.; Bacca, M.; McMeeking, R.M.; Deshpande, V.S.; Holmes, J.W. Role of boundary conditions in determining cell alignment in response to stretch. *Proc. Natl. Acad. Sci. USA* **2018**, *115*, 986–991. [CrossRef]

62. Laurence, D.; Ross, C.; Jett, S.; Johns, C.; Echols, A.; Baumwart, R.; Towner, R.; Liao, J.; Bajona, P.; Wu, Y.; et al. An investigation of regional variations in the biaxial mechanical properties and stress relaxation behaviors of porcine atrioventricular heart valve leaflets. *J. Biomech.* **2019**, *83*, 16–27. [CrossRef] [PubMed]

63. Ali, M.S.; Wang, X.; Lacerda, C.M. The effect of physiological stretch and the valvular endothelium on mitral valve proteomes. *Exp. Biol. Med.* **2019**, *244*, 241–251. [CrossRef] [PubMed]

64. Huh, D.; Matthews, B.D.; Mammoto, A.; Montoya-Zavala, M.; Hsin, H.Y.; Ingber, D.E. Reconstituting organ-level lung functions on a chip. *Science* **2010**, *328*, 1662–1668. [CrossRef]

65. Musah, S.; Mammoto, A.; Ferrante, T.C.; Jeanty, S.S.F.; Hirano-Kobayashi, M.; Mammoto, T.; Roberts, K.; Chung, S.; Novak, R.; Ingram, M.; et al. Mature induced-pluripotent-stem-cell-derived human podocytes reconstitute kidney glomerular-capillary-wall function on a chip. *Nat. Biomed. Eng.* **2017**, *1*, 1–12. [CrossRef] [PubMed]

66. Song, J.W.; Munn, L.L. Fluid forces control endothelial sprouting. *Proc. Natl. Acad. Sci. USA* **2011**, *108*, 15342–15347. [CrossRef] [PubMed]

67. Carmeliet, P. Mechanisms of angiogenesis and arteriogenesis. *Nat. Med.* **2000**, *6*, 389–395. [CrossRef] [PubMed]

68. Tzima, E.; Irani-Tehrani, M.; Kiosses, W.B.; Dejana, E.; Schultz, D.A.; Engelhardt, B.; Cao, G.; DeLisser, H.; Schwartz, M.A. A mechanosensory complex that mediates the endothelial cell response to fluid shear stress. *Nature* **2005**, *437*, 426–431. [CrossRef]

69. DuFort, C.C.; Paszek, M.J.; Weaver, V.M. Balancing forces: Architectural control of mechanotransduction. *Nat. Rev. Mol. Cell Biol.* **2011**, *12*, 308–319. [CrossRef]

70. Blackman, B.R.; García-Cardeña, G.; Gimbrone, M.A., Jr. A new in vitro model to evaluate differential responses of endothelial cells to simulated arterial shear stress waveforms. *J. Biomech. Eng.* **2002**, *124*, 397–407. [CrossRef]

71. Galie, P.A.; Nguyen, D.-H.T.; Choi, C.K.; Cohen, D.M.; Janmey, P.A.; Chen, C.S. Fluid shear stress threshold regulates angiogenic sprouting. *Proc. Natl. Acad. Sci. USA* **2014**, *111*, 7968–7973. [CrossRef]

72. Sim, W.Y.; Park, S.W.; Park, S.H.; Min, B.H.; Park, S.R.; Yang, S.S. A pneumatic micro cell chip for the differentiation of human mesenchymal stem cells under mechanical stimulation. *Lab Chip* **2007**, *7*, 1775–1782. [CrossRef] [PubMed]

73. Huang, W.; Zhang, S.; Ahmad, B.; Kawahara, T. Three-motorized-stage cyclic stretching system for cell monitoring based on chamber local displacement waveforms. *Appl. Sci.* **2019**, *9*, 1560. [CrossRef]

74. Kasahara, K.; Kurashina, Y.; Miura, S.; Miyata, S.; Onoe, H. Shape deformation analysis of single cell in 3D tissue under mechanical stimuli. In Proceedings of the 2019 20th International Conference on Solid-State Sensors, Actuators and Microsystems Eurosensors XXXIII (Transducers Eurosensors XXXIII), Berlin, Germany, 23–27 June 2019; pp. 413–416.
75. Cho, S.; Islas-Robles, A.; Nicolini, A.M.; Monks, T.J.; Yoon, J.-Y. In situ, dual-mode monitoring of organ-on-a-chip with smartphone-based fluorescence microscope. *Biosens. Bioelectron.* **2016**, *86*, 697–705. [CrossRef] [PubMed]

 *micromachines*

Review

Engineered Liver-On-A-Chip Platform to Mimic Liver Functions and Its Biomedical Applications: A Review

Jiu Deng [1,†], Wenbo Wei [1,†], Zongzheng Chen [2,†], Bingcheng Lin [1,*], Weijie Zhao [1], Yong Luo [1] and Xiuli Zhang [3,*]

1   State Key Laboratory of Fine Chemicals, Department of Chemical Engineering, Dalian University of Technology, Dalian 116024, China; dengjiu@mail.dlut.edu.cn (J.D.); wwb1004@hotmail.com (W.W.); zyzhao@dlut.edu.cn (W.Z.); yluo@dlut.edu.cn (Y.L.)
2   Integrated Chinese and Western Medicine Postdoctoral research station, Jinan University, Guangzhou 510632, China; chenmond@foxmail.com
3   College of Pharmaceutical Science, Soochow University, Suzhou 215123, China
*   Correspondence: bclin@dicp.ac.cn (B.L.); zhangxl@suda.edu.cn (X.Z.);
    Tel.: +86-0411-8437-9065 (B.L.); +86-155-0425-7723 (X.Z.)
†   These authors have equally contribution to this work.

Received: 17 September 2019; Accepted: 3 October 2019; Published: 7 October 2019

Abstract: Hepatology and drug development for liver diseases require in vitro liver models. Typical models include 2D planar primary hepatocytes, hepatocyte spheroids, hepatocyte organoids, and liver-on-a-chip. Liver-on-a-chip has emerged as the mainstream model for drug development because it recapitulates the liver microenvironment and has good assay robustness such as reproducibility. Liver-on-a-chip with human primary cells can potentially correlate clinical testing. Liver-on-a-chip can not only predict drug hepatotoxicity and drug metabolism, but also connect other artificial organs on the chip for a human-on-a-chip, which can reflect the overall effect of a drug. Engineering an effective liver-on-a-chip device requires knowledge of multiple disciplines including chemistry, fluidic mechanics, cell biology, electrics, and optics. This review first introduces the physiological microenvironments in the liver, especially the cell composition and its specialized roles, and then summarizes the strategies to build a liver-on-a-chip via microfluidic technologies and its biomedical applications. In addition, the latest advancements of liver-on-a-chip technologies are discussed, which serve as a basis for further liver-on-a-chip research.

Keywords: liver-on-a-chip; drug hepatotoxicity; drug metabolism

---

1. Introduction

The liver is the largest intracorporeal organ in the human body and plays a predominant role in several pivotal functions to maintain normal physiological activities [1] such as blood sugar and ammonia level control, synthesis of various hormones, and detoxification of endogenous and exogenous substances [2]. Normally, the liver has a tremendous regenerative capacity to cope with physical and chemical damage. However, injury caused by adverse reactions to drugs (e.g., aristolochene and ibuprofen) and chronic diseases (e.g., viral and alcoholic hepatitis) may impair its ability to perform physiological functions [3,4].

Although in vivo models are commonly established in mammals to study liver functions, especially for pharmaceutical research, the accuracy of this kind of model is still unsatisfactory [5]. For example, roughly half of the drugs found to be responsible for liver injury during clinical trials did not result in any damage in animal models in vivo [6]. In addition, as a parenchymal organ, liver cells are continuously exposed to a variety of abundant exogenous substances. Moreover, it is inconvenient to observe highly dynamic biological processes in the current in vivo animal models.

Based on these facts, it is necessary to establish a reliable liver model in vitro for in-depth understanding of the physiological/pathological processes in the liver and the development of drugs for liver diseases. Currently, the liver models used for in vitro studies commonly include bioreactors (perfusion model of an isolated liver system) [7], 2D planar primary rat hepatocytes [8,9], 3D-printed liver tissue [10,11], liver organoids [12,13], and liver-on-a-chip systems [14–16]. To date, many previous reviews have discussed the differences in these models [17–20]. However, it is well known that liver-on-a-chip technology is innovative to manage liver microenvironments in vitro, and a variety of liver chips have emerged [18,20–22]. However, there is still no comprehensive review of the strategies to fabricate liver chips or their broad applications in various fields. The purpose of this review is to summarize the strategies to build liver-on-chips via microfluidic technologies and their applications.

We first introduce the physiological microenvironment of the liver, especially the cell composition and its specialized roles in the liver. We highlight the simulation objects of a liver-on-a-chip, including the liver sinusoid, liver lobule, and zonation in the lobule. Secondly, we discuss the general strategies to replicate human liver physiology and pathology ex vivo for liver-on-a-chip fabrication, such as liver chips based on layer-by-layer deposition. Third, we summarize the current applications and future direction. Finally, challenges and bottlenecks encountered to date will be presented.

2. Physiological Microenvironment of the Liver

2.1. Cell Types and Composition

The liver is composed of many types of primary resident cells such as hepatocytes (HCs), hepatic stellate cells (HSCs), Kupffer cells (KCs), and liver sinusoid endothelial cells (LSECs), which form complex signaling and metabolic environments. These cells perform liver functions directly and are connected to each other through autocrine and paracrine signaling. Below, we review each cell type and its contributions to liver functions along with their importance in the context of toxicity. The characteristics of each cell type are summarized in Table 1.

2.1.1. Parenchymal Cells

Parenchymal cells, also called hepatocytes, are highly differentiated large epithelial cells (20–30 μm) responsible for the major liver functions [23] such as metabolism of blood sugar, decomposition of ammonia, and synthesis of bile acids. They comprise ~60% of total cells and ~80% of the total mass in the liver [24]. The main function of hepatocytes is metabolism of both internal and external substances. With a large number of mitochondria (1000–2000/cells), peroxisomes (400–700/cells), lysosomes (~250/cells), Golgi complexes (~50/cells), and endoplasmic reticulum both rough and smooth, each hepatocyte acts as a metabolism factory [23]. Nonetheless, the metabolic capacity of each hepatocyte is not exactly the same because of different oxygen pressures, nutrient supplies, and hormone concentrations in the various zones of liver lobules. Another feature of hepatocytes is polarization [25]. Physiologically, spatially adjacent hepatocytes are closely arranged in cords to form liver plates with strong polarization characteristics that allow substances to enter from the blood for excretion with bile.

Drug metabolism is the most studied function of hepatocytes, including both phase I and II, even though non-parenchymal support contributes to xenobiotic metabolism. Cytochrome P450 (CYP 450) family members, terminal oxygenases distributed on the endoplasmic reticulum and mitochondrial inner membrane, are the critical phase I metabolic enzymes in drug metabolism of the liver. They also have important effects on cytokines and thermoregulation. In particular, subtypes CYP 1A2, 2A6, 2C9, 2C19, 2D6, 2E1, and 3A4 are involved in almost all aspects of drug phase I metabolism, including oxidation, reduction, hydrolysis, and dehydrogenation. Phase II modifications are mostly carried out by cytosolic enzymes termed transferases that allow drug excretion by the kidneys after sulfation or glucuronidation.

Table 1. Main cell types of the liver and their features.

Cell	Type	Diameter (µm)	Proportion (number)	Features
Parenchymal	-	-	-	-
hepatocytes	Epithelial	20–30	60%–65%	Large in size, abundant glycogen, mostly double nuclei.
Non-parenchymal	-	-	-	-
Kupffer cells	Macrophages	10–13	~15%	Irregularly shaped, mobile cells, secretion of mediators.
liver sinusoid endothelial cells	Epithelial	6.5–11	16%	SE-1, CD31, fenestrations, none basement membrane.
hepatic stellate cells	Fibroblastic	10.7–11.5	8%	Vitamin-storing, Distinct basement membrane.
Biliary Epithelial Cells	Epithelial	~10	Little	Containing unique proteoglycans, adhesion glycoproteins.

2.1.2. Hepatic Stellate Cells

HSCs, also called fat-storing cells, perisinusoidal cells, and lipocytes, reside within the space of Disse formed between hepatic cords and sinusoidal endothelial cells, which represent approximately 8% of hepatic cells. They have the same characteristics as fibroblasts and mainly play roles in vitamin A storage. Stellate cells play major roles in maintaining the morphology of LSECs and the progression of liver fibrosis. They exhibit various forms in different physiological environments, i.e., resting and activated states. Physiologically, stellate cells are in the resting state under normal conditions. However, when stellate cells are activated by changes in the microenvironment, such as alcohol intake and viral infections, the cells proliferate and migrate. The synthesis of collagen I and α-smooth muscle actin increases rapidly, leading to extensive extracellular matrix (ECM) deposition. Recent studies have shown that stellate cells are also involved in immune-mediated liver injury, causing secondary damage to the liver.

2.1.3. Hepatic Sinusoidal Endothelial Cells

LSECs are long and slender endothelial cells in direct contact with liver blood flow [26,27]. They represent a major fraction of non-parenchymal cells (~48%) with extended processes. In addition, LSECs express SE-1 and CD31 proteins abundantly and thus can be identified easily. Unlike vascular endothelial cells, LSECs not only acts as a physical barrier for blood circulation, they also have their own characteristics by lacking a basement membrane and rich in fenestrations [27]. However, these features are lost when a lesion occurs. LSECs express endothelial nitric oxide synthase (eNOS) protein, which are affected by blood flow shear and vascular endothelial growth factor (VEGF), thus adapting to changes in blood flow velocity and pressure in liver sinusoids. LSECs are also involved in the immune response of the liver, such as phagocytosis of particles and adhesion of immune cells. In addition, LSECs express toll-like receptors that detect exogenous substances and self-apoptotic products that trigger the inflammation pathway.

2.1.4. Kupffer Cells

Kupffer cells are macrophages that reside in the liver, accounting for approximately 80%–90% of all fixed macrophages in the body and about 15% of total liver cells [26]. They are predominantly localized in the lumen of hepatic sinusoids and are anchored to the surface of LSECs by long extended processes. KCs are irregular in shape and about 10–13 µm in size. The main function of KCs is to remove particulates and foreign matter from the portal vein by phagocytosis. In addition, KCs are closely related to homeostasis of the liver environment. They release many cytokines, such as TNF-α, IL-1, and IL-6, which are involved in immunomodulation. For example, tumor-necrosis factor alpha (TNF-α) released by KCs acts on LSECs, leading to fibrin deposition in liver tissue, which may cause ischemia and hypoxia [28,29]. Recent studies have found that KCs also participate in antigen presentation [30].

2.1.5. Biliary Epithelial Cells

Biliary epithelial cells are the main epithelial cells located in the bile duct wall with a diameter of about 10 µm. They are one of the few cell types rarely studied in the liver because of their small effect on liver functions. However, recent studies have shown that the bile excretion pathway of drugs is related to hepatotoxicity [31]. Moreover, these cells express multiple bile receptors, which may be of interest to study cholestasis-induced liver disease.

2.1.6. Other Non-Parenchymal Cells

In addition to the above five kinds of cells, there are many other kinds of cells in the liver, such as neutrophils, natural killer (NK) cells, and infiltrating macrophages [24]. Physiologically, the numbers of such cells are small, but when the liver develops inflammation, these cells enter the liver rapidly in large amounts, which should be considered under pathological conditions. The main function of these cells is to release a large number of cytokines and chemokines to regulate the liver. Increasing evidence has revealed that these cells are closely related to immune-mediated hepatotoxicity [32].

2.2. Simulation Objects of a Liver-on-a-Chip

2.2.1. Liver Sinusoid

Liver sinusoid, a lacuna between adjacent liver plates (Figure 1C), is the physiological microenvironment of the liver with strong permeability to exchange materials between liver cells and blood flow [33]. In addition to containing the main liver cell types, liver sinusoid has its own specific structure in which HCs and LSECs are separated by the sinusoidal space, and hepatic stellate cells and extracellular matrix fill the gaps between HCs and LSECs. KCs are not attached to the lumen of blood vessels. The vertices of HCs fuse with each other to form bile canaliculi. The characteristics of liver sinusoids are such that cells in the liver can be approximately seen as assembled layer-by-layer. There are a large number of liver sinusoid models in vitro to reconstruct the physiologically relevant and controlled environment of liver. To date, the most used strategies employ additional polycarbonate (PC) membranes or layering by a laminar.

2.2.2. Liver Lobule

The liver lobule, which is considered as the smallest functional unit of the liver, appears as a polygon of approximately 1.1 mm in diameter and 1.7 mm in length [23]. Each lobule consists of hepatocytes radiating from the central vein and are separated by vascular endothelial cells. There are about 1×10^3 lobules in each human liver. At the center of each hexagon, there is a large vein called the central vein. The corners of the hexagon contain three conduits, the hepatic portal vein, hepatic artery, and bile duct. They are characterized by blood concentrating from six corners to the center, while bile moves from the center to the outside. Cell capturing using traps and micropatterning methods are the conventional methods to reconstruct hepatic lobules. Recently, 3D printing has become another rapid and simple method [35].

2.2.3. Zonation in the Lobule

Liver zonation is an evolutionary optimized segregation of the broad liver functions into spatial, temporarily defined, and highly specialized zones [36]. In liver zonation, different pathways are carried out in different zones—even in single cells. As shown in Figure 1C, cells in the periphery of liver lobules are relatively rich in oxygen and glucose, resulting in relatively higher albumin and urea synthesis. In contrast, internal cells possess relatively higher glycolysis than cells in peripheral zones. Liver zoning also leads to differences in hepatotoxicity. As shown in Figure 1D, because of the lower

CYP activity in zone 1, less cells are damaged, while the oxygen content in zone 2 is lower and CYP activity is therefore enhanced, thereby showing greater hepatotoxicity [34]. Such heterogeneity and functional plasticity of the liver are survival strategies for each cell to perform simultaneously without affecting each other and to use resources efficiently.

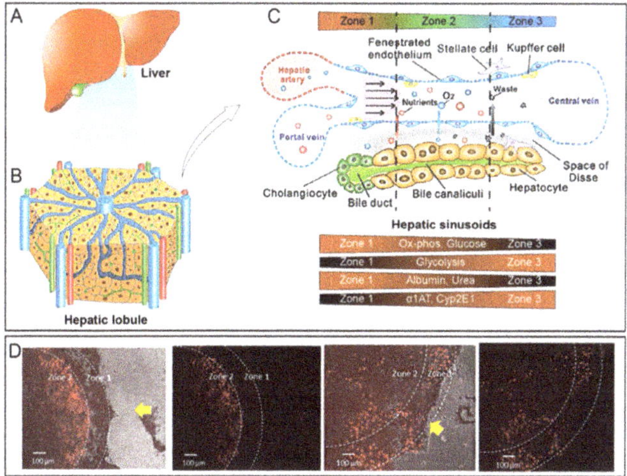

Figure 1. Cellular composition and anatomical microstructure of the liver. (**A**) Shape of the liver. It is a red-brown V-shaped organ divided into right and left parts by the hepatic artery, portal vein, hepatic vein, and bile ducts. (**B**) The liver lobule has a hexagonal shape with a diameter of about 1 mm and thickness of about 2 mm. (**C**) Zonation in the lobule. Reproduced with permission from [33]. (**D**) Zonal heterogeneity of acetaminophen-induced hepatotoxicity. The yellow arrow indicates the flow direction. Reproduced with permission from [34].

3. General Strategies for in Vitro Liver Models

As shown in Figure 2, the currently used in vitro liver models commonly include 2D planar primary hepatocytes [8,9], bioreactors (perfusion model of an isolated liver system) [7], 3D-printed liver tissue [37,38], 3D liver spheroids [39,40], and liver-on-a-chip. Table 2 compares the advantages and disadvantages of each model.

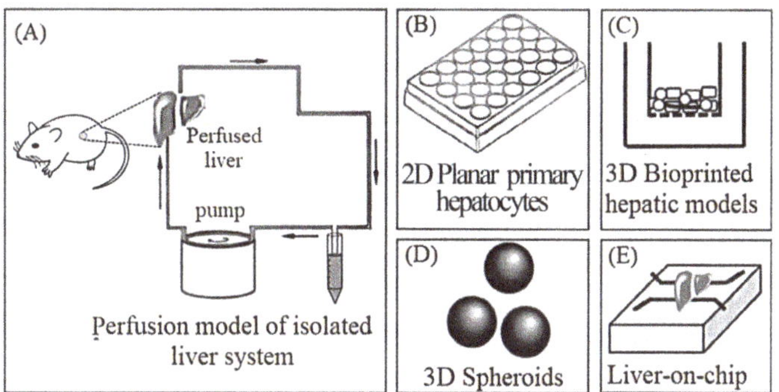

Figure 2. Liver models used commonly in vitro. (**A**) Perfusion model of an isolated liver system; (**B**) 2D planar primary rat hepatocytes; (**C**) 3D-printed liver tissue; (**D**) 3D spheroids; (**E**) liver-on-chip.

Because of the convenience and ease of handling, the conventional 2D culture of hepatocytes has been widely used as an in vitro liver model to study drug metabolism and cytotoxicity. However, most 2D-cultured hepatocytes lose their intrinsic biochemical cues and cell-cell communications necessary to maintain the physiological phenotype and cannot fully recapitulate liver-specific functions [34]. The perfusion model of an isolated liver system employs blood filtration, which is used to assist treatment of liver dysfunction and related diseases, and rarely used as a liver model to study pathology and physiology in vitro. Rapid development of 3D printing technology has provided a promising approach for in vitro liver models, which precisely controls the placement of cells, allowing the formation of separate hepatocyte and nonparenchymal cell (NPC) compartments. However, defects of the bulky dimension without flow mobility make it difficult to use for rapid, high throughput drug and toxicology evaluations. Furthermore, current printing accuracy cannot always allow placement of individual cells, which makes it impossible to reproduce physiological cues faithfully.

Primary hepatocyte aggregation culture forms 3D spheroids as a representative liver model to mimic early liver development. Many studies have shown that culturing hepatocytes within a 3D ECM-like matrix not only mimics the architecture of the liver, but also improves cell-to-cell and cell-to-matrix interactions and supports intrinsic liver functions, including production of albumin and urea as well as phase I and II drug metabolism [41]. Construction of liver organoids as a model system is an appealing experimental approach to exploit the physiological mechanisms that occur during organ development and regeneration [39]. The main techniques to generate organoids are the formation of spheroids by aggregation of cells and extracellular matrix components [40]. Such 3D liver organoid structures can meet the needs of the pharmacological and toxicological industry for drug screening [42]. The disadvantage of spheroids is that the cells are distributed randomly without formation of spatial organization i.e., liver spheroids neither possess the typical hepatic cord-like alignment of polarized hepatocytes nor sinusoids lined with endothelial cells reflecting the in vivo condition [17]. Furthermore, the sizes of spheroids are difficult to unify, and necrosis can occur in the center of large spheroids.

Recently, microfluidics-based cell culture devices have gained the most interest for biological and biomedical applications. The hepatocytes cultured within these devices under flow conditions allow for more frequent nutrient and waste exchange compared with conventional models. In addition, the hepatocytes better recapitulate liver-specific functions. For example, gradients of oxygen/hormones can be created to model zonal liver phenotypes. However, liver-on-chip fabrication requires many trivial operations, which will be described in detail in Aection 4.

In vitro liver models are critical for hepatology studies and drug development for liver diseases. An important aspect of these models is the cell source. There are three major types of cells to reconstruct liver tissue in vitro: primary human hepatocytes (PHHs), hepatic-derived cell lines, and stem cell-derived hepatocytes [43]. Table 3 compares the advantages and disadvantages of each cell source in detail. PHHs are considered as the gold standard of liver models in vitro. Isolated PHHs exhibit many intrinsic liver characteristics, including phase I and II metabolic enzyme activities, glucose metabolism, and ammonia detoxification. However, culturing PHHs on dishes has several issues such as lose of liver-specific functions, unsuitability for long-term studies, high costs, and donor variation.

Table 2. Advantages and limitations of in vitro liver models (note: these methods may have crossovers).

In Vitro Approaches	References	Advantages	Limitations
Monolayer	[8,9]	Easily manipulated, low-cost, good repeatability.	Cannot recapitulate in-vivo like cellular morphology and 3D microenvironment, loss of cell polarity, poor function.
Co-culture	[44–46]	Multi-cellular environment, cell-cell interaction, improve functions and longevity, cellular polarity.	Difficult isolation of NPCs, variations of NPCs, differentiation status and viability are varied depending on culture conditions.
3D culture	[11,47–50]	Recapitulation of 3D microenvironment and ECM properties, improve gene and protein expression, improve functions and longevity, cellular polarity.	Complicated methods of culture. Necrotic regions within 3D cellular models caused by oxygen diffusion.
Spheroids	[41,46,51]	In vivo-like microenvironment, cellular interaction, maintain liver-specific functionality over long term culture, enhanced CYP 450 and transporter expression, formation of secondary structure (e.g., bile canalicular-like structure).	Spheroid size limitation (~200 μm) and variations, necrotic cores, Oxygen and nutrient diffusion through cellular aggregates.
Liver-on-a-chip	[35,52–58]	Dynamic microenvironment, suitable for co-culture, 3D culture, and spheroid, improve liver-specific, functionality, enhanced CYP 450 and transporter expression formation of secondary structure, pattern cells spatially, high through put and low cost.	Complicated methods of operate chip and culture cell in the chip, required perfusion systems, non-specific binding of drugs to chip materials, may wash away molecules in the chamber under perfusion, no standard yet.

Hepatic cell lines, such as HepG2 (derived from human hepatocellular carcinoma of a 15-year-old male) and HepaRG (terminally differentiated hepatic cells derived from a hepatic progenitor cell line of a female hepatocarcinoma), have been widely used in toxicological investigations, because they have a stable phenotype, are essential for drug metabolism and toxicity responses, and are easily manipulated with unlimited proliferation [59,60]. For example, HepG2 cells have been used for toxicological and pharmacological research since the 1970s. However, compared with PHHs, the cells lines cannot represent the phenotype of in vivo hepatocytes and their drug reactions are inaccurate because of the low activities of CYP450 and transporters such as organic anion transporting polypeptide and sodium taurocholate co-transporting polypeptide [18]. Therefore, hepatic-derived cell lines are only suitable for the early stages of drug safety and screening assessment.

Stem cell-derived hepatocytes have become a new alternative liver cell source [61]. Du et al. employed induced pluripotent stem cells (iPSCs) differentiated into hepatocytes and endothelial cells, and then encapsulated them in fibers to form liver tissue-like constructs [51]. In addition, the 3D-aggregated stem cells can be differentiated into liver organoids with stable functions including albumin secretion, liver-specific gene expression, urea production, and metabolic activities [33,34,62]. However, manipulation of stem cells is not as simple as that of cell lines and requires specific induction factors during a >15-day culture period to obtain differentiated liver cells. Moreover, hepatocytes differentiated from iPSCs show less albumin secretion, exhibit dramatically reduced CYP450 activity, and express immature markers such as alpha-feto protein. Therefore, the maturation of organoids is hardly comparable with that of organoids formed by PHHs [18,63,64].

It has been demonstrated that culturing hepatocytes with non-parenchymal cells increases functionality and longevity. In addition, coculture provides multiple types cellular interactions and recovers cellular polarity, which more similar to in vivo conditions. Moreover, coculture of multiple types of cells forms many typical liver structures such as sinusoid [17], lobule [31], and biliary systems [34].

Table 3. Advantages and limitations of cells used in liver-on-chips.

Cell Type	Advantages	Limitations
Primary hepatocytes (human, rat)	Liver intrinsic characteristics, including phase I and II metabolic enzyme activity, glucose metabolism, ammonia detoxification	Losing liver specific function; unsuitable for long-term; high cost; donor variation, difficult isolation
Hepatic-derived cell lines (HepG2, HepaRG, C3A)	Unlimited lifespan; easily manipulated; stable phenotype; essential for drug metabolism and toxicity response.	Drug reaction are inaccurate; low metabolic competence and rapid loss of expression of liver-specific enzymes/transporters.
Stem cell induces hepatocytes	A stable source of hepatocytes; liver organoid; stable functions including albumin secretion, liver-specific gene expression, urea production and metabolic activity.	Hardly manipulated; required specific induce factor; high cost; insufficient maturate.

4. Liver-On-A-Chip Technology

Mimicking the liver in vitro remains a great challenge. Even coculture systems hardly simulate the complexity of the liver, because different types of cells are mixed and seeded randomly in coculture, which cannot form the complex cellular architecture or manipulate cell-to-cell interactions. However, rapid development of microfabrication and microfluidic technologies has provided a promising approach to establish microscale functional liver constructs on a chip. Moreover, microfluidic devices have many attractive advantages compared with conventional culture. For example, a microfluidic device can easily generate a concentration gradient, control cellular spatial distribution, and provide a flow environment. Therefore, researchers have developed various strategies to build a liver-on-a-chip via microfluidic technology for biological and biomedical applications. Table 4 summarizes the current general methods to build liver-on-chips and their advantages and disadvantages.

4.1. Liver Chips Based on 2D Planar Culture

Monolayer culture, also called 2D planar culture, has been widely used in the early stages of drug screening because it is easy to handle and amenable to screening large numbers of compounds in a short amount of time. However, mounting evidence indicates that 2D planar culture of hepatocytes results in rapid loss of hepatic marker expression and phenotypes within hours and is unsuitable for long-term culture. Moreover, cell-cell communication between hepatocytes and non-parenchymal cells improves liver functions, whereas these interactions are weak in 2D planar culture. To mimic the complexity of interactions of each cell type of the liver, researchers have developed methods for patterning and coculturing hepatocytes with other cell types on 2D substrates. Micropatterned substrates not only alter the distribution of different types of cells in a controllable manner, but also provide suitable biochemical cues for both parenchymal and non-parenchymal cells. For example, Ho et al. designed hepatic lobule arrays to pattern and coculture HepG2 cells and human umbilical vein endothelial cells (HUVECs) [65]. By manipulating dielectrophoresis, the original randomly distributed cells in the microfluidic chamber were able to form the desired pattern, which mimicked the morphology of a lobule. In addition, they found that the activity of the CYP450 enzyme was 80% higher compared with non-patterned HegG2 cells after two days of culture. However, the surface characteristics of 2D materials, such as stiffness and hydrophilicity, also affect the phenotype and function of hepatocytes. Therefore, the development of suitable materials has become the direction of 2D planar culture.

4.2. Liver Chips Based on Matrixless 3D Spheroid Culture

Aggregating hepatocytes into a 3D spheroid is another conventional method and promising in vitro model for hepatic metabolism and cytotoxicity research. Hanging drop technology facilitates cell aggregation, resulting in spheroids [53,66]. Aeby et al. used hanging hydrogel drops to form primary human liver microtissues for more than nine days [67]. In addition, Boos et al. combined primary human liver microtissues with embryoid bodies in the same hanging drop platform and found that the metabolites of primary human liver microtissues were directly transported to the EBs [68].

The cell-repellent plate method is commonly used to self-assemble liver cell spheroids [59–61,69]. Desai et al. used magnetic nanoparticles to modify PHHs and then applied magnetic force to rapidly assemble and easily handle the spheroids [59]. Moreover, in toxicological applications, by integrating microsensors with a liver-on-a-chip, the metabolic parameters and cell viability of a single spheroid could be monitored without microscopy [60]. Furthermore, the established human liver spheroid model could be used to study other liver diseases. For example, 3D spheroids of cocultured HepaRG and HSCs enabled testing of drug-induced liver fibrosis in vitro. After applying drugs, these cocultured spheroids presented HSC-specific gene expression and fibrotic features, including HC activation, and collagen secretion and deposition [69]. In addition to static culture, Tostoes et al. showed that human hepatocyte spheroids maintained liver-specific protein synthesis, CYP450 activity, and phase II and III drug-metabolizing enzyme gene expression and activity in a perfusion bioreactor system for two to four weeks [41].

Alternatively, a microwell array can also be applied to generate 3D spheroids in a high-throughput and controllable manner [33,51]. For example, Miyamoto et al. showed that hundreds of uniform HepG2 spheroids were generated with a TASCL (tapered stencil for cluster culture) device [51]. The size of the spheroids relied on the size of the microwell. In addition, the spheroids showed high viability and an albumin secretion activity. Moreover, the liver chip could be fabricated with a perfusion chamber that provided a fluidic shear stress biomimetic microenvironment. Subsequently, Ma et al. designed a reversible concave microwell-based polydimethylsiloxane (PDMS)-membrane-PDMS sandwich multilayer chip [33]. Their results showed that 1080 HepG2/C3A cell spheroids could be perfused in parallel in long-term culture on the chip, which significantly improved the establishment of cell polarity and enhanced liver-specific functions and metabolic activities.

Inspired by lattice growth mechanisms in materials science, Weng et al. designed a liver chip with a hexagonal culture chamber to mimic physiology and control the assembly of liver cells into an organotypic hierarchy [34]. First, they deposited primary liver cells (PLCs) onto a micropatterned hydrophobic PDMS membrane. The cell-coated membrane was then enclosed within the hexagonal culture chamber. The medium was introduced by flow in the chamber from each corner of the hexagonal chamber, which simulated the portal vein function. They found that the PLCs formed a scaffold-free hierarchical tissue and represented complex functional dynamics at the tissue level, such as dose effects of acetaminophen-induced hepatotoxicity.

4.3. Liver Chips Based on Matrix-Dependent 3D Culture

In liver tissue engineering, an ECM-like scaffold is required to facilitate cell adhesion, support cell growth, and enhance cell-matrix interactions. Therefore, various ECM components (natural and synthetic) have been applied to liver chips to improve their liver functions for pharmaceutical and cytotoxicity applications. Researchers have used an ECM within a liver chip to maintain and mimic the native microenvironment. For example, Toh et al. developed a multiplexed microfluidic 3D hepatocyte chip for in vitro drug toxicity testing [62]. The chip was coated with collagen, a natural ECM component, to support hepatocyte adhesion and growth (Figure 3a). They found that hepatocytes cultured in the coated chip showed both cell-cell and cell-ECM interactions, and maintained hepatocyte synthetic and metabolic functions.

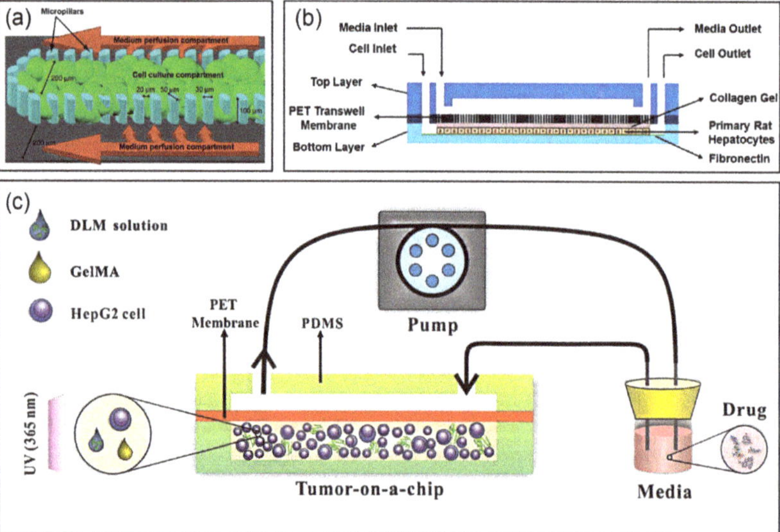

Figure 3. Strategy to build a liver-on-a-chip in matrix-dependent 3D culture: (**a**) liver cell culture channel of the multiplexed cell culture chip. Reproduced with permission from [62]. (**b**) Cross-sectional view of the assembled device showing that hepatocytes in collagen gel are introduced and cultured in the bottom layer and growth medium is introduced through the top layer. Reproduced with permission from [70]. (**c**) Schematic of HepG2-laden decellularized liver matrix with gelatin methacryloyl (DLM-GelMA) in a microfluidic device. Hydrogel precursors are injected into a microfluidic device using a pipette and photopolymerized by UV exposure to form HepG2-laden DLM-GelMA for subsequent drug screening. Reproduced with permission from [71].

Hepatocytes have also been introduced into liver chips with a pre-gel ECM component solution [54,70,72]. Hegde et al. demonstrated a method to culture hepatocytes in a collagen sandwich configuration in which hepatocytes were immobilized at the bottom chamber and the top chamber remained open for perfusion, as shown in Figure 3b [70]. Their results demonstrated that hepatocytes within the chip exhibited not only higher albumin and urea secretion, but also higher collagen secretion under perfusion. Interestingly, hepatocytes showed a well-connected cellular network with bile canaliculus formation over two weeks of perfusion culture. Similarly, Jang et al. also showed that Matrigel-embedded HepG2 cells formed a bile canaliculus structure over 14 days of perfusion culture [72]. Moreover, they reported HepG2 cell cultivation with an indirect flow but without a physical barrier. These data demonstrated that a microfluidic chip improves and stabilizes hepatic functions by mimicking an in vivo hepatocyte environment. Furthermore, Bavli et al. extended liver chip applications for real-time monitoring of mitochondrial respiration by co-encapsulating oxygen-sensing beads in collagen [54]. The HepG2/C3A cells formed 3D aggregates in the microfluidic chip, which provided native-like physiological shear forces and a stable oxygen gradient. These results showed that their system permitted detection of minute shifts from oxidative phosphorylation to glycolysis or glutaminolysis, which demonstrated the unique advantage of organ-on-chip technology.

Hydrogels have various intrinsic critical features to mimic native mechanical and structural cues that promote cell adhesion, proliferation, and differentiation. For example, Christoffersson et al. developed a modular and flexible hyaluronan and poly(ethylene glycol) (HA-PEG) hydrogel in which the mechanical properties and hydrogelation kinetics could be conveniently tuned by modulating the degree of crosslinking and temperature, respectively [73]. HepG2 cells or iPSC-derived hepatocytes (hiPS-HEPs) were encapsulated in the HA-PEG hydrogel in a perfusion device to enable implementation to a liver-on-a-chip. In addition, they compared the HA-PEG hydrogel with agarose and alginate. HepG2 cells encapsulated within all hydrogels formed spheroids with high viability, and the albumin and urea secretion were the highest in alginate hydrogels. Furthermore, hiPS-HEPs migrated and grew in 3D within Arg-Gly-Asp (RGD) peptide-modified HA-PEG hydrogels and showed increased viability and higher albumin secretion compared with other hydrogels. Zhu et al. synthesized state-specific liver microtumors using thermal-sensitive hydrogels with tailored stiffness and 3D scaffolds [74]. Their results indicated a close relationship between tissue biomechanics and drug efficacy, which provided a powerful tool for discovery and optimization of tissue-specific stroma-reprogrammed combinatorial therapy.

The bottom-up tissue engineering approach creates relatively bionic 3D tissues containing multiple types of hierarchically assembled cells. 3D liver spheroids can be generated easily by various methods, as described above. However, the major limitation of spherical morphology is a uniform supply of oxygen and nutrients, where cells located at the center are prone to die or lose their functions because of the hypoxic environment. Therefore, researchers have developed a cell-laden microfiber strategy for effective delivery of oxygen and nutrients via molecular diffusion. Yajima et al. packed HepG2 cells into the core of sandwich-type anisotropic microfibers. In addition, vascular endothelial cells were seeded on the fiber surface to form vascular network-like conduits between fibers. These cell-laden microfibers were further cultivated in a perfusion chamber. HepG2 cells within the microfiber not only exhibited high viability and functions, but also mimicked the structure of the hepatic lobule and sinusoidal in vitro [75].

Although hydrogel scaffolds resemble both structural and mechanical cues of the ECM, the artificial 3D systems still lack the appropriate native growth factors that promote cell growth and sustained cell functions. Therefore, a decellularized liver matrix (DLM) has become the most promising candidate for engineering native-like liver tissue. Lu et al. developed a biomimetic 3D liver tumor-on-a-chip with a DLM and gelatin metharcyloyl (GelMA) in a microfluidic 3D dynamic cell culture system (Figure 3c) [75]. As expected, the liver-on-a-chip integrated with DLM-GelMA better recapitulated the tumor microenvironment, such as essential scaffold proteins, growth factors, stiffness, and shear stress. Moreover, it demonstrated dose-dependent responses to the toxicity of acetaminophen and sorafenib.

4.4. Liver Chips Based on Layer-by-Layer Deposition

As shown in Figure 1C, the liver sinusoid, a functional repetitive microvascular unit that is formed by the sinusoidal wall composed of endothelial cells connected to the portal vein and hepatic artery, has unique structural characteristics. In detail, HCs and LSECs are separated by the hepatic sinusoidal space, HSCs and extracellular matrix fill the gap, and mainly KCs are free in the vascular lumen. This layered structure of each type of cell allows building liver-on-chips by layer-by-layer deposition. Therefore, researchers have developed a layer-by-layer cell-coating technique to create in vitro 3D vascularized tissue. For example, Sasaki et al. coated cells with layer-by-layer nanofilms of fibronectin and gelatin, and the coated cells reconstructed homogenous, dense, well-vascularized liver tissue with high a cellular function (albumin production) and cytochrome P450 activity [76]. Ahmed et al. designed an approach to seed hepatocytes, endothelial cells, and stellate cells sequentially on hollow fiber membranes [44]. The cells attached on the fiber surface, self-assembled, and formed tissue-like structures around and between fibers, which synthesized albumin and urea for 28 days.

Owing to the flexibility of microfabrication and the microfluidic technique, a multichannel microfluidic chip has become a promising platform to recapitulate the critical features of the liver sinusoid. Mi et al. constructed a liver sinusoid based on the laminar flow on a chip [77]. Specifically, HepG2 cell- and HUVEC-laden collagens were synchronously injected into the microfluidic chip. Then, by taking advantage of laminar flow, the two collagen layers formed a clear borderline, and the HUVECs in the collagen self-assembled into a monolayer by controlling cell density and injection of a growth factor.

Kang et al. reported a dual channel microfluidic platform, where primary hepatocytes and endothelial cells were cocultured within the channel to mimic the architecture of the liver sinusoid [45]. Hepatocytes maintained their normal morphology under continuous perfusion and produced urea for at least 30 days. In addition, the sinusoid-on-a-chip could be used to analyze replication of the hepatitis B virus. Rennert et al. established a liver organoid integrating all major types of cells (hepatocytes, stellate cells, endothelial cells, and macrophages) in a perfused biochip that was used a porous membrane to mimic the space of Disse. Endothelial cells and macrophages were seeded on the top side of the membrane, and hepatocyte-like HepaRG and stellate cells were cocultured on the opposite side of the membrane [78]. The liver organoid displayed clear differentiation and structural reorganization. Moreover, perfusion increased hepatobiliary secretion of 5(6)-carboxy-2′,7′-dichlorofluorescein and enhanced hepatocyte microvillus formation. More evidence demonstrated that the artificial liver sinusoid maintained high viability and in vivo-like morphology in long-term perfusion culture. In addition, the flow rate was not only related to albumin and urea responses, but also enhanced HGF production and CYP450 metabolism [79–81]. Furthermore, Deng et al. used all cell lines (HepG2, LX-2, EAhy926, and U937 cells) to build a liver sinusoid-on-a-chip with artificial liver blood flow and biliary efflux flowing in the opposite direction (Figure 4a). The all cell line liver chip was used to test the hepatoxicity of acetaminophen with other drugs. The results were similar to the "gold standard" primary hepatocyte plate model, indicating that the all cell line liver chip provided an alternative approach to investigate drug hepatoxicity and drug-drug interactions [31].

Another advantage of an organ-on-a-chip is that it can achieve real-time monitoring by integrating with sensors. As shown in Figure 4b, Moya et al. used an inkjet-printed oxygen sensor with a liver-on-a-chip to evaluate metabolic activity in real-time [82].

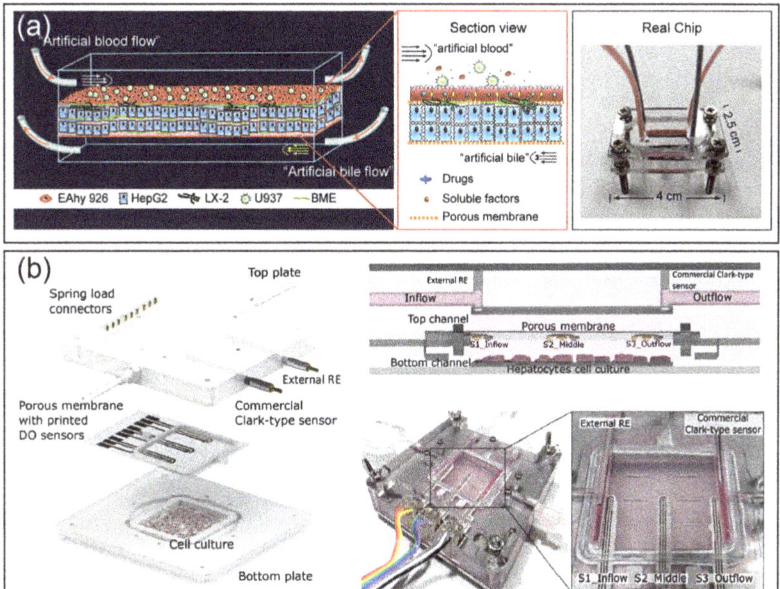

Figure 4. Establishment of a liver-on-a-chip using layer-by-layer deposition. (**a**) Schematic of the liver sinusoid structure and LSOC microdevice. Reproduced with permission from [31]. (**b**) Schematic, cross-section, and real image of an oxygen sensor-integrated liver chip. Reproduced with permission from [82].

4.5. Liver Chips Based on 3D Bioprinting

Rapid development of 3D printing technology has provided a promising approach for liver tissue engineering, which facilitates automated and high-throughput fabrication of precisely controlled 3D architectures. 3D bioprinting can produce anatomically accurate liver anatomy including the specific spatial structure and vascular network of the liver. The unique aspects of 3D bioprinting techniques are making them increasingly popular tools to manufacture in vitro liver models to study liver diseases and screen drugs. For example, Noroma et al. bioprinted human liver constructs comprising primary hepatocytes, hepatic stellate cells, and endothelial cells to model methotrexate- and thioacetamide-induced liver injury leading to fibrosis [83]. After exposure to these compounds, liver injury was detected, including hepatocellular damage, and deposition and accumulation of fibrillar collagens, which indicated that the 3D-bioprinted liver recapitulated compound-induced liver injury responses. Similarly, Nguyen et al. found that 3D-bioprinted liver tissue not only effectively modeled drug-induced liver injury, but also distinguished between highly related compounds with differential profiles [84].

Table 4. Summary of the strategies used for liver-on-chip fabrication.

Strategies	References	Characteristics	Culture Period	Advantages	Disadvantages
Liver chip based on 2D planar culture	[65]	Pattern or capture hepatocytes in 2D form; co-culture with non-parenchymal cells.	Short term	Relatively easy and fast; suitable for high throughput screening.	No polarization; low cell-cell communication; depended on the nature of substrate.
Liver chip based on matrixless 3D spheroid culture	[33,41,51,59–61]	Hepatocytes form spheroid spontaneously, due to gravity or modification of material surface; also suitable for co-culture.	Medium to long term	Scaffold-free; easy to achieve mass production of uniform size; good part form for stem cell differentiation	Needs special technology, such as cell-repellent plate and hanging drop technique.
Liver chip based on matrix-dependent 3D culture	[54,62,70–75]	Encapsulate cells within a three-dimensional (3D) matrix, such as hydrogel, BME and collagen, which replicates the supportive functions of the extracellular matrix.	Long term	Provide support and fixation for cells; enhanced cell-cell and cell-matrix interaction; conducive to cell adhesion and regulate dynamic cue of cells	Dependent on matrix, such as stability, stiffness; batch-to-batch variability; potential immunogenicity and presence of biological contaminants; unpredictable effects on signaling pathways.
Liver chip based on layer-by-layer deposition	[33,44,45,76,80,85]	Pattern hepatocytes and nonparenchymal cells lay by lay by porous membrane or 3D printing technology, etc.	Long term	Easy to control the position of cell layers to mimic the distribution of liver cells; forming tightly connected endotheliocytes for perfusion; hepatocyte polarization and angiogenesis	Not suitable for organs with unclear cell stratification; depends on other auxiliary tool, such as membrane and bio-ink.
Liver chip based on 3D bioprinting	[10,11,84,86–88]	Cells and extracellular matrix are laid out according to a preset path through a 3D printer in the form of additive manufacturing.	Long term	Easy to construct complex 3D biological microscale structures with various cell types and biomaterials; time save and high throughput	Limited by printing accuracy, it is difficult to control individual cells; the properties of printed materials are not optimized enough.
Liver chip-based cell microarrays such as microwell systems	[74,89,90]	Seed cells in an array of well plates.	Medium to long term	High throughput; miniaturize and parallelize.	Lack of spatial distribution and cellular interactions of cells in vivo.
Liver chip-based hanging drops	[68,91]	Form 3D micro-tissues of cells (one type or multi-types) by hanging cells in drop.	Medium term	Controllable and reproducible spheroid formation; no need to use scaffold; each drop served as a culture compartment for a single microtissue that was suitable for high throughput screening.	Not suitable for long-term culture for chronic toxicity and chronic liver disease.

Even though 3D-bioprinted liver tissue has been considered as an ideal liver model in vitro, a printed 3D liver analogue in a conventional plate or transwell cannot provide the dynamic perfusion condition. To overcome this limitation, a bioprinted liver was further integrated with microfluidic and perfusion devices to form a liver-on-a-chip. For example, HepG2/C3A spheroids were printed with GelMA ink and perfusion cultured in a bioreactor chamber for 30 days. The printed cell spheroid presented liver-specific functions including secretion of albumin, alpha-1, antitrypsin, transferrin, and ceruloplasmin [86]. Massa et al. developed a perfusable vascularized 3D liver construct via a sacrificial bioprinting technique [92]. Agarose fibers were printed in per-polymerized GelMA, which encapsulated HepG2/C3A cells. After GelMA polymerization, the agarose fibers were removed and HUVECs were seeded within the hollow channel to form a vascularized construct. The encapsulated HepG2/C3A cells exhibited high viability within the vascularized construct, demonstrating a protective role of the introduced endothelial cell layer. In another study, Grix et al. printed a complex cell-laden lobule with a hollow channel system using a stereolithographic bioprinting approach [88]. This printed liver organoid also showed higher albumin and cytochrome P450 expression compared with the monolayer control over 14 days of cultivation. Furthermore, Lee et al. generated a 3D liver-on-a-chip with multiple cell types using decellularized ECM bio ink and integrated it with a microfluidic device containing vascular and biliary fluidic channels [10]. Their results demonstrated that the liver functionalities were significantly enhanced by the formation of the biliary system on a chip, which becomes an effective potential candidate for drug discovery.

4.6. Liver Chips Based on Other Technologies

In addition to the above techniques, other methods can be applied to the manufacture of liver-on-chips, such as cell microarrays, microwell systems, and hanging drops [90]. The most notable feature of cell microarrays and microwell systems is the high throughput for large-scale screening of drugs at the early stage. Micropillars and traps are common methods used to form a cell microarray. For example, Lee et al. [89] designed a microchip platform for 3D culture of Hep3B human hepatic cells. This device features 532 micropillars and corresponding microwells that combine for high-throughput assessment of compound hepatotoxicity. The device was demonstrated to be suitable for 3D cell encapsulation, gene expression, and rapid toxicity assessment. The hanging drop method is another approach for liver chip fabrication. Frey et al. [68] reported a microfluidic hanging drop platform that seamlessly integrated liver metabolism into an embryonic stem cell test. This device was operated by gravity-driven flow to ensure constant inter-tissue communication as well as rapid and efficient exchange of metabolites.

5. Applications of a Liver-on-a-Chip

In the past decades, liver-on-chips have been a useful tool for drug screening and toxicity testing, prediction of metabolism, establishment of liver disease models, and studying the interactions of multiple organs. Because a microfluidic approach aims at mimicking the physiological and pathological conditions, various liver chips have been described for 2D and 3D culture of hepatocytes alone or in coculture with other types of cells in the liver for long-term culture and to study toxicity, metabolism, and disease. Table 5 summarizes the typical applications of liver-on-a-chip systems.

5.1. Drug Screening and Toxicity Testing

Drug-induced liver injury remains a significant source of clinical attrition, restrictive drug labeling, and post-market withdrawal of therapeutics [93,94]. Because of the ability to mimic in vivo physiological parameters, organ-on-a-chip devices enhance hepatocyte functions to assess the cellular behavior responses to drugs, which offer an alternative to animal experimentation [94]. Toh and colleagues developed multichannel microfluidics with a 3D engineered microenvironment to maintain the synthetic and metabolic functions of hepatocytes [62]. The multiplexed channels allow simultaneous management of drug doses at a concentration gradient to hepatocytes, enabling prediction of hepatotoxicity in vitro. Yu and colleagues designed a perfusion-incubator-liver-chip for 3D spheroid culture with a tangential flow, which maintained cell viability for over 24 days [95]. Then, chronic drug responses to repeated dosing of diclofenac and acetaminophen were evaluated by this device. Based on the channel modification technique, a 3D liver sinusoid-mimicking model was established [55]. Three hepatotoxins were evaluated by this liver chip. IC50 values were closely aligned with the LD50 values in mice, thus demonstrating the hepatotoxicity testing effectiveness of the proposed liver model. Except for endpoint assays to assess drug toxicity, dynamic information can be used as well to assess a drug's mechanism of action. Bavli and colleges constructed a liver-on-chip device capable of being maintained for more than a month in vitro under physiological conditions to allow real-time analysis of minute shifts from oxidative phosphorylation to anaerobic glycolysis in the process of mitochondrial stress [54]. Through this microfluidic platform, the dynamics and strategies of cellular adaptation to mitochondrial damage induced by rotenone and troglitazone were revealed.

A liver-on-chip provides a credible tool for studying relatively complete drug action pathways. Prot and colleagues integrated transcriptomic, proteomic and metabolomics profiles by cultivating liver cells in microfluidic biochips treated with or without acetaminophen (APAP), which allowed a more complete reconstruction of APAP-induced injury pathways. In addition, the validity of the results was confirmed by comparisons with in vivo studies [96,97]. However, other than the liver, drug hepatotoxicity metabolism is also associated with other organs. Thus, an organ-on-chip makes it possible to study the interactions between multiple organs. A series of multiple organ chips have been established to mimic the physiological environmental systems of multiple organs, such as intestines with the liver [97], nephridium with the liver, and lungs with the liver [98].

5.2. Prediction of Metabolism

A goal of a liver-on-chip is to establish a hepatocyte microenvironment in vitro consistent with that in vivo. Compared with cell culture techniques, the comprehensive metabolic capacity of liver cells cultured on organ chips is enhanced greatly. For example, primary hepatocytes hosted in 3D heparin-coated microtrenches secreted high levels of albumin and urea for over four weeks [55]. In addition, a liver-on-chip provides a dynamic flow environment for hepatocytes to perform higher albumin synthesis and urea excretion (detoxification) compared with static cultures [79]. Precision-cut liver slices can also be incubated directly on a microfluidic-based biochip, and the metabolic rate was significantly improved by embedding slices in Matrigel-based microfluidic chips [99]. Another important function of the liver is participation in drug metabolism. A liver-organ-chip can be used to evaluate phase I and II metabolisms in the liver [79] and study the first pass metabolism of drugs by integrating a gut-like structure in the front end of the liver chip [100], which is difficult to reproduce in vitro by conventional cell culture systems. Zhou et al. developed a five-chamber microsystem—two for coculturing hepatocytes with HSCs and three other chambers integrating aptamer-modified electrodes to monitor secretion of transforming growth factor-β [101]. This microsystem is capable of monitoring paracrine crosstalk between two cell types communicating via signaling molecules.

In conclusion, liver-on-chips based on microfabrication technology make it feasible to establish a liver model closer to the actual physiological environment in vivo, which is characterized by coculturing multiple types of cells under physiological flow conditions, high metabolic activity of hepatocytes, and establishment of complex and reliable cellular microenvironments.

Table 5. Typical applications of liver-on-a-chip systems.

Application	Reference	Cells Used	Description	Experimental Specifications
Drug screening and toxicity testing	[95]	Primary rat hepatocytes	A perfusion-incubator-liver-chip (PIC) was designed for 3D rat hepatocyte spheroids culture; chronic drug response to repeated dosing of Diclofenac and Acetaminophen were evaluated in PIC.	PIC system structure, functionality and optimization; Maintenance of cell function in PIC; application of PIC-cultured hepatocytes in drug safety testing.
Prediction of metabolism	[100]	Caco-2; HepG2	A microfluidic chip consists of two separate layers for Caco-2 and HepG2 was designed; first pass metabolism of a flavonoid, apigenin was evaluated as a model compound.	Gut-liver chip design for cells proliferation and differentiation; Paracellular permeability of intestinal barrier; first pass metabolism of apigenin.
Establishment of liver disease models	[52]	HepDE19; cryopreserved PHH; HepG2	A 3D microfluidic PHH system permissive to HBV infection; This system enables the recapitulation of all steps of the HBV life cycle, replication of patient-derived HBV and the maintenance of HBV cccDNA.	HBV patient-derived viruses and infections; exogenous stimulation of KC suppresses HBV replication.
Fabrication of multi-organ on a chip	[102]	HepaRG; human primary hepatic stellate cells; prepuce	A system for the co-culture of human 3D liver spheroids with human gut barrier and skin toward systemic repeated dose substance testing.	Fourteen-day performance of liver-intestinal co-cultures; 14-day performance of liver-skin co-cultures; repeated dose substance exposure.

5.3. Establishment of Liver Disease Models

Based on the characteristic of organ-on-chips controlling the external microenvironment of cells, a series of models for liver diseases—including alcoholic liver disease [103], fatty liver disease, liver fibrosis, acute liver injury, and patient-specific liver diseases—were established in the form of a liver-on-chip.

In Lee's study, rat primary hepatocytes and HSCs were cocultured in a fluid activity chip to observe structural changes, which exhibited a decrease in hepatic functions with the increase in ethanol concentration [17]. To develop a human in vitro model of non-alcoholic fatty liver disease, HepG2 cells [104] and primary hepatocytes [105] were cultured under 2D and 3D perfused dynamic conditions with free fatty acid supplementation, respectively. The models allowed for sustained culture of hepatocytes in vitro, which were used to investigate FFA-induced intrahepatic triglyceride accumulation (steatosis), which initially leads to a benign condition but can progress to more advanced conditions of steatohepatitis and fibrosis. Primary hepatocytes and LSECs, which were isolated from control and cirrhotic humans, were cocultured on the chip to mimic the in vivo physiological sinusoidal environment [81]. A human, 3D, four-cell, sequentially layered, and self-assembled microfluidic liver model demonstrated the development and characteristics of early fibrotic activation induced by 30 nM methotrexate as indicated by the expression of alpha-smooth muscle actin and collagen, and increased stellate cell migration [106].

A liver-on-a-chip device has also been used to study host/pathogen interactions. A 3D microfluidic liver culture system was constructed to provide a valuable preclinical platform for hepatitis B virus (HBV) research, which is capable of recapitulating all steps of the HBV life cycle, including the replication of patient-derived HBV and maintenance of HBV cccDNA [52]. The pathological process may reflect the specificity and genetics of the patient. Therefore, the principle of disease therapy based on other non-specific models cannot be applied to everyone. In these examples, a liver-on-chip provides an advanced model that better preserves the liver phenotype and can employ different cell types critical to facilitate the development of personalized/targeted medicine. Primary hepatocytes and LSECs, which were isolated from control and cirrhotic humans, have also been cocultured on a chip to mimic the in vivo-specific physiological and pathology of the sinusoidal environment [81]. Schepers and colleges described perfusable liver chip-cultured 3D organoids consisting of primary and iPSC-derived cells from a patient of interest. In their study, organoids were encapsulated and cultured in C-trap architecture for at least 28 days [58]. This strategy can be applied to other microfluidic organ models, which provides an opportunity to query patient-specific liver responses in vitro. Another hiPSC-induced liver organoid-on-a-chip system also demonstrated a promising organoid-based liver chip platform with applications in precision medicine and disease modeling [107].

5.4. Fabrication of Multiple Organs on a Chip

As an organ, the liver does not perform its function alone in vivo, but undertakes higher physiological functions together with other organs or tissues. An organ-on-chip allows for crosstalk between multiple organs to be studied by connecting a liver chip with organotypic models. Another important application of liver-on-chips is the combination with other organ chips, called "body-on-a-chip" or "human-on-a-chip", used to study complex mechanisms in disease and drug screening. The most common organ-organ interaction studies are based on the intestines and liver, which are physically closest to each other.

Lee and colleagues built a gut-liver coculture chip with a PK model to predict first-pass metabolism by comparing the PK profile of paracetamol obtained by this chip with the known profile in humans. The clearance of the drug in this chip was significantly slow, and the gap was closed by improving the absorption surface area and metabolic capacity of the chip [97]. In addition, the gut-liver chip mimicked the absorption and accumulation of fatty acids in the gut and liver. Moreover, the effects of TNF-α, butyrate, and α-lipoic on hepatic steatosis via different mechanisms were evaluated by this chip [108]. To evaluate nanoparticle interactions with human tissues, a gastrointestinal tract and liver tissue system was constructed by combining the human intestinal epithelium and liver represented by coculture of enterocytes (Caco-2) with mucin-producing cells (TH29-MTX) and HepG2/C3A cells. High doses of nanoparticles induced aspartate aminotransferase release, indicating liver cell injury. Therefore, this device successfully simulated the uptake, metabolism, and toxicity of acetaminophen in vitro [109] (Figure 5A).

In addition to the intestines, the liver is associated with tumors, lungs, and skin, and can be used to study how substances are metabolized in multiple organs or tissues. For example, a liver and tumor-combined organ-on-chip was used as a PK-PD model to interpret drug actions in multiple organs [110]. The anti-cancer activity of luteolin was evaluated in this study, which was significantly weaker than that in 2D culture. These results revealed that simultaneous metabolism and tumor-killing actions likely resulted in a decreased anti-cancer effect. This study demonstrates that multiple organs on a chip established by combining the liver with a tumor is a useful tool for gaining insights into the mechanisms of drugs by interactions among multiple organs. Moreover, a lung/liver-on-a-chip has been reported (Figure 5B). Liver spheroids were connected in a single circuit, and normal human bronchial epithelial cells were cultured at the air-liquid interface. Aflatoxin B1 (AFB1) toxicity in lung tissues decreased when liver spheroids were present in the same chip circuit, indicating that the liver-mediated detoxification protected lung tissues [98]. The lung/liver-on-a-chip platform presented here offers new opportunities to study the toxicity of inhaled aerosols or to demonstrate the safety and efficacy of new drug candidates targeting the human lung. A simplified liver-kidney-on-chip model has been reported, which was used to investigate the biotransformation and toxicity of aflatoxin B1 (AFB1) and benzoalphapyrene (BαP) [111]. Coculture of human artificial liver microtissues and skin biopsies in multi-organ-chip was performed to emulate the systemic organ complexity of the human body, and each tissue had 1/100,000 of the biomass of their original human organ counterparts. After 14 days of coculture in a fluid flow environment, crosstalk between the liver and skin tissues was observed. This model facilitates exposure of skin at the air-liquid interface and provides a potential new tool for systemic substance testing [112].

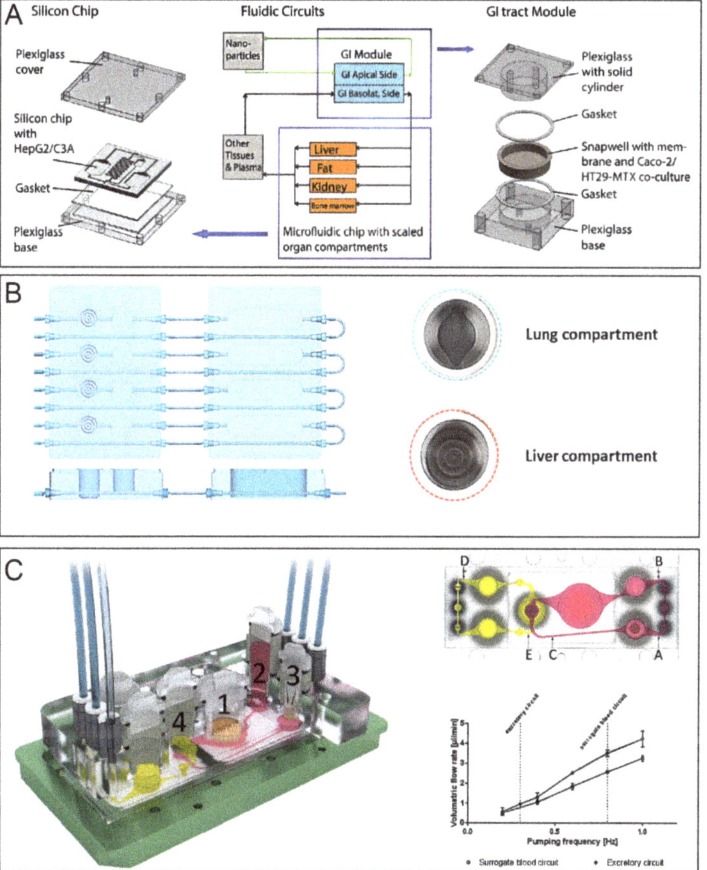

Figure 5. (**A**) Gastrointestinal tract and liver tissue system construct by combining the human intestinal epithelium and liver represented by coculture of enterocytes (Caco-2) with mucin-producing cells (TH29-MTX) and HepG2/C3A cells. Reproduced with permission from [109]. (**B**) Lung/liver-on-a-chip, in which liver spheroids were connected in a single circuit and normal human bronchial epithelial cells were cultured at the air-liquid interface. Reproduced with permission from [98]. (**C**) The microfluidic four-organ-chip device. (i) 3D view of the device comprising two polycarbonate cover plates, a PDMS-glass chip accommodating a surrogate blood flow circuit (pink), and an excretory flow circuit (yellow). Numbers represent the four tissue culture compartments for the intestines (1), liver (2), skin (3), and kidney (4). (ii) Central cross-section of each tissue culture compartment aligned along the interconnecting microchannel. (iii) Average volumetric flow rate plotted against the pumping frequency of the flow circuit. Reproduced with permission from [113].

In addition to interactions between the liver and a single organ, the crosstalk between the liver and more than one organ/tissue has been studied by connecting multiple organs in one microfluidic system. A homeostatic long-term coculture of human liver equivalents with either a reconstructed human intestinal barrier model or human skin biopsy platform has been reported. This platform provides pulsatile fluid flow within physiological ranges at low medium-to-tissue ratios and supports submersed cultivation of an intact intestinal barrier model and an air-liquid interface for the skin model during their coculture with liver equivalents [102]. Moreover, a four-organ-chip was employed to evaluate systemic absorption and metabolism of drugs in the small intestines, as shown in Figure 5C. Metabolism by the liver and excretion by the kidney are key determinants of the efficacy and safety of therapeutic candidates. Within two to four days, establishment of reproducible homeostasis among the co-cultures appeared at fluid-to-tissue ratios near to those in the physiological environment, and this homeostasis was sustained for at least 28 days [113].

These results clearly support the importance of advanced interconnected multi-organs in microfluidic devices for application to in vitro toxicity testing as well as optimized tissue culture systems for in vitro drug screening.

6. Challenges and Future Directions

The microfluidic technique offers substantial benefits to generate a functional liver on a chip. Accumulating evidence shows that liver-on-a-chip technology has achieved significant success in biomanufacturing for recreating key aspects of the liver, drug development in hepatotoxicity testing, and investigation of fundamental mechanisms in liver disease. However, the development of liver-on-a-chip technology is still in the early stage and many challenges remain. A critical issue is whether the results from liver-on-chips can replace those from animal experiments. To resolve this issue, the following points are worthy of attention.

As the building material of liver chips, a sustainable and reliable liver cell source is one of the key limitations. Primary hepatocytes and multiple types of non-parenchymal cells—rather than animal cells—are the best candidates, even though dedifferentiation hepatocytes in vitro remain to be overcome. Recently, mutable human embryonic stem cells and iPSCs have been considered as the most promising alternative sources. In particular, liver organoids have shown a similar spatial organization as the liver, which are able to reproduce some of the functions of the liver. However, to generate and control physiologically relevant structural, mechanical and biochemical cues that instruct directional differentiation remain great challenges. In addition, even though recent studies have begun to simulate bile tubes and shown hepatocyte polarization, most existing liver models cannot allow for the bile canaliculus and bile acid secretion, which may have a significant effect on liver functions and subsequent applications.

Another critical technical challenge arises from the stability of readouts and excessive complexity of operations. There have been a variety of liver chip devices to date, but none of them have received FDA approval, which is mainly due to the lack of uniform testing standards. The readouts of recent liver chips are often end point results, lacking real-time dynamic monitoring, which ignores many important physiological processes. Moreover, the throughput of liver chips is relatively low, which is unsuitable for the rapid high throughput of industrial applications. Biosensors integration will make up for the drawbacks of liver chip devices in terms of readouts and throughput. It is foreseeable that future organ chips will integrate many biosensors to meet the requirements for the automatization and monitorization. In addition, the use of high-tech detection technologies, such as live cell imaging and super resolution microscopy, are beneficial as well to the future commercial applications of liver chips.

Finally, a liver-on-a-chip itself cannot recapitulate the communication between different organs, thereby lacking the pharmacokinetic properties and toxic effects between other organs during hepatotoxicity testing. The immune system also plays a critical role in liver infection, disease progression, and drug-induced hepatotoxicity. To our knowledge, there is still no liver chip containing the immune system. In addition, to achieve crosstalk of a liver-on-a-chip with other organs, studies are anticipated to develop the next generation of multi-organs-on-chips such as a liver-kidney chip, liver-intestine chip, liver-immune chip, and finally a human-on-a-chip. In the near future, with the development of microengineering and microfluidic technologies, there may emerge a large number of fast, high-throughput liver-on-a-chip devices and liver organoids that can be widely used in pathological studies and environmental toxicology.

Author Contributions: X.Z., B.L. and J.D. conceived this review. J.D., W.W., Z.C., Y.L., W.Z. and X.Z. wrote the manuscript.

Funding: This work was supported by the National Natural Science Foundation of China (NNSFC) (grant no. 21675017) and the National Key Research and Development Program of China (grant no. 2017YFC1702001).

References

1. Wiśniewski, J.R.; Vildhede, A.; Norén, A.; Artursson, P. In-depth quantitative analysis and comparison of the human hepatocyte and hepatoma cell line HepG2 proteomes. *J. Proteomics* **2016**, *136*, 234–247. [CrossRef] [PubMed]
2. Casotti, V.; Antiga, L.D. Basic principles of liver physiology. In *Pediatric Hepatology and Liver Transplantation*; D'Antiga, L., Ed.; Springer International Publishing: New York City, NY, USA, 2019; pp. 21–39.
3. De Boer, Y.S.; Kosinski, A.S.; Urban, T.J.; Zhao, Z.; Long, N.; Chalasani, N.; Kleiner, D.E.; Hoofnagle, J.H. Features of Autoimmune Hepatitis in Patients with Drug-induced Liver Injury. *Clin. Gastroenterol. H* **2017**, *15*, 103–112. [CrossRef] [PubMed]
4. Chen, M.; Suzuki, A.; Borlak, J.; Andrade, R.J.; Lucena, M.I. Drug-induced liver injury: Interactions between drug properties and host factors. *J. Hepatol.* **2015**, *63*, 503–514. [CrossRef] [PubMed]
5. Van Midwoud, P.M.; Verpoorte, E.; Groothuis, G.M.M. Microfluidic devices for in vitro studies on liver drug metabolism and toxicity. *Integr. Biol. Uk* **2011**, *3*, 509. [CrossRef] [PubMed]
6. Olson, H.; Betton, G.; Robinson, D.; Thomas, K.; Monro, A.; Kolaja, G.; Lilly, P.; Sanders, J.; Sipes, G.; Bracken, W.; et al. Concordance of the Toxicity of Pharmaceuticals in Humans and in Animals. *Regul. Toxicol. Pharm.* **2000**, *32*, 56–67. [CrossRef]
7. Choi, W.; Eun, H.S.; Lee, Y.; Kim, S.J.; Kim, M.; Lee, J.; Shim, Y.; Kim, H.; Kim, Y.E.; Yi, H.; et al. Experimental Applications of in situ Liver Perfusion Machinery for the Study of Liver Disease. *Mol. Cells* **2019**, *42*, 45–55.
8. Tomlinson, L.; Hyndman, L.; Firman, J.W.; Bentley, R.; Kyffin, J.A.; Webb, S.D.; McGinty, S.; Sharma, P. In vitro Liver Zonation of Primary Rat Hepatocytes. *Front. Bioeng. Biotechnol.* **2019**, *7*, 17. [CrossRef]
9. Moravcova, A.; Cervinkova, Z.; Kucera, O.; Mezera, V.; Rychtrmoc, D.; Lotkova, H. The effect of oleic and palmitic acid on induction of steatosis and cytotoxicity on rat hepatocytes in primary culture. *Physiol. Res.* **2015**, *64*, S627–S636.
10. Lee, H.; Chae, S.; Kim, J.Y.; Han, W.; Kim, J.; Choi, Y.; Cho, D. Cell-printed 3D liver-on-a-chip possessing a liver microenvironment and biliary system. *Biofabrication* **2019**, *11*, 25001. [CrossRef]
11. Kim, Y.; Kang, K.; Jeong, J.; Paik, S.S.; Kim, J.S.; Park, S.A.; Kim, W.D.; Park, J.; Choi, D. Three-dimensional (3D) printing of mouse primary hepatocytes to generate 3D hepatic structure. *Ann. Surg Treat. Res.* **2017**, *92*, 67. [CrossRef]
12. Broutier, L.; Mastrogiovanni, G.; Verstegen, M.M.; Francies, H.E.; Gavarró, L.M.; Bradshaw, C.R.; Allen, G.E.; Arnes-Benito, R.; Sidorova, O.; Gaspersz, M.P.; et al. Human primary liver cancer–derived organoid cultures for disease modeling and drug screening. *Nat. Med.* **2017**, *23*, 1424–1435. [CrossRef] [PubMed]
13. Lancaster, M.A.; Knoblich, J.A. Organogenesis in a dish: Modeling development and disease using organoid technologies. *Science* **2014**, *345*, 1247125. [CrossRef]
14. Li, L.; Gokduman, K.; Gokaltun, A.; Yarmush, M.L.; Usta, O.B. A microfluidic 3D hepatocyte chip for hepatotoxicity testing of nanoparticles. *Nanomedicine* **2019**, *14*, 16. [CrossRef]
15. Haque, A.; Gheibi, P.; Stybayeva, G.; Gao, Y.; Torok, N.; Revzin, A. Ductular reaction-on-a-chip: Microfluidic co-cultures to study stem cell fate selection during liver injury. *Sci. Rep. UK* **2016**, *6*, 36077. [CrossRef] [PubMed]
16. Esch, M.B.; Ueno, H.; Applegate, D.R.; Shuler, M.L. Modular, pumpless body-on-a-chip platform for the co-culture of GI tract epithelium and 3D primary liver tissue. *Lab Chip* **2016**, *16*, 2719–2729. [CrossRef] [PubMed]
17. Underhill, G.H.; Khetani, S.R. Bioengineered Liver Models for Drug Testing and Cell Differentiation Studies. *Cellul. Mol. Gastroenterol. Hepatol.* **2018**, *5*, 426–439. [CrossRef]
18. Beckwitt, C.H.; Clark, A.M.; Wheeler, S.; Taylor, D.L.; Stolz, D.B.; Griffith, L.; Wells, A. Liver 'organ on a chip'. *Exp. Cell Res.* **2018**, *363*, 15–25. [CrossRef]

19. Lauschke, V.M.; Hendriks, D.F.G.; Bell, C.C.; Andersson, T.B.; Ingelman-Sundberg, M. Novel 3D Culture Systems for Studies of Human Liver Function and Assessments of the Hepatotoxicity of Drugs and Drug Candidates. *Chem. Res. Toxicol.* **2016**, *29*, 1936–1955. [CrossRef]

20. Yoon No, D.; Lee, K.; Lee, J.; Lee, S. 3D liver models on a microplatform: well-defined culture, engineering of liver tissue and liver-on-a-chip. *Lab Chip* **2015**, *15*, 3822–3837. [CrossRef]

21. Bhatia, S.N.; Ingber, D.E. Microfluidic organs-on-chips. *Nat. Biotechnol.* **2014**, *32*, 760–772. [CrossRef]

22. Zhang, B.; Korolj, A.; Lai, B.F.L.; Radisic, M. Advances in organ-on-a-chip engineering. *Nat. Rev. Mater.* **2018**, *3*, 257–278. [CrossRef]

23. Usta, O.B.; McCarty, W.J.; Bale, S.; Hegde, M.; Jindal, R.; Bhushan, A.; Golberg, I.; Yarmush, M.L. Microengineered cell and tissue systems for drug screening and toxicology applications: Evolution ofin-vitro liver technologies. *TECHNOLOGY* **2015**, *3*, 1–26. [CrossRef] [PubMed]

24. Godoy, P.; Hewitt, N.J.; Albrecht, U.; Andersen, M.E.; Ansari, N.; Bhattacharya, S.; Bode, J.G.; Bolleyn, J.; Borner, C.; Böttger, J.; et al. Recent advances in 2D and 3D in vitro systems using primary hepatocytes, alternative hepatocyte sources and non-parenchymal liver cells and their use in investigating mechanisms of hepatotoxicity, cell signaling and ADME. *Arch. Toxicol.* **2013**, *87*, 1315–1530. [CrossRef] [PubMed]

25. Treyer, A.; Müsch, A. Hepatocyte polarity. *Compr. Physiol.* **2013**, *3*, 243. [PubMed]

26. Laskin, D. Nonparenchymal Cells and Hepatotoxicity. *Semin. Liver Dis.* **1990**, *10*, 293–304. [CrossRef] [PubMed]

27. Poisson, J.; Lemoinne, S.; Boulanger, C.; Durand, F.; Moreau, R.; Valla, D.; Rautou, P. Liver sinusoidal endothelial cells: Physiology and role in liver diseases. *J. Hepatol.* **2017**, *66*, 212–227. [CrossRef] [PubMed]

28. Li, J.; Li, P.; He, K.; Liu, Z.; Gong, J. The role of Kupffer cells in hepatic diseases. *Mol. Immunol.* **2017**, *85*, 222–229. [CrossRef] [PubMed]

29. Boltjes, A.; Movita, D.; Boonstra, A.; Woltman, A.M. The role of Kupffer cells in hepatitis B and hepatitis C virus infections. *J. Hepatol.* **2014**, *61*, 660–671. [CrossRef]

30. Ringelhan, M.; Pfister, D.; O Connor, T.; Pikarsky, E.; Heikenwalder, M. The immunology of hepatocellular carcinoma. *Nat. Immunol.* **2018**, *19*, 222–232. [CrossRef]

31. Deng, J.; Zhang, X.; Chen, Z.; Luo, Y.; Lu, Y.; Liu, T.; Wu, Z.; Jin, Y.; Zhao, W.; Lin, B. A cell lines derived microfluidic liver model for investigation of hepatotoxicity induced by drug-drug interaction. *Biomicrofluidics* **2019**, *13*, 24101. [CrossRef]

32. Adams, D.H.; Ju, C.; Ramaiah, S.K.; Uetrecht, J.; Jaeschke, H. Mechanisms of Immune-Mediated Liver Injury. *Toxicol. Sci.* **2010**, *115*, 307–321. [CrossRef] [PubMed]

33. Ma, L.; Wang, Y.; Wang, J.; Wu, J.; Meng, X.; Hu, P.; Mu, X.; Liang, Q.; Luo, G. Design and fabrication of a liver-on-a-chip platform for convenient, highly efficient, and safe in situ perfusion culture of 3D hepatic spheroids. *Lab Chip* **2018**, *18*, 2547–2562. [CrossRef] [PubMed]

34. Weng, Y.; Chang, S.; Shih, M.; Tseng, S.; Lai, C. Scaffold-Free Liver-On-A-Chip with Multiscale Organotypic Cultures. *Adv. Mater.* **2017**, *29*, 1701545. [CrossRef] [PubMed]

35. Ma, X.; Qu, X.; Zhu, W.; Li, Y.; Yuan, S.; Zhang, H.; Liu, J.; Wang, P.; Lai, C.S.E.; Zanella, F.; et al. Deterministically patterned biomimetic human iPSC-derived hepatic model via rapid 3D bioprinting. *Proc. Natl. Acad. Sci. USA* **2016**, *113*, 2206–2211. [CrossRef] [PubMed]

36. Kang, Y.B.; Eo, J.; Mert, S.; Yarmush, M.L.; Usta, O.B. Metabolic Patterning on a Chip: Towards in vitro Liver Zonation of Primary Rat and Human Hepatocytes. *Sci. Rep. UK* **2018**, *8*, 8951. [CrossRef] [PubMed]

37. Madurska, M.J.; Poyade, M.; Eason, D.; Rea, P.; Watson, A.J.M. Development of a Patient-Specific 3D-Printed Liver Model for Preoperative Planning. *Surg. Innov.* **2017**, *24*, 145–150. [CrossRef]

38. Witowski, J.S.; Pędziwiatr, M.; Major, P.; Budzyński, A. Cost-effective, personalized, 3D-printed liver model for preoperative planning before laparoscopic liver hemihepatectomy for colorectal cancer metastases. *Int. J. Comput. Ass. Rad.* **2017**, *12*, 2047–2054. [CrossRef] [PubMed]

39. Ramachandran, S.D.; Schirmer, K.; Münst, B.; Heinz, S.; Ghafoory, S.; Wölfl, S.; Simon-Keller, K.; Marx, A.; Øie, C.I.; Ebert, M.P.; et al. In Vitro Generation of Functional Liver Organoid-Like Structures Using Adult Human Cells. *PLoS ONE* **2015**, *10*, e139345. [CrossRef]

40. Matsusaki, M.; Case, C.P.; Akashi, M. Three-dimensional cell culture technique and pathophysiology. *Adv. Drug Deliv. Rev.* **2014**, *74*, 95–103. [CrossRef]

41. Tostões, R.M.; Leite, S.B.; Serra, M.; Jensen, J.; Björquist, P.; Carrondo, M.J.T.; Brito, C.; Alves, P.M. Human liver cell spheroids in extended perfusion bioreactor culture for repeated-dose drug testing. *Hepatology* **2012**, *55*, 1227–1236. [CrossRef]

42. Schanz, S.; Schmalzing, M.; Guenova, E.; Metzler, G.; Ulmer, A.; K Ötter, I.; Fierlbeck, G. Interstitial Granulomatous Dermatitis with Arthritis Responding to Tocilizumab. *Arch. Dermatol.* **2012**, *148*, 17–20. [CrossRef]

43. Kyffin, J.A.; Sharma, P.; Leedale, J.; Colley, H.E.; Murdoch, C.; Mistry, P.; Webb, S.D. Impact of cell types and culture methods on the functionality of in vitro liver systems-A review of cell systems for hepatotoxicity assessment. *Toxicol. Vitro* **2018**, *48*, 262–275. [CrossRef] [PubMed]

44. Ahmed, H.M.M.; Salerno, S.; Morelli, S.; Giorno, L.; De Bartolo, L. 3D liver membrane system by co-culturing human hepatocytes, sinusoidal endothelial and stellate cells. *Biofabrication* **2017**, *9*, 25022. [CrossRef] [PubMed]

45. Kang, Y.B.A.; Sodunke, T.R.; Lamontagne, J.; Cirillo, J.; Rajiv, C.; Bouchard, M.J.; Noh, M. Liver sinusoid on a chip: Long-term layered co-culture of primary rat hepatocytes and endothelial cells in microfluidic platforms. *Biotechnol. Bioeng.* **2015**, *112*, 2571–2582. [CrossRef] [PubMed]

46. Lee, S.; No, D.Y.; Kang, E.; Ju, J.; Kim, D.; Lee, S. Spheroid-based three-dimensional liver-on-a-chip to investigate hepatocyte–hepatic stellate cell interactions and flow effects. *Lab Chip* **2013**, *13*, 3529. [CrossRef] [PubMed]

47. Lee, J.; Choi, B.; No, D.Y.; Lee, G.; Lee, S.; Oh, H.; Lee, S. A 3D alcoholic liver disease model on a chip. *Integr. Biol. UK* **2016**, *8*, 302–308. [CrossRef] [PubMed]

48. Lauschke, V.M.; Shafagh, R.Z.; Hendriks, D.F.G.; Ingelman Sundberg, M. 3D Primary Hepatocyte Culture Systems for Analyses of Liver Diseases, Drug Metabolism, and Toxicity: Emerging Culture Paradigms and Applications. *Biotechnol. J.* **2019**, *14*, 1800347. [CrossRef]

49. Meyer, K.; Ostrenko, O.; Bourantas, G.; Morales-Navarrete, H.; Porat-Shliom, N.; Segovia-Miranda, F.; Nonaka, H.; Ghaemi, A.; Verbavatz, J.; Brusch, L.; et al. A Predictive 3D Multi-Scale Model of Biliary Fluid Dynamics in the Liver Lobule. *Cell Syst.* **2017**, *4*, 277–290. [CrossRef]

50. Van Grunsven, L.A. 3D in vitro models of liver fibrosis. *Adv. Drug Deliv. Rev.* **2017**, *121*, 133–146. [CrossRef]

51. Miyamoto, Y.; Ikeuchi, M.; Noguchi, H.; Yagi, T.; Hayashi, S. Spheroid Formation and Evaluation of Hepatic Cells in a Three-Dimensional Culture Device. *Cell Med.* **2015**, *8*, 47–56. [CrossRef]

52. Ortega-Prieto, A.M.; Skelton, J.K.; Wai, S.N.; Large, E.; Lussignol, M.; Vizcay-Barrena, G.; Hughes, D.; Fleck, R.A.; Thursz, M.; Catanese, M.T.; et al. 3D microfluidic liver cultures as a physiological preclinical tool for hepatitis B virus infection. *Nat. Commun.* **2018**, *9*, 682. [CrossRef] [PubMed]

53. Frey, O.; Misun, P.M.; Fluri, D.A.; Hengstler, J.G.; Hierlemann, A. Reconfigurable microfluidic hanging drop network for multi-tissue interaction and analysis. *Nat. Commun.* **2014**, *5*, 4250. [CrossRef] [PubMed]

54. Bavli, D.; Prill, S.; Ezra, E.; Levy, G.; Cohen, M.; Vinken, M.; Vanfleteren, J.; Jaeger, M.; Nahmias, Y. Real-time monitoring of metabolic function in liver-on-chip microdevices tracks the dynamics of mitochondrial dysfunction. *Proc. Natl. Acad. Sci. USA* **2016**, *113*, E2231–E2240. [CrossRef] [PubMed]

55. Delalat, B.; Cozzi, C.; Rasi Ghaemi, S.; Polito, G.; Kriel, F.H.; Michl, T.D.; Harding, F.J.; Priest, C.; Barillaro, G.; Voelcker, N.H. Microengineered Bioartificial Liver Chip for Drug Toxicity Screening. *Adv. Funct. Mater.* **2018**, *28*, 1801825. [CrossRef]

56. Ma, C.; Zhao, L.; Zhou, E.; Xu, J.; Shen, S.; Wang, J. On-Chip Construction of Liver Lobule-like Microtissue and Its Application for Adverse Drug Reaction Assay. *Anal. Chem.* **2016**, *88*, 1719–1727. [CrossRef] [PubMed]

57. Li, X.; George, S.M.; Vernetti, L.; Gough, A.H.; Taylor, D.L. A glass-based, continuously zonated and vascularized human liver acinus microphysiological system (vLAMPS) designed for experimental modeling of diseases and ADME/TOX. *Lab Chip* **2018**, *18*, 2614–2631. [CrossRef] [PubMed]

58. Schepers, A.; Li, C.; Chhabra, A.; Seney, B.T.; Bhatia, S. Engineering a perfusable 3D human liver platform from iPS cells. *Lab Chip* **2016**, *16*, 2644–2653. [CrossRef] [PubMed]

59. Desai, P.; Tseng, H.; Souza, G. Assembly of Hepatocyte Spheroids Using Magnetic 3D Cell Culture for CYP450 Inhibition/Induction. *Int. J. Mol. Sci.* **2017**, *18*, 1085. [CrossRef]

60. Weltin, A.; Hammer, S.; Noor, F.; Kaminski, Y.; Kieninger, J.; Urban, G.A. Accessing 3D microtissue metabolism: Lactate and oxygen monitoring in hepatocyte spheroids. *Biosens. Bioelectron.* **2017**, *87*, 941–948. [CrossRef]

61. Foster, A.J.; Chouhan, B.; Regan, S.L.; Rollison, H.; Amberntsson, S.; Andersson, L.C.; Srivastava, A.; Darnell, M.; Cairns, J.; Lazic, S.E.; et al. Integrated in vitro models for hepatic safety and metabolism: evaluation of a human Liver-Chip and liver spheroid. *Arch. Toxicol.* **2019**, *93*, 1021–1037. [CrossRef]

62. Toh, Y.; Lim, T.C.; Tai, D.; Xiao, G.; van Noort, D.; Yu, H. A microfluidic 3D hepatocyte chip for drug toxicity testing. *Lab Chip* **2009**, *9*, 2026. [CrossRef] [PubMed]

63. Schwartz, R.E.; Fleming, H.E.; Khetani, S.R.; Bhatia, S.N. Pluripotent stem cell-derived hepatocyte-like cells. *Biotechnol. Adv.* **2014**, *32*, 504–513. [CrossRef] [PubMed]

64. Yi, F.; Liu, G.; Belmonte, J.C.I. Human induced pluripotent stem cells derived hepatocytes: rising promise for disease modeling, drug development and cell therapy. *Protein Cell* **2012**, *3*, 246–250. [CrossRef] [PubMed]

65. Ho, C.; Lin, R.; Chen, R.; Chin, C.; Gong, S.; Chang, H.; Peng, H.; Hsu, L.; Yew, T.; Chang, S.; et al. Liver-cell patterning Lab Chip: mimicking the morphology of liver lobule tissue. *Lab Chip* **2013**, *13*, 3578. [CrossRef] [PubMed]

66. Misun, P.M.; Rothe, J.; Schmid, Y.R.F.; Hierlemann, A.; Frey, O. Multi-analyte biosensor interface for real-time monitoring of 3D microtissue spheroids in hanging-drop networks. *Microsyst. Nanoeng.* **2016**, *2*, 16022. [CrossRef] [PubMed]

67. Aeby, E.A.; Misun, P.M.; Hierlemann, A.; Frey, O. Microfluidic Hydrogel Hanging-Drop Network for Long-Term Culturing of 3D Microtissues and Simultaneous High-Resolution Imaging. *Adv. Biosyst.* **2018**, *2*, 1800054. [CrossRef]

68. Boos, J.A.; Misun, P.M.; Michlmayr, A.; Hierlemann, A.; Frey, O. Microfluidic Multitissue Platform for Advanced Embryotoxicity Testing In Vitro. *Adv. Sci.* **2019**, *6*, 1900294. [CrossRef] [PubMed]

69. Leite, S.B.; Roosens, T.; El Taghdouini, A.; Mannaerts, I.; Smout, A.J.; Najimi, M.; Sokal, E.; Noor, F.; Chesne, C.; van Grunsven, L.A. Novel human hepatic organoid model enables testing of drug-induced liver fibrosis in vitro. *Biomaterials* **2016**, *78*, 1–10. [CrossRef]

70. Hegde, M.; Jindal, R.; Bhushan, A.; Bale, S.S.; McCarty, W.J.; Golberg, I.; Usta, O.B.; Yarmush, M.L. Dynamic interplay of flow and collagen stabilizes primary hepatocytes culture in a microfluidic platform. *Lab Chip* **2014**, *14*, 2033–2039. [CrossRef]

71. Lu, S.; Cuzzucoli, F.; Jiang, J.; Liang, L.; Wang, Y.; Kong, M.; Zhao, X.; Cui, W.; Li, J.; Wang, S. Development of a biomimetic liver tumor-on-a-chip model based on decellularized liver matrix for toxicity testing. *Lab Chip* **2018**, *18*, 3379–3392. [CrossRef]

72. Jang, M.; Neuzil, P.; Volk, T.; Manz, A.; Kleber, A. On-chip three-dimensional cell culture in phaseguides improves hepatocyte functionsin vitro. *Biomicrofluidics* **2015**, *9*, 34113. [CrossRef] [PubMed]

73. Christoffersson, J.; Aronsson, C.; Jury, M.; Selegård, R.; Aili, D.; Mandenius, C. Fabrication of modular hyaluronan-PEG hydrogels to support 3D cultures of hepatocytes in a perfused liver-on-a-chip device. *Biofabrication* **2019**, *11*, 15013. [CrossRef] [PubMed]

74. Zhu, L.; Fan, X.; Wang, B.; Liu, L.; Yan, X.; Zhou, L.; Zeng, Y.; Poznansky, M.C.; Wang, L.; Chen, H.; et al. Biomechanically primed liver microtumor array as a high-throughput mechanopharmacological screening platform for stroma-reprogrammed combinatorial therapy. *Biomaterials* **2017**, *124*, 12–24. [CrossRef] [PubMed]

75. Yajima, Y.; Lee, C.N.; Yamada, M.; Utoh, R.; Seki, M. Development of a perfusable 3D liver cell cultivation system via bundling-up assembly of cell-laden microfibers. *J. Biosci. Bioeng.* **2018**, *126*, 111–118. [CrossRef] [PubMed]

76. Sasaki, K.; Akagi, T.; Asaoka, T.; Eguchi, H.; Fukuda, Y.; Iwagami, Y.; Yamada, D.; Noda, T.; Wada, H.; Gotoh, K.; et al. Construction of three-dimensional vascularized functional human liver tissue using a layer-by-layer cell coating technique. *Biomaterials* **2017**, *133*, 263–274. [CrossRef] [PubMed]

77. Mi, S.; Yi, X.; Du, Z.; Xu, Y.; Sun, W. Construction of a liver sinusoid based on the laminar flow on chip and self-assembly of endothelial cells. *Biofabrication* **2018**, *10*, 25010. [CrossRef]

78. Rennert, K.; Steinborn, S.; Gröger, M.; Ungerböck, B.; Jank, A.; Ehgartner, J.; Nietzsche, S.; Dinger, J.; Kiehntopf, M.; Funke, H.; et al. A microfluidically perfused three dimensional human liver model. *Biomaterials* **2015**, *71*, 119–131. [CrossRef]

79. Prodanov, L.; Jindal, R.; Bale, S.S.; Hegde, M.; McCarty, W.J.; Golberg, I.; Bhushan, A.; Yarmush, M.L.; Usta, O.B. Long-term maintenance of a microfluidic 3D human liver sinusoid. *Biotechnol. Bioeng.* **2016**, *113*, 241–246. [CrossRef]

80. Du, Y.; Li, N.; Yang, H.; Luo, C.; Gong, Y.; Tong, C.; Gao, Y.; Lü, S.; Long, M. Mimicking liver sinusoidal structures and functions using a 3D-configured microfluidic chip. *Lab Chip* **2017**, *17*, 782–794. [CrossRef]

81. Ortega-Ribera, M.; Fernández-Iglesias, A.; Illa, X.; Moya, A.; Molina, V.; Maeso-Díaz, R.; Fondevila, C.; Peralta, C.; Bosch, J.; Villa, R.; et al. Resemblance of the human liver sinusoid in a fluidic device with biomedical and pharmaceutical applications. *Biotechnol. Bioeng.* **2018**, *115*, 2585–2594. [CrossRef]

82. Moya, A.; Ortega-Ribera, M.; Guimerà, X.; Sowade, E.; Zea, M.; Illa, X.; Ramon, E.; Villa, R.; Gracia-Sancho, J.; Gabriel, G. Online oxygen monitoring using integrated inkjet-printed sensors in a liver-on-a-chip system. *Lab Chip* **2018**, *18*, 2023–2035. [CrossRef]

83. Norona, L.M.; Nguyen, D.G.; Gerber, D.A.; Presnell, S.C.; LeCluyse, E.L. Editor's Highlight: Modeling Compound-Induced FibrogenesisIn Vitro Using Three-Dimensional Bioprinted Human Liver Tissues. *Toxicol. Sci.* **2016**, *154*, 354–367. [CrossRef] [PubMed]

84. Nguyen, D.G.; Funk, J.; Robbins, J.B.; Crogan-Grundy, C.; Presnell, S.C.; Singer, T.; Roth, A.B. Bioprinted 3D Primary Liver Tissues Allow Assessment of Organ-Level Response to Clinical Drug Induced Toxicity In Vitro. *PLoS ONE* **2016**, *11*, e158674. [CrossRef] [PubMed]

85. Itai, S.; Tajima, H.; Onoe, H. Double-layer perfusable collagen microtube device for heterogeneous cell culture. *Biofabrication* **2018**, *11*, 15010. [CrossRef] [PubMed]

86. Bhise, N.S.; Manoharan, V.; Massa, S.; Tamayol, A.; Ghaderi, M.; Miscuglio, M.; Lang, Q.; Shrike Zhang, Y.; Shin, S.R.; Calzone, G.; et al. A liver-on-a-chip platform with bioprinted hepatic spheroids. *Biofabrication* **2016**, *8*, 14101. [CrossRef] [PubMed]

87. Arai, K.; Yoshida, T.; Okabe, M.; Goto, M.; Mir, T.A.; Soko, C.; Tsukamoto, Y.; Akaike, T.; Nikaido, T.; Zhou, K.; et al. Fabrication of 3D-culture platform with sandwich architecture for preserving liver-specific functions of hepatocytes using 3D bioprinter. *J. Biomed. Mater. Res. A* **2017**, *105*, 1583–1592. [CrossRef]

88. Grix, T.; Ruppelt, A.; Thomas, A.; Amler, A.; Noichl, B.; Lauster, R.; Kloke, L. Bioprinting Perfusion-Enabled Liver Equivalents for Advanced Organ-on-a-Chip Applications. *Genes Basel* **2018**, *9*, 176. [CrossRef]

89. Roth, A.D.; Lama, P.; Dunn, S.; Hong, S.; Lee, M. Polymer coating on a micropillar chip for robust attachment of PuraMatrix peptide hydrogel for 3D hepatic cell culture. *Mater. Sci. Eng. C* **2018**, *90*, 634–644. [CrossRef]

90. Khetani, S.R.; Bhatia, S.N. Microscale culture of human liver cells for drug development. *Nat. Biotechnol.* **2008**, *26*, 120–126. [CrossRef]

91. Messner, S.; Agarkova, I.; Moritz, W.; Kelm, J.M. Multi-cell type human liver microtissues for hepatotoxicity testing. *Arch. Toxicol.* **2013**, *87*, 209–213. [CrossRef]

92. Massa, S.; Sakr, M.A.; Seo, J.; Bandaru, P.; Arneri, A.; Bersini, S.; Zare-Eelanjegh, E.; Jalilian, E.; Cha, B.; Antona, S.; et al. Bioprinted 3D vascularized tissue model for drug toxicity analysis. *Biomicrofluidics* **2017**, *11*, 44109. [CrossRef] [PubMed]

93. Ware, B.R.; Khetani, S.R. Engineered Liver Platforms for Different Phases of Drug Development. *Trends Biotechnol.* **2016**, *35*, 172–183. [CrossRef] [PubMed]

94. Deng, J.; Qu, Y.; Liu, T.; Jing, B.; Zhang, X.; Chen, Z.; Luo, Y.; Zhao, W.; Lu, Y.; Lin, B. Recent organ-on-a-chip advances toward drug toxicity testing. *Microphysiol. Syst.* **2018**, *2*, 4798. [CrossRef]

95. Yu, F.; Deng, R.; Hao Tong, W.; Huan, L.; Chan Way, N.; IslamBadhan, A.; Iliescu, C.; Yu, H. A perfusion incubator liver chip for 3D cell culture with application on chronic hepatotoxicity testing. *Sci. Rep. UK* **2017**, *7*, 14528. [CrossRef] [PubMed]

96. Prot, J.; Bunescu, A.; Elena-Herrmann, B.; Aninat, C.; Snouber, L.C.; Griscom, L.; Razan, F.; Bois, F.Y.; Legallais, C.; Brochot, C.; et al. Predictive toxicology using systemic biology and liver microfluidic "on chip" approaches: Application to acetaminophen injury. *Toxicol. Appl. Pharm.* **2012**, *259*, 270–280. [CrossRef] [PubMed]

97. Lee, D.W.; Ha, S.K.; Choi, I.; Sung, J.H. 3D gut-liver chip with a PK model for prediction of first-pass metabolism. *Biomed. Microdevices* **2017**, *19*, 100. [CrossRef] [PubMed]

98. Bovard, D.; Sandoz, A.; Luettich, K.; Frentzel, S.; Iskandar, A.; Marescotti, D.; Trivedi, K.; Guedj, E.; Dutertre, Q.; Peitsch, M.C.; et al. A lung/liver-on-a-chip platform for acute and chronic toxicity studies. *Lab Chip* **2018**, *18*, 3814–3829. [CrossRef]

99. Midwoud, P.M.V. An Alternative Approach Based on Microfluidics to Study Drug Metabolism and Toxicity Using Liver and Intestinal Tissue. Ph.D. Thesis, Faculty of Mathematics and Natural Sciences, University of Groningen, Groningen, The Netherlands, 2010.

100. Choe, A.; Ha, S.K.; Choi, I.; Choi, N.; Sung, J.H. Microfluidic Gut-liver chip for reproducing the first pass metabolism. *Biomed. Microdevices* **2017**, *19*, 4. [CrossRef]
101. Zhou, Q.; Patel, D.; Kwa, T.; Haque, A.; Matharu, Z.; Stybayeva, G.; Gao, Y.; Diehl, A.M.; Revzin, A. Liver injury-on-a-chip: microfluidic co-cultures with integrated biosensors for monitoring liver cell signaling during injury. *Lab Chip* **2015**, *15*, 4467–4478. [CrossRef]
102. Maschmeyer, I.; Hasenberg, T.; Jaenicke, A.; Lindner, M.; Lorenz, A.K.; Zech, J.; Garbe, L.; Sonntag, F.; Hayden, P.; Ayehunie, S.; et al. Chip-based human liver-intestine and liver-skin co-cultures - A first step toward systemic repeated dose substance testing in vitro. *Eur. J. Pharm. Biopharm.* **2015**, *95*, 77–87. [CrossRef]
103. Deng, J.; Chen, Z.; Zhang, X.; Luo, Y.; Wu, Z.; Lu, Y.; Liu, T.; Zhao, W.; Lin, B. A liver-chip-based alcoholic liver disease model featuring multi-non-parenchymal cells. *Biomed. Microdevices* **2019**, *21*, 57. [CrossRef] [PubMed]
104. Gori, M.; Simonelli, M.C.; Giannitelli, S.M.; Businaro, L.; Trombetta, M.; Rainer, A. Investigating Nonalcoholic Fatty Liver Disease in a Liver-on-a-Chip Microfluidic Device. *PLoS ONE* **2016**, *11*, e159729. [CrossRef] [PubMed]
105. Kostrzewski, T.; Cornforth, T.; Snow, S.A.; Ouro-Gnao, L.; Rowe, C.; Large, E.M.; Hughes, D.J. Three-dimensional perfused human in vitro model of non-alcoholic fatty liver disease. *World J. Gastroenterol.* **2017**, *23*, 204. [CrossRef] [PubMed]
106. Vernetti, L.A.; Senutovitch, N.; Boltz, R.; DeBiasio, R.; Ying Shun, T.; Gough, A.; Taylor, D.L. A human liver microphysiology platform for investigating physiology, drug safety, and disease models. *Exp. Biol. Med.* **2015**, *241*, 101–114. [CrossRef] [PubMed]
107. Wang, Y.; Wang, H.; Deng, P.; Chen, W.; Guo, Y.; Tao, T.; Qin, J. In situ differentiation and generation of functional liver organoids from human iPSCs in a 3D perfusable chip system. *Lab Chip* **2018**, *18*, 3606–3616. [CrossRef] [PubMed]
108. Lee, S.Y.; Sung, J.H. Gut-liver on a chip toward an in vitro model of hepatic steatosis. *Biotechnol. Bioeng.* **2018**, *115*, 2817–2827. [CrossRef]
109. Esch, M.B.; Mahler, G.J.; Stokor, T.; Shuler, M.L. Body-on-a-chip simulation with gastrointestinal tract and liver tissues suggests that ingested nanoparticles have the potential to cause liver injury. *Lab Chip* **2014**, *14*, 3081–3092. [CrossRef]
110. Lee, H.; Kim, D.S.; Ha, S.K.; Choi, I.; Lee, J.M.; Sung, J.H. A pumpless multi-organ-on-a-chip (MOC) combined with a pharmacokinetic-pharmacodynamic (PK-PD) model. *Biotechnol. Bioeng.* **2017**, *114*, 432–443. [CrossRef]
111. Theobald, J.; Ghanem, A.; Wallisch, P.; Banaeiyan, A.A.; Andrade-Navarro, M.A.; Taskova, K.; Haltmeier, M.; Kurtz, A.; Becker, H.; Reuter, S.; et al. Liver-Kidney-on-Chip to Study Toxicity of Drug Metabolites. *ACS Biomater. Sci. Eng.* **2018**, *4*, 78–89. [CrossRef]
112. Wagner, I.; Materne, E.; Brincker, S.; S Bier, U.; Fr Drich, C.; Busek, M.; Sonntag, F.; Sakharov, D.A.; Trushkin, E.V.; Tonevitsky, A.G.; et al. A dynamic multi-organ-chip for long-term cultivation and substance testing proven by 3D human liver and skin tissue co-culture. *Lab Chip* **2013**, *13*, 3538. [CrossRef]
113. Maschmeyer, I.; Lorenz, A.K.; Schimek, K.; Hasenberg, T.; Ramme, A.P.; Hubner, J.; Lindner, M.; Drewell, C.; Bauer, S.; Thomas, A.; et al. A four-organ-chip for interconnected long-term co-culture of human intestine, liver, skin and kidney equivalents. *Lab Chip* **2015**, *15*, 2688–2699. [CrossRef] [PubMed]

 *micromachines*

Review

Microfluidic-Based 3D Engineered Microvascular Networks and Their Applications in Vascularized Microtumor Models

Xiaolin Wang [1,2,3,*], Qiyue Sun [1] and Jianghua Pei [1]

[1] Department of Micro/Nano Electronics, School of Electronic Information and Electrical Engineering, Shanghai Jiao Tong University, Shanghai 200240, China; kizluy@sjtu.edu.cn (Q.S.); Peijianghua@sjtu.edu.cn (J.P.)

[2] National Key Laboratory of Science and Technology on Micro/Nano Fabrication, Department of Micro/Nano Electronics, School of Electronic Information and Electrical Engineering, Shanghai Jiao Tong University, Shanghai 200240, China

[3] Key Laboratory for Thin Film and Microfabrication Technology (Ministry of Education), Department of Micro/Nano Electronics, School of Electronic Information and Electrical Engineering, Shanghai Jiao Tong University, Shanghai 200240, China

[*] Correspondence: xlwang83@sjtu.edu.cn; Tel.: +86-021-3420-6683

Received: 21 August 2018; Accepted: 25 September 2018; Published: 27 September 2018

Abstract: The microvasculature plays a critical role in human physiology and is closely associated to various human diseases. By combining advanced microfluidic-based techniques, the engineered 3D microvascular network model provides a precise and reproducible platform to study the microvasculature in vitro, which is an essential and primary component to engineer organ-on-chips and achieve greater biological relevance. In this review, we discuss current strategies to engineer microvessels in vitro, which can be broadly classified into endothelial cell lining-based methods, vasculogenesis and angiogenesis-based methods, and hybrid methods. By closely simulating relevant factors found in vivo such as biomechanical, biochemical, and biological microenvironment, it is possible to create more accurate organ-specific models, including both healthy and pathological vascularized microtissue with their respective vascular barrier properties. We further discuss the integration of tumor cells/spheroids into the engineered microvascular to model the vascularized microtumor tissue, and their potential application in the study of cancer metastasis and anti-cancer drug screening. Finally, we conclude with our commentaries on current progress and future perspective of on-chip vascularization techniques for fundamental and clinical/translational research.

Keywords: microfluidics; vascularization; organ-on-a-chip; vascularized tumor model; tissue engineering

---

1. Introduction

The circulatory system plays a vital role to maintain homeostasis in the human body. It comprises a closed network of arteries, veins, and capillaries that allow blood to circulate throughout the body, not only for waste product removal, but also for gas exchange and nutrient transportation, all of which are essential for organ viability. Besides participating in metabolic function, microvasculature in different organ microenvironment has unique biological functions and physical properties, such as maintaining solute and water balance between the blood and tissue compartments, or responding to different deformations and stress fluctuations [1]. Recently, the concept of "organ-on-a-chip" has been proposed to establish in vitro models that can mimic the microphysiological function and three-dimensional (3D) microstructure of human organ more accurately and specifically compared to the traditional two-dimensional (2D) cultures and animal models [2]. In addition to supplying nutrient and oxygen to the cultured tissue by perfusing the culture medium, vascularization of organ-on-a-chip can also contribute to the establishment of organ-specific microenvironments and microphysiological function by constructing the microvascular with selective barrier function similar to that in vivo. In other words, to better mimic the characteristics and functions of specific human organs in vitro, it is necessary to integrate a perfusable and functional 3D microvasculature to different organ-on-a-chip systems. Microfluidic technologies have emerged as useful tools for the development of organ-on-a-chip, which can offer precise control over various aspects of the cellular microenvironment such as a different profile of fluid flow, gradient of various growth factors, and mechanical properties of versatile biomaterials. All these advantages can facilitate the formation of biomimicking in vitro vascularized microtissue models.

Furthermore, besides the substantial physiological research on constructing and characterizing the microvasculature in homeostatic conditions in these systems, there is a great potential to model pathological conditions to study vascular-related diseases [3]. Especially for cancer biology, the tumor vasculature plays a critical role in several key events in the metastatic cascade, such as intravasation and extravasation. Engineered microvessels can be well suited to the study of mechanisms of tumor growth and metastasis, drug screening, and cancer therapies by establishing the vascularized microtumor models in vitro.

In this review, we focus on the generation of microvascular networks in 3D engineered tissue constructs and their integration into vascularized microtumor models by combining microfluidics, microfabrication, biomaterials, and tissue engineering technologies. We first discuss the current strategies for tissue vascularization. Next, we highlight relevant factors that induce vascularization inside microfluidic systems. We then provide a brief introduction of selective vascular barrier properties in different human organs and methods to construct and characterize these properties. Then, we review the current, state-of-the-art in vitro vascularized tumor-on-a-chip models in various disease stages, and their potential applications for anti-cancer drug screening. Finally, we conclude with our visions to improve the current approaches to create vascularized microtissues. This review will provide a better understanding of the vascularization process for organ-on-a-chip systems and its applications in cancer biology.

2. In Vitro Vascularization Strategies

In many early studies to understand the basis of vascular biology, 2D models were constructed by plating endothelial cells (ECs) on a flat surface such as Petri dish [4], porous membrane [5], or patterned hydrogel [6] to form a confluent monolayer to mimic the blood vessel wall. However, these 2D models cannot replicate well the proper physical structure of blood vessel in vivo, and of note is its circular shape and polarized luminal/abluminal surfaces. It is known that the blood vessel in human body is with circular cross-section. In order to mimic the microvascular in vitro, it is necessary to replicate not only its microphysiological function, but also its 3D microstructure (i.e., circular cross-section). In addition, blood vessel polarization in the apical (luminal) to basal (abluminal) axis is important for directed secretion of proteins, which can induce cord hollowing by changing the cell shape to form a lumen [7]. Thus, it is challenging to use these 2D models to replicate luminal flow in vivo and study its effect on vasculature. Due to the advantages of highly regulated spatial and temporal control over cell patterning, chemical gradients, and mechanical stimuli, microfluidics-based techniques have been widely utilized to create 3D models in vitro [8,9]. Besides that, microfluidics also enable the continuous perfusion of cell culture medium to supply the oxygen and nutrient as well as the removal of waste product, which is the necessary condition to realize the long-term survival of these vascularized microtissue. In addition, with the combination of versatile biomaterials and tissue engineering techniques, many 3D in vitro vascularization strategies have been developed, which can be broadly categorized into three main types: (1) EC lining-based methods, (2) vasculogenesis and angiogenesis-based methods, and (3) hybrid methods.

2.1. EC Lining-Based Methods

Briefly, the EC lining methods are achieved by allowing EC to form a monolayer on the inner walls of the microfluidic channel. The major advantage of these methods is that the microvasculature geometry and dimensions can be easily controlled based on different microfabrication techniques with high flexibility. Furthermore, shear stress imposed on the microvessels can be precisely controlled and calculated based on the channel dimensions and the applied flow rate. However, these methods cannot mimic vascular formation in vivo that mainly relies on the natural process of vasculogenesis and angiogenesis to form the complex microvascular networks. Since it is difficult to evenly distribute a high density of ECs inside the microfluidic channels with small diameter due to the blockage, EC lining methods are only suitable to construct large blood vessels with diameter greater than 50 μm [10]. The hollow microstructure (i.e., a single channel or a channel network) can be made of either hydrogel or polydimethylsiloxane (PDMS). For hydrogel-based microchannel, various micro-molding methods reported in the literature so far can be grouped into three main types: (1) Microneedle-based removable method for a single microchannel, (2) micropatterned, planar hydrogel slab bonding method for single-layer microchannel network, and (3) dissolvable material-based sacrificial micromolding method for multi-layer microchannel network. For EC lining inside a PDMS-based microfluidic channel, since cells might not adhere tightly to PDMS surface, a thin coating layer of basement membrane proteins (e.g., laminin, fibronectin, collagen IV, etc.) onto the microchannel inner walls is needed to enhance cell adherence.

2.1.1. Microneedle-Based Removable Method

The basic operation procedure of this method is to insert a cylindrical object (e.g., microneedle, wire, capillary tubes, etc.) into a hydrogel solution and remove it after the hydrogel is fully polymerized to create a single microchannel. Then ECs are seeded onto the inner surface of the microchannel to form a confluent EC monolayer (Figure 1A) [11–13]. Briefly, this method typically requires two separate steps: Creating a microchannel, and lining it with ECs. Furthermore, Sadr et al. proposed a new method to realize a rapid one-step engineering of microtubular constructs that combined self-assembled monolayer-based cell transfer and hydrogel photocrosslinking techniques [14]. After UV crosslinking, the gold sputtered rod modified with oligopeptide was removed from the polymerized hydrogel by applying electrical stimulation, which would transfer the layer of human umbilical vein endothelial cells (HUVECs) to 3D geometrically defined vascular-like structures in hydrogels. Moreover, a bilayer vascular structure of smooth muscle cells (SMCs) and HUVECs could be formed by a sequential deposition of SMCs, fibronectin as the cell adhesive layer and HUVECs, which exhibited stronger barrier function than EC monolayer [15]. However, since the cylindrical objects need to be removed from hydrogel, this method is limited to a simple, single straight microvascular channel [12,16].

2.1.2. Micropatterned Planar Hydrogel Slab Bonding Method

Rather than a single microchannel, a single-layer microchannel network can also be fabricated by using standard lithographic patterning method. Zheng et al. developed an interconnected microvascular array by using the additive bonding between a micropatterned collagen gel slab and a flat hydrogel layer (Figure 1B). Briefly, a micropatterned collagen gel slab was imprinted from a microstructured PDMS stamp, and bonded to a flat collagen substrate to form a collagen-based microchannel network. Then, ECs were seeded onto the inner wall of these collagen microchannels. Using this model, several vascular biology studies were performed, including vessel sprouting, interaction between ECs and mural cells, endothelial barrier function under different flow conditions, and EC response to various biochemical signals [17]. Compared with the microneedle-based removable method, although 2D planar networks can be easily formed, they usually have the rectangular cross-section due to the inherent characteristics of templates produced by lithographic methods.

2.1.3. Dissolvable Materials-Based Sacrificial Micromolding Method

In this method, a multi-layer microchannel network can be fabricated by dissolving or melting the dissolvable gel or solid material embedded in a gel matrix. With the development of 3D printing technology, more complex multi-layer microchannel networks can be interconnected using a wide range of template materials, such as carbohydrate glass [18], Pluronic F127 [19], agarose [20], gelatin [21], sodium alginate [22], and synthetic polyethylene glycol (PEG) [23]. Miller et al. created a multi-layer microvessel network from the carbohydrate glass template that can provide the sufficient mechanical support in an open lattice (Figure 1C) [18]. After dissolving the carbohydrate glass with cell culture medium, the perfusable, hollow, cylindrical network structure was left inside fibrin gel. HUVECs are then seeded into these hollow microstructures to form a 3D vascular network. While this strategy has the potential to create a complex 3D microvascular network, it is time-intensive and still limited to larger vessels determined by the resolution of 3D printer.

2.1.4. EC Lining inside a PDMS-Based Microfluidic Channel

The PDMS-based microfluidic chips enable a single-layer microchannel network design with the diameter ranging from 60–200 µm, and different microenvironment factors (e.g., flow profile, shear stress, etc.) can be precisely controlled with the sophisticated microfluidic technologies [24]. In vivo, ECs are attached to a basement membrane with the thickness of 40–120 nm [25]. Therefore, for ECs to adhere to a PDMS-based microchannel, extracellular matrix (ECM) proteins like laminin, fibronectin, or collagen IV are required to be deposited onto the microchannel inner wall prior to lining it with ECs (Figure 1(Di)) [26,27]. Moreover, microchannels fabricated through soft lithography technology have the inherent rectangular or square cross section, which cannot closely mimic real microvessels in vivo with a circular cross section. However, Esch et al. showed that ECs could be grown within both square and semicircular microchannels [28]. It was found that shear force has a greater impact on these endothelial structures than geometry, and the channel geometry only influences the final cross-sectional profile of the vascular lining.

To solve the problem of non-circular cross section, Bischel et al. developed a "viscous finger patterning" method that took advantage of basic fluidic principles to create 3D lumens with circular cross section inside a PDMS-based microchannel [29,30]. Briefly, culture medium was pumped into a partially polymerized hydrogel at the central position of the microchannel, creating a circular channel at the center while filling the rectangular corner with hydrogel. Similarly, circular PDMS microchannel can also be fabricated by introducing a pressurized air stream inside a partially solidified rectangular microchannel filled with liquid PDMS or liquid silicone oligomer, followed by baking the device to fully cure the coated layer [31,32]. Furthermore, Zhang et al. developed a hollow PDMS tube with adjustable diameters and wall thickness to mimic the elastomeric free-form blood vessels after EC lining, which could potentially replace inert plastic tubes to integrate multiple organoids into a single microfluidic circuitry (Figure 1(Dii)) [33].

2.2. Vasculogenesis and Angiogenesis-Based Methods

Vasculogenesis and angiogenesis are the fundamental processes of vascular development in vivo [34,35], as shown in Figure 2A. Vasculogenesis gives rise to formation of the first primitive vascular plexus during early embryogenesis, while angiogenesis is responsible for the remodeling and expansion of the vascular network [36]. In adults, formation of new blood vessels may occur via vasculogenesis through recruitment of ECs from bone marrow [37], as well as through angiogenesis stimulated by distress signals from the parenchymal tissue under certain physiological and pathological conditions such as wound healing, exercise, and tumor growth [38]. Different from EC lining-based methods, vasculogenesis, and angiogenesis-based methods do not require additional microstructures to guide ECs into forming vascular structure. It allows the seeded cells inside ECM to self-assemble into a 3D microvascular network *de novo*. Therefore, these methods can create a more natural vascular structure than the EC lining methods.

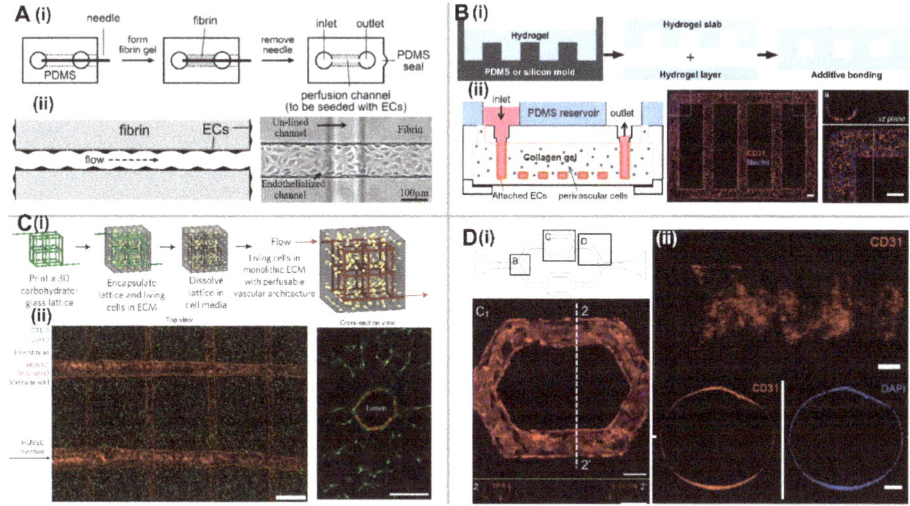

Figure 1. EC lining-based methods for in vitro vascularization. (**A**) Microneedle-based removable method. (**i**) Schematic of the simple needle-molding technique to create fluidic channels inside hydrogels; (**ii**) Endothelialized microchannel under the effect of fluidic flow. Adapted by permission from Reference [13], copyright Wiley Periodicals, Inc. 2012. (**B**) Micropatterned planar hydrogel slab bonding method. (**i**) Schematic of fluidic hydrogels fabricated by micromolding from a PDMS or silicon mold followed by a bonding process; (**ii**) Microvessel network shape inside collagen gel constructed from a micropatterned silicon stamp, and Z-stack projection of horizontal confocal sections of endothelialized microfluidic vessels immunostained with CD 31 from the view of overall network and the corner. Scale bar, 100 µm. Adapted by permission from Reference [17], copyright National Academy of Sciences, USA 2012. (**C**) Dissolvable materials-based sacrificial micromolding method. (**i**) Schematic of a 3D interconnected microvessel network formed by rapid casting of carbohydrate glass lattice as the sacrificial element with a 3D printer. The lattice is encapsulated in ECM along with living cells and dissolved in minutes in cell media without damage to nearby cells, which can generate a monolithic tissue construct with a vascular architecture. (**ii**) The micrograph (left) shows HUVECs expressing mCherry attached to the hydrogel wall to generate the microvessel network. Scale bar, 1mm. The micrograph (right) shows endothelial monolayer lined vascular lumen surrounded by 10T1/2 cells after 9 days in culture. Scale bar, 200 µm. Adapted by permission from Reference [18], copyright Nature Publishing Group 2012. (**D**) EC lining inside a PDMS-based microfluidic channel. (**i**) The schematic of PDMS-based microfluidic channel, and the confocal image of endothelial cell at location C of the channel with rectangle cross section. Scale bar, 100 µm. Adapted by permission from Reference [27], under the Creative Commons Attribution License. (**ii**) Confocal reconstruction image showing the complete lumen formed by the HUVECs inside the PDMS tube, and confocal fluorescence micrographs of the cross-sectional views of an endothelialized PDMS tube stained with CD31/nuclei. Scale bar, 100 µm (top) or 200 µm (bottom). Adapted by permission from Reference [33], copyright Royal Society of Chemistry 2016.

2.2.1. Vasculogenesis

Vasculogenesis is a process in which new blood vessels are formed from endothelial progenitor cells (EPCs) *de novo*. Within a microfluidic device, Hsu et al. demonstrated formation of microvascular networks using endothelial colony forming cell-derived endothelial cells (ECFC-ECs) and normal human lung fibroblast (NHLF) mixed in fibrin gel, and embedded inside a series of diamond-shaped tissue chambers (Figure 2B). Under a physiological level of interstitial flow, a continuous and perfusable microvessel network formed after three weeks by perfusing the cell culture medium in absence of vascular endothelial growth factor (VEGF) and basic fibroblast growth factor (bFGF) [39–41]. It was reported that the vascularized microtissue could be cultured and remained viable for up to 40 days and showed disruption and shedding over time [41]. Furthermore, it was found that combination of fibroblast-derived proteins would promote EC sprouting and was necessary for EC lumen formation [42]. VEGF and bFGF can also be utilized as chemical factors to stimulate the vasculogenesis. Raghavan et al. presented a method to control the microvessel geometry by spatially patterning ECs within micromolded collagen gels, and demonstrated that the lumenized microvessels formed within 24–48 h by stimulating with VEGF and bFGF [43]. With this method, various microvessel size and shape can be achieved by modifying the collagen gel concentration and the seeded cell density. It was found that both the average branch length and effective diameter decreased with the increase in fibrinogen concentrations, while the branch length, diameter, and area fraction of the vascularized region all increased with the increase of EC seeding density [44]. Beyond these, paracrine signaling by co-culturing with other types of cells can also influence vascular morphology, such as the number of branches, average branch length, percent vascularized area, and average vessel diameter. It was found that the larger coverage area and network stability can be achieved by co-culturing with fibroblasts, while the diameter and average vessel length are reduced under the effect of angiogenic growth factors [45]. Furthermore, besides the co-culture with fibroblast, the other stromal cells like pericytes, SMCs, or mesenchymal stem cells (MSCs) can contribute the vascular stabilization [46]. These findings will provide mechanisms to fine-tune microvascular network in vitro with specified morphological properties.

2.2.2. Angiogenesis

Angiogenesis is the formation of new blood vessel from pre-existing vessels, which is an important mechanism for vascular network remodeling. Angiogenesis occurs through a series of defined steps, including ECM degradation by matrix metalloproteinases (MMPs), angiogenic stimulus, sprouting, elongation and branching, lumen formation, anastomosis, and finally stabilization or regression [47]. Within microfluidic device, angiogenesis can be induced from either EC-lined microvessel or vasculogenesis-derived microvascular network. After applying the transendothelial flow, the sprouting angiogenesis could be induced from an EC monolayer. It was found that focal adhesion kinase-mediated signaling accompanied by extensive remodeling of cell–cell junctions and redistribution of the actin cytoskeleton contributed to the effect of transendothelial flow on vascular sprouting (Figure 2C) [48]. Furthermore, angiogenic sprouting from the EC monolayer can also be induced by VEGF gradient [49,50]. In addition, Kim et al. developed a 3D microfluidic in vitro model to investigate angiogenic sprouting from vasculogenesis-derived microvascular networks in the presence of interstitial flow (Figure 2D) [51]. It was found that angiogenic sprouting was only promoted in the direction opposite to that of interstitial flow and suppressed in the direction of flow.

To form a perfusable microvascular network, these sprouts need to connect with each other through a process called anastomosis. Yeon et al. established an array of perfusable microvessels from EC monolayer lined on both sidewalls of fibrin gel confined inside a microfluidic channel (Figure 2E) [52]. By co-culturing with lung fibroblasts, the angiogenic sprouts were stimulated and fused together after 3–4 days. The spacing of the angiogenic sprouts can be well controlled by adjusting the spacing of microstructures that confine the ECM during the microfabrication process. Similarly, Song et al. demonstrated a microfluidic device that could accurately reproduce the dynamics of vascular anastomosis under treatment of VEGF [53]. All these devices will enable a new generation of studies to investigate the mechanisms of angiogenesis, and provide a novel and practical platform for drug screening.

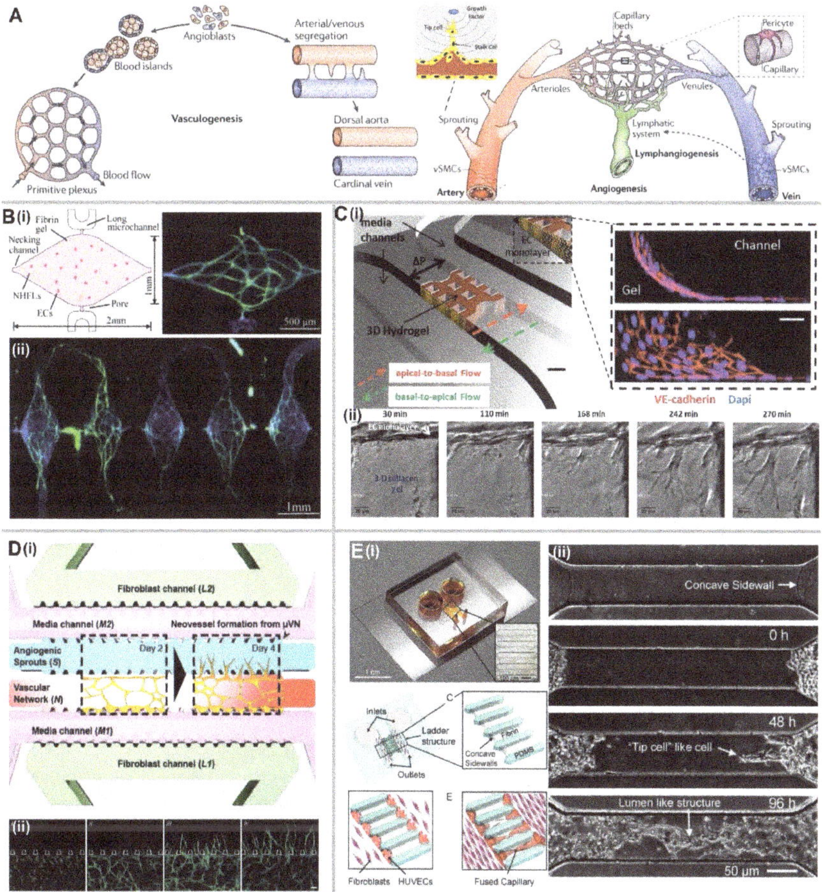

Figure 2. Vasculogenesis and angiogenesis-based methods for in vitro vascularization. (**A**) Schematic of vasculogenesis and angiogenesis in vivo. Adapted by permission from Reference [36], copyright Nature Publishing Group 2011. (**B**) 3D microvascular network formation by vasculogenesis. (**i**) Schematic of 3D microtissue chamber seeded with ECs and NHLFs in fibrin gel, and formed microvascular network labeled with CD31 (green) and 4′,6-diamidino-2-phenylindole (DAPI) (blue) for nuclei. (**ii**) Fluorescent image of 5 interconnected microvascular networks. Adapted by permission from Reference [39], copyright Royal Society of Chemistry 2013. (**C**) Angiogenesis from EC monolayer. (**i**) Schematic of microfluidic-based 3D cell culture system, and the confocal images of confluent EC monolayer showing coverage on gel and channel surfaces immunostained with anti-VE-cadherin (red) and nuclei (blue). Scale bar, 500 μm. (**ii**) Time-lapse images of sprouting angiogenesis induced by transendothelial flow. Scale bar, 20 μm. Adapted by permission from Reference [48], copyright Royal Society of Chemistry 2012. (**D**) Angiogenesis from 3D microvascular network. (**i**) Configuration of the microfluidic device and schematic of the angiogenesis assay from microvascular network. (**ii**) Temporal sequence of interconnected vascular network formation and angiogenic sprouting from the formed network inside the adjacent microchannel. Scale bar, 100 μm. Adapted by permission from Reference [51], copyright Royal Society of Chemistry 2016. (**E**) Anastomosis of sprouts into interconnected and perfusable blood vessels. (**i**) Schematic of microfluidic device design and self-organized capillary networks formed by HUVECs attached on the concave sidewalls from both sides. (**ii**) Images of sprouting HUVECs from both sides and fused capillary formation over time. Adapted by permission from Reference [52], copyright Royal Society of Chemistry 2012.

2.3. Hybrid Methods

By controlling different factors within the microenvironment, different mechanisms to stimulate vascularization can be achieved on a single microfluidic device. Kim et al. presented a microfluidic device with five parallel microchannels that could generate perfusable microvessel network through either vasculogenesis or angiogenesis by seeding the cells and fibrin gel into different channels (Figure 3A) [54]. For the vasculogenesis process, HUVECs and NHLFs were respectively mixed in fibrinogen solution and introduced separately into the central channel and the stromal cell culture channels. For angiogenesis process, the central channel and stromal cell culture channels were solely filled with fibrin gel, and HUVECs were attached onto the gel wall inside one medium channel contralateral to the NHLF seeding. Recently, Wang et al. developed a versatile microfluidic device design with robust construction methodology to establish an interconnected, perfused vascular network from artery to capillary beds to vein (Figure 3B) [26]. This method combined multiple strategies and vascular development processes such as vasculogenesis, EC lining, angiogenesis, and anastomosis to take place on a single platform in proper sequence. After the formation of a capillary network inside the tissue chamber via vasculogenesis, EC lining along the microfluidic channels adjacent to the tissue chamber was performed to serve as artery and vein. To promote a tight interconnection between the artery/vein and the capillary network, sprouting angiogenesis was induced, leading to anastomosis of the microvascular network inside the tissue chamber and the EC lining along the microfluidic channels. More importantly, this work demonstrated that non-physiologic leakage from the vasculature into interstitial space could be prevented, which was validated by 70 kDa fluorescein isothiocyanate (FITC)-labeled dextran perfusion.

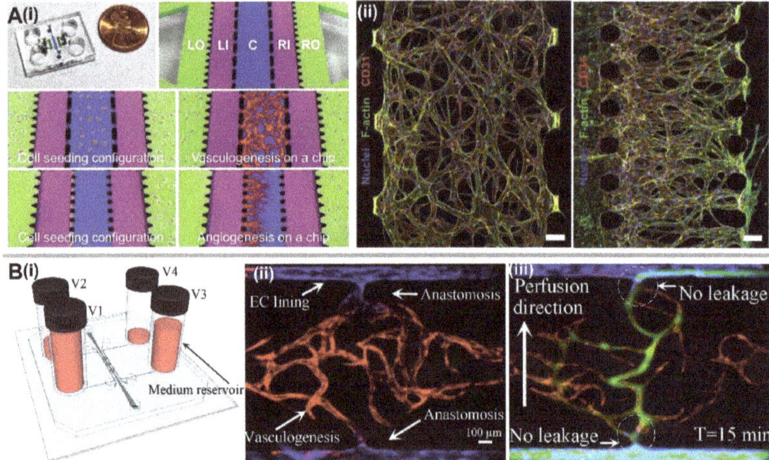

Figure 3. Hybrid methods for in vitro vascularization. (**A**) Vasculogenesis or angiogenesis realized on a single microfluidic device. (**i**) Schematic of microfluidic chip design and cell seeding configurations for vasculogenesis and angiogenesis at different microfluidic channels. (**ii**) Confocal micrographs showing the overall architectures of vascular networks established by vasculogenic and angiogenic processes at day 4. Scale bar, 100 μm. Adapted by permission from Reference [54], copyright Royal Society of Chemistry 2013. (**B**) Advanced vascularization model for generating an intact and perfusable 3D microvascular network. (**i**) Schematic of microfluidic device design by using four medium reservoirs with different hydrostatic pressures. (**ii**) Intact microvascular network formation that incorporates different stages of vascular development including vasculogenesis, EC lining, sprouting angiogenesis, and anastomosis in sequential order. (**iii**) 70 kDa FITC-dextran perfusion confirms physiologic tightness of the EC junctions and completeness of the interconnections between artery/vein and the capillary network. Adapted by permission from Reference [26], copyright Royal Society of Chemistry 2016.

3. Vascular Inducing Factors

Microfluidic system has become an emerging tool to develop microvasculature in vitro due to its advantages in precise control of factors within the microenvironment such as fluid flow at the physiological levels, distribution of different chemical factors, and different ECM properties (e.g., stiffness, orientation, etc.). These inducing factors can be divided into three main types: Biomechanical factors, extracellular (or diffusible) signaling molecules, as well as cell source and cell-cell interaction. Table 1 shows the summary of main factors on the influence of different vessel parameters.

3.1. Biomechanical Factors

In microfluidic device, biomechanical factors are various forces generated by fluid flow through the microfluidic channel or the ECM. Shear stress is the dominant biomechanical factor that highly depends on the fluid flow. It was found that shear stress would enhance barrier functions by decreasing vascular permeability and narrowing vascular wall to improve stability [12,55–57]. For EC lining, ECs were elongated and aligned with the direction of the flow under the effect of shear stress (Figure 4(Aii)). In addition, shear stress has been shown to attenuate HUVEC invasion through nitric oxide (NO) signaling irrespective of interstitial flow direction and the VEGF gradient [58,59]. Galie et al. further concluded that intraluminal shear stress and transmural flow through the endothelium above 10 dyn/cm^2 can trigger ECs to sprout and invade into the underlying matrix, independent of cell-cell junction maturation or pressure gradient across the monolayer [60].

Interstitial flow is another important regulator of various cell behaviors both in vitro and in vivo [61], and its force imposed on ECs can promote capillary formation. In vivo, interstitial flow in the range of 0.1–1 μm/s can modulate vessel formation [62]. Inside the microfluidic device, the convective velocity in the range of 1.7–11 μm/s with high Péclet number (Pe > 10), defined as the ratio of convective to diffusive transport, can stimulate vasculogenesis regardless of the interstitial flow direction (Figure 4(Ai)) [40]. However, for angiogenesis, new sprouting from an existing vascular network is more active in the reverse direction of interstitial flow [51]. Moreover, it was found that the basal-to-apical transendothelial flow could trigger the transition of ECs from a quiescent to an invasive phenotype to undergo angiogenesis [48].

In addition to fluid forces, hydrogel/ECM mechanical properties such as gel permeability, pore size, fiber diameter and stiffness, and degradation speed also plays an important role in regulating vascular formation [45]. Collagen and fibrin are the most commonly used natural hydrogels, and their mechanical properties can be flexibly fine-tuned. Collagen mechanical properties can be adjusted by changes in temperature or pH while fibrin mechanical properties can be adjusted by changes in fibrinogen and thrombin concentrations. It was found that stiffer gels would generate vessel lumens with small diameters [63], and the invasion of sprouting tubular structures into the stiff collagen gel was significantly reduced compared with the soft gel (Figure 4(Aiii)) [64]. Furthermore, fiber arrangements inside the ECM can also affect vessel alignment in response to contact guidance [65]. In addition to these passive mechanical forces generated by ECM with different mechanical properties, the active mechanical force imposed on ECM will have a profound effect on the microvasculature response. For example, the free-floating scaffolds resulted in randomly oriented vessels. However, the cyclic stretching applied on uniaxially fixated seeded constructs would result in diagonal vessels, whereas the static stretching resulted in vertical vessels [66].

3.2. Extracellular (or Diffusible) Signaling Molecules

Growth factors are cell-secreted substances capable of stimulating cellular growth, proliferation, and differentiation. VEGF is the most important growth factor that regulates EC migration and proliferation through VEGF receptor-2 signaling [67]. It was found that VEGF at lower concentrations (2.5–5 ng/ml) induces endothelial sprouting rather than at higher concentrations (15–35 ng/ml) [68].

In addition, negative VEGF gradients result in a process resembling vessel dilation, whereas positive VEGF gradients induce sprouting [59]. Besides VEGF, other angiogenic growth factors such as Angiopoietin 1 (Ang-1), Transforming Growth Factor Beta 1 (TGF-β1), and bFGF can direct angiogenic sprouting from a low-to-high concentration gradient via endothelial tip cell filopodia, and align them into well-organized structures [64,69,70]. Therefore, establishing a concentration gradient of growth factors is crucial for tissue vascularization through either vasculogenesis or angiogenesis in vitro (Figure 4B). This can be achieved by embedding the fibroblast-contained alginate spheroids or the growth factor encapsulated microparticles within the hydrogel to create a local concentration gradient, or by perfusing cell culture media containing growth factors through the microfluidic device [71,72]. Compared to single growth factors acting on the microvascular network, a combination of multiple growth factors can highly enhance vascular density and stability [73,74]. Growth factors with different gradient profiles can also be adjusted by tuning flow rate and designing different microfluidic configurations [75]. For example, a simple device with one central gel channel positioned between two medium channels can generate a linear gradient profile across the gel channel [76]. Besides a linear gradient profile in one direction, two gradient profiles can also be generated orthogonally, which could be used to examine cellular morphogenesis under the influence of multiple well-defined gradients [49]. In addition to the spatial effect of growth factors, temporal change of VEGF concentration could also affect angiogenesis. It was found that an initial gradient of a high VEGF concentration, with a programmed concentration decrease over time, yielded optimal angiogenic sprouting, as compared to a constant dose of VEGF [77]. Therefore, controlled release of growth factors, optimal combination of growth factors, dosage, gradient profile, and appropriate exposure time will facilitate proper vascular formation and stability.

In addition to the direct influence of different growth factors, the oxygen tension may be considered as an indirect method that would affect the secretion of VEGF. It is known that hypoxia inducible factor (HIF)-1 secreted in hypoxic environment can initiate the transcription of VEGF, which plays a critical role in inducing the expression of anti-apoptotic protein, especially in these pathological angiogenesis models such as tumor angiogenesis [78]. In microfluidics, the desired oxygen microenvironment can be generated through different methods. For example, due to the high gas permeability of PDMS used in microfluidics, both hypoxia microenvironment and oxygen tension gradient can be flexibly created by flowing different gases with certain concentration, perfusing gas-equilibrated medium, or loading oxygen scavenging or generation chemicals [79,80]. Furthermore, with the combination of poor gas-permeable thermoplastic, the interference from the external microenvironment can be inhibited, which can improve the oxygen control inside the PDMS-thermoplastic hybrid microfluidic devices [81,82].

3.3. Cell Source and Cell-Cell Interaction

ECs are the main cellular component for tissue vascularization. HUVECs have been widely used for microvascular construction in vitro due to their relatively simple isolation process and well-established literature references. However, other EC sources may be used instead of HUVECs for better outcomes or for certain applications that require tissue-specific ECs. For example, it was found that EPCs exhibited a substantial proliferative capacity, which is different from HUVECs with limited proliferation in the absence of pro-angiogenic factors [83,84]. Recently, induced pluripotent stem cells (iPSCs)-derived ECs have become a new EC source for personalized medicine research [85–87]. To create a mature and stable microvascular network in vitro, it is necessary to co-culture ECs with mural cells, such as fibroblasts, pericytes or MSCs. Fibroblasts can facilitate EC sprouting and lumen formation by synthesizing and maintaining essential matrix proteins (e.g. Collagen I), as well as secreting various angiogenic growth factors [40,52]. In addition, pericytes and SMCs wrapping around capillary and larger vessels (e.g. arterioles, arteries, venioles, and veins) play an important role in stabilizing newly formed vessels (Figure 4C) [88–91]. MSCs also exhibit a functional role in vascular stabilization. For example, a co-culture of HUVECs and MSCs derived from either bone marrow or human embryonic stem cells (hESCs) can increase sprouting branches and maintain structural integrity of microvascular network in vitro [76,92].

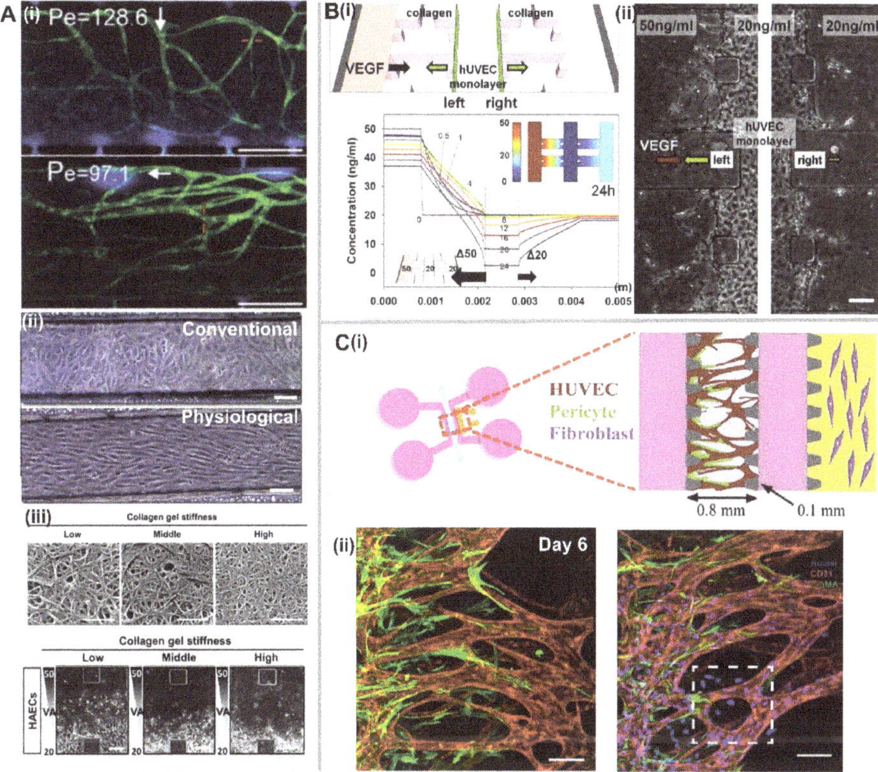

Figure 4. Three major vascular inducing factors in vascularization on chip. (**A**) Biomechanical factors. (**i**) The influence of interstitial flow with high Péclet number on vasculogenesis, and the orientation of microvascular network under the effect of interstitial flow in transverse and longitudinal directions. Scale bar, 500 μm. Adapted by permission from Reference [40], copyright Royal Society of Chemistry 2013. (**ii**) The effect of shear stress on EC lining, and ECs were elongated and aligned with the direction of the flow under the physiological conditions compared with the conventional conditions. Scale bar, 100 μm. Adapted by permission from Reference [57], copyright Nature Publishing Group 2014. (**iii**) The effect of gel stiffness on angiogenesis, and invasion of tubular structures into the stiff collagen gel was significantly reduced compared with the soft gel. Scale bar, 1 μm (top) or 150 μm (bottom). Adapted by permission from Reference [64], copyright Nature Publishing Group 2016. (**B**) Extracellular (or diffusible) signaling molecules. (**i**) Schematic of chip design where HUVECs were cultured in the center channel, and simulation result of VEGF gradient. (**ii**) Apparent sprouting angiogenesis on the left-hand side with a steeper gradient compared with the right-hand side with a gentle gradient. Scale bar, 150 μm. Adapted by permission from Reference [50], copyright American Chemical Society 2011. (**C**) Cell source and cell-cell interactions. (**i**) Schematic of microfluidic device composed of a central vessel channel, two adjacent media channels, and the outermost fibroblast channel. (**ii**) Confocal images showing matured pericytes covered the perfusable EC network on day 6. Scale bar, 100 μm. Adapted by permission from Reference [91], under the Creative Commons Attribution License 2015.

Table 1. Main factors on the influence of different vessel parameters.

Factors	Vessel Parameters	Effect
Shear stress	Barrier function	Decreasing permeability [12,55–57]
	Cell orientation	Elongated and aligned along flow direction in EC lining [57]
	EC invasion	Attenuate invasion [58,59]
Interstitial flow	Vasculogenesis	Promote capillary formation regardless of flow direction [40]
	Angiogenesis	More active in reverse direction of interstitial flow [51]
Stiff ECM	Vessel lumen	Small diameter [63]
	Sprouting invasion	Significantly reduced [70]
Fiber arrangements of ECM	Vessel alignment	Along the fiber orientation [64]
VEGF concentration	Endothelial sprouting	Induce sprouting at low concentration (2.5–5 ng/mL) [67]
VEGF gradient	Vessel morphology	Positive gradient induces sprouting, and negative gradient induces vessel dilation [59]
Hypoxia	Angiogenesis	Promote angiogenesis, such as tumor angiogenesis [78]
Co-culture with fibroblast	Perfusable vascular formation	Promote EC sprouting and lumen formation [40,52]
Co-culture with mural cells	Vessel stabilization	More stable [88–91]

4. Selective Vascular Barrier

ECs exhibit different cellular structure and barrier properties depending on specific organs in the human body. For example, endothelial cells at the blood-brain barrier (BBB) has very tight cell-cell junctions, resulting in very low permeability to protect the central nervous system from toxins and plasma fluctuations (Figure 5(Ai)) [93–95]. In kidney, ECs and podocytes form a glomerular filtration barrier that can retain 99.9% of large proteins (Figure 5(Aii)) [96,97]. In contrast, liver sinusoidal ECs have large openings that facilitate bidirectional macromolecular exchange, liposomal transportation, and nitric oxide synthesis (Figure 5(Aiii)) [98,99]. Therefore, to develop organ-on-a-chip systems with tissue-specific characteristics, it is necessary to incorporate a microvascular network with tissue-specific barrier functions. Besides improving the barrier function by imposing shear stress, an increase in circumferential stretch on the vessel wall perpendicular to the direction of flow will enhance endothelial permeability and vascular tone due to the overloaded pressure that is associated with hypertension [100]. In addition to the stimuli with various mechanical factors, the microvascular would react with the loss of barrier function upon the simulation with inflammatory cytokines [11]. Therefore, factors including cell phenotype, perfusion condition, vessel geometry, as well as growing factors and cytokines, will have a significant effect on the vessel permeability.

In versatile organ-on-chips, most designs consist of a vascular barrier made from porous membrane that sandwiched between the blood vessel compartment containing ECs and the other tissue/organ compartment containing other cell types [101,102]. This porous membrane could facilitate the creation of cellular/tissue interface as well as the crosstalk between these two compartments. The porous membrane can be fabricated from a variety of materials with different methods, like the PDMS through the template-based soft lithography [103], polymeric track etched membranes [104], etc. The pore size of porous membrane is likely the most important parameter to affect the cell transmigration, physical contact with the other cells and the paracrine signaling. In addition, the mechanical properties of the porous membrane like stiffness and strain would affect the organization and function of organs as successfully demonstrated in lung-on-a-chip [105] and gut-on-a-chip (Figure 5B) [106]. Furthermore, the surface properties of the porous membrane, functioned as an equivalent to ECM and more specifically the cell basement membrane, would directly stimulate the cell response. It was found that the porous membrane with higher roughness would increase cell adhesion due to more adsorbed protein, while a porous membrane with moderate hydrophilicity would lead to the highest cell adhesion, spreading, and growth. Therefore in order to better mimic the organ-specific vascular barrier properties, it is necessary to integrate a porous membrane with suitable microstructures (pore size, porosity, etc.), mechanical properties (substrate stiffness, strain, etc.), and surface properties (surface roughness and topography, surface chemistry, etc.).

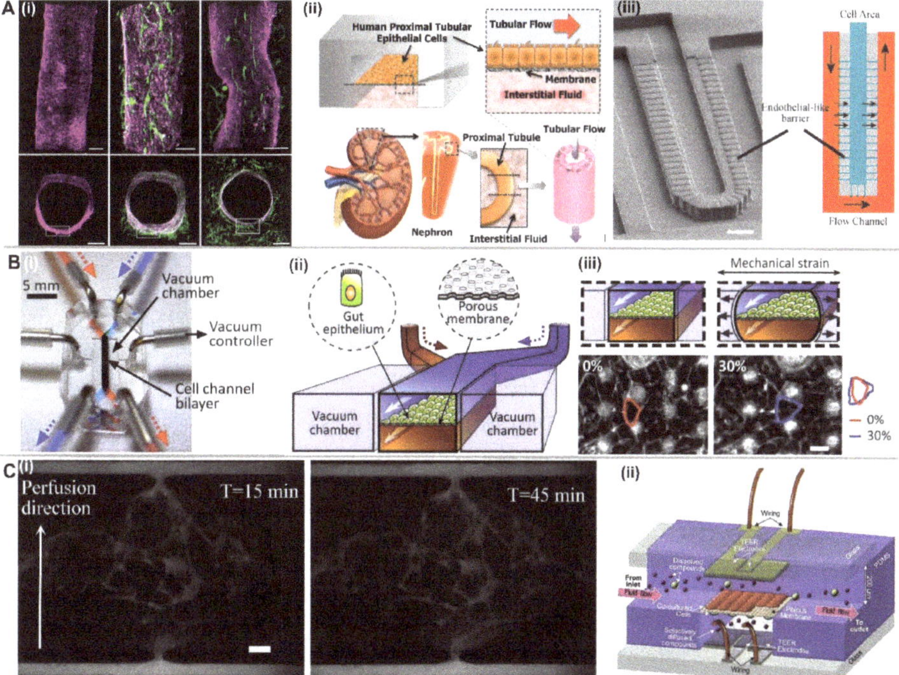

Figure 5. Selective vascular barrier property of different organ chips, construction with porous membrane and permeability characterization methods. (**A**) Microvascular with different structure and selective barrier function in different organ-on-a-chip. (**i**) BBB with tight cell-cell junctions. Fluorescence confocal micrographs of the engineered brain microvessel from human brain microvascular endothelial cells (left), and co-culture with pericytes (middle) and astrocytes (right). Scale bar, 200 μm. Adapted by permission from Reference [95], under the Creative Commons Attribution License 2016. (**ii**) Schematic of a human kidney proximal tubule-on-a-chip in an attempt to mimic the natural architecture, tissue–tissue interface, and dynamically active mechanical microenvironment of the living kidney proximal tubule. Adapted by permission from Reference [96], copyright Royal Society of Chemistry 2013. (**iii**) Schematic of the liver-on-a-chip design that resembles a liver sinusoid, including the endothelial-like barrier layer. Scale bar, 50 μm. Adapted by permission from Reference [99], copyright John Wiley and Sons, Inc. 2007. (**B**) Human gut-on-a-chip. (**i**) Photographic image of the chip prototype. (**ii**) A monolayer of gut epithelial cells is cultured on the flexible porous ECM-coated membrane in the middle of the central microchannel. Vacuum chambers are designed parallel to the channels that can apply mechanical strain to the cell layer. (**iii**) Schematics (top) and phase contrast images (bottom) of intestinal monolayers cultured within the gut-on-a-chip in the absence (left) or presence (right) of mechanical strain (30%; arrow indicated direction) exerted by applying suction to the vacuum chambers. Scale bar, 20 μm. Adapted by permission from Reference [106], copyright Royal Society of Chemistry 2012. (**C**) Vascular permeability characterization. (**i**) Permeability coefficient calculation by measuring perfusion fluorescent solutes across vessel lumen wall over time. Scale bar, 100 μm. Adapted by permission from Ref. [26], copyright Royal Society of Chemistry 2016. (**ii**) Schematic of a multi-layered microfluidic device design with incorporated electrode for internal TEER measurement on a BBB model. Adapted by permission from Reference [93], copyright Royal Society of Chemistry 2012.

As is known, the tighter endothelial junctions of established tissue barrier correspond to the lower solute permeability. Immunostaining of junctional markers such as platelet endothelial cell adhesion molecule-1 (PECAM-1) and vascular endothelial (VE)-cadherin, as well as the deposition of basement membrane proteins such as laminin and collagen IV are commonly used to validate vascular integrity qualitatively [66]. For quantitative analysis, fluorescent-tagged dextran or other biologically relevant molecules are typically utilized to characterize barrier function by measuring their permeability coefficient across the vessel wall (Figure 5(Ci)) [26,107,108]. Since albumin (MW = 66.5 kDa) is the most common protein in blood plasma, fluorescent-tagged bovine serum albumin (BSA) or 70 kDa dextran are commonly used [108]. Vascular permeability coefficient of microvessels in vitro is typically in the 10^{-6} cm s^{-1} range, which is comparable to the value of tumor microvasculature in vivo [11,17]. In addition to permeability measurement, transendothelial electrical resistance (TEER) is a widely accepted method to quantify vascular integrity by characterizing electric impedance across the endothelium (Figure 5(Cii)) [93,109]. Compared to the molecular permeability test, TEER can be performed in real-time with ease of implementation. However, TEER measurement in self-assembled microvascular models remains a challenge compared to EC monolayer or single-vessel models.

5. Application of Engineered Microvascular Networks to Cancer Biology

Besides a better understanding of the mechanisms of vascularization, the other promising application of engineered microvessels is in modeling human diseases in vitro. Endothelial dysfunction is a major physiological mechanism that can cause various vascular diseases such as thrombosis, atherosclerosis, and inflammation [110]. Especially within the field of cancer biology, in vitro tumor models have provided important tools for cancer research and served as low-cost anti-cancer drug screening platforms. Based on the tumor formation type and vascular integration, these models can be broadly classified into four categories: Transwell-based [111], spheroid-based [112], hydrogel droplet embedded culture [113], and vascularized tumor models [114]. More than 90% of cancer-related mortality is attributed to cancer metastasis [115], which often involves multiple steps closely associated with vascular pathology such as tumor angiogenesis [116], intravasation [117], and extravasation [118]. Therefore, engineered vascularized tumor models play an important role in studying cancer metastasis.

5.1. Tumor Angiogenesis

Since the tumor growth can be suppressed by cutting out their nutrient and oxygen supply through tumor vessels, it is essential to have a complete understanding of tumor angiogenesis to develop new cancer therapies. It is known that the rapidly growing angiogenic vasculature around the tumor consequently leads to irregular sprouting, tortuous microstructure, dysfunctional or absent perivascular cells, and leaky barrier properties, all of which are thought to promote tumor metastasis [119]. Several in vitro studies have been performed to study tumor angiogenesis by co-culturing cancer cells and ECs in microfluidic devices. It is normally performed by establishing a vertical 2D endothelial monolayers on the side walls, which can facilitate the better imaging of angiogenic sprouting into the 3D ECM on the axial plane. By attaching ECs to the side wall of fibrin gel, 3D sprouting was promoted by the factors secreted by the highly malignant human glioblastoma, which exhibited aberrant morphology compared to the NHLF-induced sprouts [54]. Similarly, Chung et al. engineered a platform to evaluate and quantify capillary growth and endothelial cell migration from an intact EC monolayer attached onto a collagen gel wall by co-culturing with MTLn3 cancer cells [63]. They observed that cancer cells could either attract ECs and induce capillary formation or have minimal effect, while SMCs suppress endothelial activity. Furthermore, Lee et al. presented a tumor angiogenesis model to quantify angiogenic sprouting from vasculogenesis-derived microvessels (Figure 6A) [120]. It was found that low flow velocities commonly exist in tumor vasculature, suggesting high shear stress regulation of angiogenic activity was absent in these vessels, thereby driving tumor angiogenesis due to the increased tumor-expressed angiogenic factors and endothelial permeability [121]. To further study the complex multicellular interactions during the tumor angiogenesis, a pre-vascularized tumor spheroid model was developed by co-culturing of tumor cells, ECs, and fibroblasts embedded in fibrin gel, which exhibited the robust sprouting angiogenesis after seven days of culture [122]. In addition, different from physiological angiogenesis, tumor angiogenesis is mainly stimulated by the release of HIF-1 due to the excessive cell proliferation and dysfunctional vasculature to initiate transcription of VEGF to stimulate vascular outgrowth toward the tumor tissue [78]. DelNero et al. developed the oxygen-controlled 3D alginate-based tumor model to study the tumor angiogenesis in hypoxic conditions, which demonstrated that the pro-inflammatory pathways were a critical regulator for tumor angiogenesis under the hypoxia microenvironment [123].

5.2. Tumor Intravasation

Tumor intravasation is the invasion of cancer cells through the basal membrane into blood or lymphatic vessels near the tumor stroma that can be trigged by chemotactic gradient, oxygen tension, and impaired endothelial barrier function [117]. Zervantonakis et al. demonstrated a model to explore the relationship between cancer cell intravasation and endothelial permeability in the context of cytokine-induced endothelial cell activation and paracrine signaling loops involving macrophages and tumor cells (Figure 6B). They found that macrophage-secreted tumor necrosis factor alpha (TNF-α) could enhance breast cancer cell intravasation due to impaired endothelial barrier function [124]. Similarly, Wong et al. presented a platform that positioned metastatic cancer cells next to artificial, EC-lined microvessels embedded in the ECM. Using real-time live cell fluorescence microscopy, they observed a variety of tumor-endothelium interactions, including invasion, intravasation, angiogenesis, and tumor cell dormancy [125].

5.3. Tumor Extravasation

Tumor extravasation refers to circulating tumor cells (CTCs) transmigrating through the endothelium and lodge at the secondary organs. Chen et al. developed a microfluidic platform to study tumor cell extravasation from vasculogenesis-derived microvascular networks (Figure 6C) [126]. Using high resolution time-lapse microscopy, they observed a highly dynamic nature of extravasation events, beginning with thin cancer cell protrusions across the endothelium and followed by extrusion of the remaining cell body through small openings in the endothelial barrier, which grow in size to allow nuclear transmigration. They also found that tumor transendothelial migration efficiency was significantly higher in trapped cells compared to non-trapped cells, and in cell clusters versus single cancer cells. Furthermore, Jeon et al. developed a microfluidic 3D in vitro model to analyze organ-specific human breast cancer cell extravasation into bone and muscle-mimicking microenvironments through a microvascular network concentrically wrapped by mural cells [127]. They found that the extravasation rates of breast cancer cells were significantly higher in the bone-mimicking microenvironment compared to the control matrices without stromal cells, or the muscle-mimicking microenvironments. This result validated a widely accepted phenomenon that specific tumor types preferentially metastasize to specific organ systems.

5.4. Tumor Microenvironment

Tumor microenvironment (TME) plays an important role during the tumor metastasis process, which can be broadly classified into biophysical cues (e.g., intramural flow, interstitial flow, and ECM) and biochemical cues (e.g., molecular gradient of growth factors, cytokines, oxygen, and metabolites). For intramural flow in the bioengineered microvessel models, a critical method is to develop a perfusable and functional microvascular network with either EC lining or vasculogenesis/angiogenesis-based methods. The tumor intravasation/extravasation processes are mainly regulated by the alteration of vessel wall permeability and integrity under the effect of shear stress induced by the intramural flow inside vessel lumen. Therefore, the tumor invasion and metastasis events are unlikely to occur in large vessels like arteries and veins due to the high shear stress along the tight and thick vessel wall. In contrast, the capillaries with a smaller size would easily trap the tumor cells, which are more likely to penetrate through the capillaries due to the thinner walls and lower shear stresses [128]. The interstitial flow in TME can lead to the spatial gradient of secreted cytokines, which would contribute to the directional tumor cell migration along/against the interstitial flow direction [129]. Besides the biological flow, the properties of ECM in TME are the other major biophysical factors to affect the tumor cell behavior, including stiffness, pore size, and fiber alignment [130]. It was found that the ECM with both stiff property and large pore size would facilitate the tumor cell migration to promote its invasion events. Furthermore, the hollow microchannel inside ECM remodeled by the MMP can serve as a highway to speed up the tumor cell migration [131]. Similarly, the well aligned hydrogel fibers can also be regarded as the hollow micro-sized tunnels, which can be realized by using the shear stress on the un-polymerized hydrogel inside a narrow microfluidic microstructure or directly applying the mechanical stretching force on the polymerized hydrogel.

Chemokines play a paramount role in the tumor progression. Both tumor cells and stromal cells elaborate chemokines and cytokines. These act either by autocrine or paracrine mechanisms to sustain tumor cell growth, induce angiogenesis and immune response to the tumor [132]. By using microfluidic systems, well-controlled biochemical gradient profiles can be established. Furthermore, due to the diminished supply of nutrients and oxygen caused by the defective vascular with heterogeneous perfusion in fast-growing solid tumor tissue, the TME is often hypoxic. Therefore, the oxygen tension is another critical chemical factor in TME.

5.5. Application of Vascularized Tumor-on-a-Chip

The main application for in vitro vascularized tumor models is anti-cancer drug screening. Current drug screening methods still heavily rely on 2D cell culture assays or animal models. While 2D cell culture assays fail to replicate the 3D structure and complexity of tissues in vivo, animal models fail to capture the human-specific response to drugs due to intrinsic differences in anatomy and physiology [133]. While 3D tumor spheroids can address some of these shortcomings, many tumor cell types, especially those with a highly invasive phenotype, cannot readily form spheroids. Since in vivo most drugs are delivered to the targeted tissue through the circulatory system rather than directly imposed on the cultured tissue, incorporating the vasculature into tumor models can better mimic drug delivery to provide the precise information like drug dosage as reference in further clinical trials. Besides the vascularized tumor tissue, co-culture with other stromal cells is another important aspect to mimic the drug delivery microenvironment. It was found that the co-culture of stromal cells and tumor cells showed a significantly higher drug resistance, as compared to the culture only with tumor cells [134]. Sobrino et al. developed a 3D microtumor model supported by perfused human microvasculature for anti-cancer drug screening [135]. They found that the microvessels in this platform are sensitive to the effects of both anti-angiogenic (Pazopanib, Linifanib, Cabozantinib, Axitinib, etc.) and vascular disrupting agents (Vincristine, Taxol). Therefore, this vascularized tumor model can be used to identify drugs that target cancer cells directly, or indirectly through effects on the vasculature to starve off tumor growth. Based on this work, Phan et al. further developed an arrayed vascularized microtumor platform for large-scale drug screening applications and successfully demonstrated its capacity to identify both anti-angiogenic and anti-cancer drugs from a small library of compounds (Figure 6D) [136].

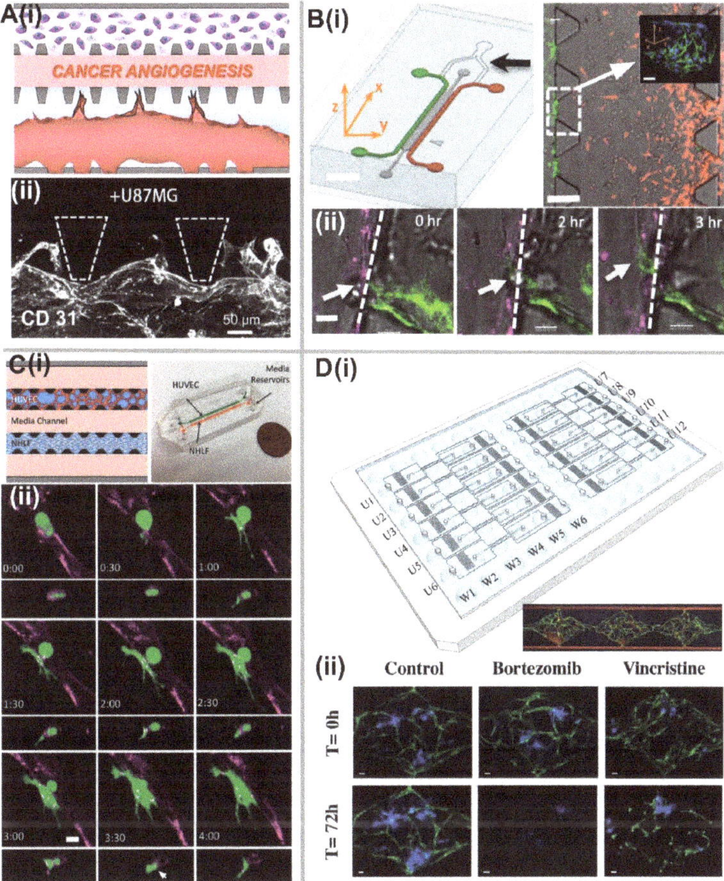

Figure 6. Vascularized tumor-on-a-chip for studying tumor metastasis and anti-cancer drug screening. (**A**) Tumor angiogenesis. (**i**) Schematic of chip design for tumor angiogenesis. (**ii**) 3D confocal image (CD31) shows the angiogenic sprouts stemmed from the microvessel wall grow toward the upper channel stimulated by the pro-angiogenic factors from U87MG cancer cell lines. Adapted by permission from Reference [120], copyright American Institute of Physics 2014. (**B**) Tumor intravasation. (**i**) Schematic of microfluidic tumor-vascular interface model, and the immunostaining image showing confluency of the endothelial monolayer on the 3D ECM. Scale bar, 2 mm (left) or 300 µm (right). (**ii**) Time sequence of a single confocal slice showing a breast carcinoma cell (white arrow) in the process of intravasaton across a HUVEC monolayer. Scale bar, 30 µm. Adapted by permission from Reference [124], copyright National Academy of Sciences, USA 2012. (**C**) Tumor extravasation. (**i**) Schematic and prototype of microfluidic device and cell-seeding configuration. (**ii**) Time-lapse confocal images showing the extravasation process of entrapped MDA-MB-231 (green) from vessel lumen (purple). Scale bar, 10 µm. Adapted by permission from Ref. [126], copyright Royal Society of Chemistry 2013. (**D**) Large-scale anti-cancer drug screening. (**i**) Schematic of microfluidic platform design that custom-fitted into a standard 96-well plate format, and the formation of vascularized microtumor tissue inside three tissue chambers. (**ii**) Drug screening validation with bortezomib and vincristine characterized by the effect on the morphology microvascular network as well as the tumor size. Scale bar, 50 µm. Adapted by permission from Reference [136], copyright Royal Society of Chemistry 2017.

The widely used strategies for anti-angiogenic therapies are to destroy the tumor vasculature, thereby depriving the tumor of nutrients. However, these methods will also severely compromise the delivery of oxygen to the solid tumor, which induces hypoxia microenvironment to make many chemotherapeutics and radiation less effective. In addition, the interstitial hypertension and nonuniform blood flow induced by the abnormalities and leakiness of tumor vasculature will also accumulate the drugs and oxygen at the concentrated region far from the solid tumors, no matter how much therapeutics and oxygenation are pumped into it. Therefore, another concept for tumor anti-angiogenic therapy is to study the normalization of tumor vasculature, which can temporarily "normalize" the abnormal structure and function of tumor vasculature into the functionally normal blood vessels by using certain antiangiogenic agents to make it more efficient for oxygen and drug delivery [137]. The microfluidics-based vascularized tumor model can provide a better understanding of tumor vessel normalization at the molecular, cellular, and functional organ level. Besides the anti-angiogenic therapies, immunotherapies against cancer have shown enormous potential, leading to the recent FDA approval of several drugs that reduce cancer progression [138]. Therefore, more studies on cancer-immune cell interactions should be better characterized by using the vascularized tumor model perfused with immune cells [139].

6. Conclusions and Future Perspectives

In this review, we have provided an overview of different vascularization strategies based on microfluidic technology. These strategies can be classified into three categories: EC lining-based methods, vasculogenesis and angiogenesis-based methods, and hybrid methods. It is apparent that each strategy carries both advantages and limitations. While the EC lining-based methods enable immediate media perfusion with well-controlled biomechanical factors, it is inefficient and only suitable to create larger blood vessels based on the predefined microchannel dimensions. In contrast, while vasculogenesis and angiogenesis-based methods provide very limited microenvironment control, they are suitable to create a capillary network down to a few micrometers in diameter by utilizing natural morphogenic properties of ECs. Therefore, an ideal strategy for proper tissue vascularization may lie in the combination of more than one technique to realize the hybrid methods. Moreover, multiple vascular inducing factors, such as biomechanical factors, extracellular (or diffusible) signaling molecules, cell source, and cell-cell interaction should be integrated into more complex microfluidic designs simultaneously to closely mimic the native microenvironment in different organs with specific vascular barrier functions. Furthermore, engineered microvessels also play an important role in modeling various human diseases, especially for the cancer research. The construction of vascularized tumor-on-a-chip can serve as a powerful tool to study the mechanisms of cancer metastasis at different stages like tumor angiogenesis, intravasation, and extravasation as well as their potential applications in anti-cancer drug screening and normalization of tumor vessels.

While many proof-of-principles studies have demonstrated versatile vascularization strategies and its applications in tumor models, challenges still remain to be solved. For example, there is an urgent need to establish more precise and geometrically controlled methods with more dynamic and flexible on-chip microenvironment control to recapitulate all aspects of vasculature in vivo. To construct a functional vascularized microtissue, interactions between ECs, supporting mural cells, and organ-specific cells with in vivo-relevant cell densities are required. Additionally, current culture media are optimized for a specific cell type when cultured in monolayer. Once these cells are in a complex microenvironment, it is necessary to have a universal blood substitute to equally supplement multiple cell types. Another major consideration is to develop thick, implantable vascularized tissue constructs from microns to millimeters with synthetic biocompatible and degradable scaffold materials for tissue repair or regeneration, as described in the recent AngioChip design [140]. The exciting future application for engineered microvessels is their potential use in the field of precision medicine by creating patient-specific microvessel models using ECs derived from iPSCs. The future work for vascularized tumor models may lie in the better understanding of the tumor microenvironment, and how to manipulate different microenvironment factors to affect the cancer progression. Besides the vascular inducing factors, the solid tumor stress induced by the abnormally high amounts of collagen and other proteins is also the key factor to design the physiologically relevant in vitro vascularized tumor models. For drug screening applications, it is imperative to develop automated, high throughput, reproducible, cost-effective, and easy-to-use microfluidic 3D cell culture systems. In terms of device design, it is necessary to integrate multiple on-chip detection sensors and analysis components for real-time monitoring and analysis of growing tissues. Finally, while the majority of organ-on-a-chip platforms are fabricated using PDMS and its derivatives due to their low cost, easy-to-use, and biocompatibility, they can absorb small molecules with transient mechanical properties, making them ineffective for long-term cell/tissue culture and drug studies. Therefore, there has been a push towards thermoplastic materials (like polystyrene, cyclic olefin copolymer, and polycarbonate, etc.) to replace PDMS for organ-on-a-chip platforms [141].

To summarize, although significant progress has been made towards tissue vascularization and related vascularized tumor models in the past decade, numerous challenges remain and will be addressed in future studies with a combination of microfluidics, microfabrication, biomaterials, tissue engineering, and other related technologies. Thus, a multidisciplinary, collaborative approach between engineers and biologists is required to develop more novel microphysiological systems that can closely mimic organ-specific physiology in vivo.

Author Contributions: X.W. designed and wrote the manuscript. Q.S. and J.P. collected references and conceived the structure of review article.

Acknowledgments: This work was supported by grants from the National Natural Science Foundation of China (No. 31600781), Science and Technology Commission of Shanghai Municipality (17JC1400200) and the Fundamental Research Funds for the Central Universities.

Conflicts of Interest: The authors declare no conflict of interest.

Abbreviations

The following abbreviations are used in this manuscript:

3D	Three-dimensional
2D	Two-dimensional
ECs	Endothelial cells
PDMS	Polydimethylsiloxane
HUVECs	Human umbilical vein endothelial cells
PEG	Polyethylene glycol
SMCs	Smooth muscle cells
ECM	Extracellular matrix
EPCs	Endothelial progenitor cells
ECFC-ECs	Endothelial colony forming cell-derived endothelial cells
NHLF	Normal human lung fibroblast
VEGF	Vascular endothelial growth factor
bFGF	Basic fibroblast growth factor
MMPs	Metalloproteinases
DAPI	4′,6-diamidino-2-phenylindole
FITC	Fluorescein isothiocyanate
NO	Nitric oxide
Ang-1	Angiopoietin 1
TGF-β1	Transforming growth factor beta 1
HIF	Hypoxia inducible factor
iPSCs	Induced pluripotent stem cells
MSCs	Mesenchymal stem cells
hESCs	Human embryonic stem cells
BBB	Blood-brain barrier

PECAM-1   Platelet endothelial cell adhesion molecule-1
VE   Vascular endothelial
BSA   Bovine serum albumin
TEER   Transendothelial electrical resistance
TNF-α Tumor necrosis factor alpha
CTCs   Circulating tumor cells
TME   Tumor microenvironment

References

1. Jain, R. Molecular regulation of vessel maturation. *Nat. Med.* **2003**, *9*, 685–693. [CrossRef] [PubMed]
2. Bhatia, S.; Ingber, D. Microfluidic organs-on-chips. *Nat. Biotechnol.* **2014**, *32*, 760–772. [CrossRef] [PubMed]
3. Rayner, S.; Zheng, Y. Engineered microvessels for the study of human disease. *J. Biomech. Eng.* **2016**, *138*, 110801. [CrossRef] [PubMed]
4. Kramer, R.; Nicolson, G. Interactions of tumor cells with vascular endothelial cell monolayers: A model for metastatic invasion. *Proc. Nat. Acad. Sci. USA* **1979**, *76*, 5704–5708. [CrossRef] [PubMed]
5. Young, E.; Watson, M.; Srigunapalan, S.; Wheeler, A.; Simmons, C. Technique for real-time measurements of endothelial permeability in a microfluidic membrane chip using laser-induced fluorescence detection. *Anal. Chem.* **2010**, *82*, 808–816. [CrossRef] [PubMed]
6. Wang, L.; Zhang, Z.; Wdzieczak-Bakala, J.; Pang, D.; Liu, J.; Chen, Y. Patterning cells and shear flow conditions: Convenient observation of endothelial cell remoulding, enhanced production of angiogenesis factors and drug response. *Lab Chip* **2011**, *11*, 4235–4240. [CrossRef] [PubMed]
7. Pelton, J.C.; Wright, C.E.; Leitges, M.; Bautch, V.L. Multiple endothelial cells constitute the tip of developing blood vessels and polarize to promote lumen formation. *Development* **2014**, *141*, 4121–4126. [CrossRef] [PubMed]
8. Halldorsson, S.; Lucumi, E.; Gómez-Sjöberg, R.; Fleming, R. Advantages and challenges of microfluidic cell culture in polydimethylsiloxane devices. *Biosens. Bioelectron.* **2015**, *63*, 218–231. [CrossRef] [PubMed]
9. Hasan, A.; Paul, A.; Vrana, N.E.; Zhao, X.; Memic, A.; Hwang, Y.S.; Dokmeci, M.R.; Khademhosseini, A. Microfluidic techniques for development of 3D vascularized tissue. *Biomaterials* **2014**, *35*, 7308–7325. [CrossRef] [PubMed]
10. Tien, J.; Wong, K.H.K.; Truslow, J.G. Vascularization of microfluidic hydrogels. In *Microfluidic Cell Culture Systems*; Bettinger, C., Borenstein, J.T., Tao, S.L., Eds.; Elsevier: Oxford, UK, 2013; pp. 205–221.
11. Chrobak, K.; Potter, D.; Tien, J. Formation of perfused, functional microvascular tubes in vitro. *Microvasc. Res.* **2006**, *71*, 185–196. [CrossRef] [PubMed]
12. Price, G.; Wong, K.; Truslow, J.; Leung, A.; Acharya, C.; Tien, J. Effect of mechanical factors on the function of engineered human blood microvessels in microfluidic collagen gels. *Biomaterials* **2010**, *31*, 6182–6189. [CrossRef] [PubMed]
13. Wong, K.; Truslow, J.; Khankhel, A.; Chan, K.; Tien, J. Artificial lymphatic drainage systems for vascularized microfluidic scaffolds. *J. Biomed. Mater. Res. Part A* **2012**, *101A*, 2181–2190. [CrossRef] [PubMed]
14. Sadr, N.; Zhu, M.; Osaki, T.; Kakegawa, T.; Yang, Y.; Moretti, M.; Fukuda, J.; Khademhosseini, A. SAM-based cell transfer to photopatterned hydrogels for microengineering vascular-like structures. *Biomaterials* **2011**, *32*, 7479–7490. [CrossRef] [PubMed]
15. Yoshida, H.; Matsusaki, M.; Akashi, M. Multilayered blood capillary analogs in biodegradable hydrogels for in vitro drug permeability assays. *Adv. Funct. Mater.* **2012**, *23*, 1736–1742. [CrossRef]
16. Buchanan, C.; Voigt, E.; Szot, C.; Freeman, J.; Vlachos, P.; Rylander, M. Three-dimensional microfluidic collagen hydrogels for investigating flow-mediated tumor-endothelial signaling and vascular organization. *Tissue Eng. Part C* **2014**, *20*, 64–75. [CrossRef] [PubMed]
17. Zheng, Y.; Chen, J.; Craven, M.; Choi, N.; Totorica, S.; Diaz-Santana, A.; Kermani, P.; Hempstead, B.; Fischbach-Teschl, C.; Lopez, J.; et al. In vitro microvessels for the study of angiogenesis and thrombosis. *Proc. Nat. Acad. Sci. USA* **2012**, *109*, 9342–9347. [CrossRef] [PubMed]
18. Miller, J.; Stevens, K.; Yang, M.; Baker, B.; Nguyen, D.; Cohen, D.; Toro, E.; Chen, A.; Galie, P.; Yu, X.; et al. Rapid casting of patterned vascular networks for perfusable engineered three-dimensional tissues. *Nat. Mater.* **2012**, *11*, 768–774. [CrossRef] [PubMed]

19. Kolesky, D.; Truby, R.; Gladman, A.; Busbee, T.; Homan, K.; Lewis, J. 3D bioprinting of vascularized, heterogeneous cell-laden tissue constructs. *Adv. Mater.* **2014**, *26*, 3124–3130. [CrossRef] [PubMed]

20. Bertassoni, L.; Cecconi, M.; Manoharan, V.; Nikkhah, M.; Hjortnaes, J.; Cristino, A.; Barabaschi, G.; Demarchi, D.; Dokmeci, M.; Yang, Y.; et al. Hydrogel bioprinted microchannel networks for vascularization of tissue engineering constructs. *Lab Chip* **2014**, *14*, 2202–2211. [CrossRef] [PubMed]

21. Golden, A.; Tien, J. Fabrication of microfluidic hydrogels using molded gelatin as a sacrificial element. *Lab Chip* **2007**, *7*, 720–725. [CrossRef] [PubMed]

22. Wang, X.; Jin, Z.; Gan, B.; Lv, S.; Xie, M.; Huang, W. Engineering interconnected 3D vascular networks in hydrogels using molded sodium alginate lattice as the sacrificial template. *Lab Chip* **2014**, *14*, 2709–2716. [CrossRef] [PubMed]

23. Tsang, V.; Chen, A.; Cho, L.; Jadin, K.; Sah, R.; DeLong, S.; West, J.; Bhatia, S. Fabrication of 3D hepatic tissues by additive photopatterning of cellular hydrogels. *FASEB J.* **2007**, *21*, 790–801. [CrossRef] [PubMed]

24. Xia, Y.; Whitesides, G. Soft lithography. *Annu. Rev. Mater. Sci.* **1998**, *28*, 153–184. [CrossRef]

25. McGuigan, A.; Sefton, M. The influence of biomaterials on endothelial cell thrombogenicity. *Biomaterials* **2007**, *28*, 2547–2571. [CrossRef] [PubMed]

26. Wang, X.; Phan, D.; Sobrino, A.; George, S.; Hughes, C.; Lee, A. Engineering anastomosis between living capillary networks and endothelial cell-lined microfluidic channels. *Lab Chip* **2016**, *16*, 282–290. [CrossRef] [PubMed]

27. Li, X.; Xu, S.; He, P.; Liu, Y. In vitro recapitulation of functional microvessels for the study of endothelial shear response, nitric oxide and (Ca2+)i. *PLoS ONE* **2015**, *10*, e0126797. [CrossRef] [PubMed]

28. Esch, M.; Post, D.; Shuler, M.; Stokol, T. Characterization of in vitro endothelial linings grown within microfluidic channels. *Tissue Eng. Part A* **2011**, *17*, 2965–2971. [CrossRef] [PubMed]

29. Bischel, L.; Young, E.; Mader, B.; Beebe, D. Tubeless microfluidic angiogenesis assay with three-dimensional endothelial-lined microvessels. *Biomaterials* **2013**, *34*, 1471–1477. [CrossRef] [PubMed]

30. Bischel, L.; Lee, S.; Beebe, D. A practical method for patterning lumens through ECM hydrogels via viscous fingering patterning. *J. Lab. Autom.* **2012**, *17*, 96–103. [CrossRef] [PubMed]

31. Abdelgawad, M.; Wu, C.; Chien, W.; Geddie, W.; Jewett, M.; Sun, Y. A fast and simple method to fabricate circular microchannels in polydimethylsiloxane (PDMS). *Lab Chip* **2011**, *11*, 545–551. [CrossRef] [PubMed]

32. Fiddes, L.; Raz, N.; Srigunapalan, S.; Tumarkan, E.; Simmons, C.; Wheeler, A.; Kumacheva, E. A circular cross-section PDMS microfluidics system for replication of cardiovascular flow conditions. *Biomaterials* **2010**, *31*, 3459–3464. [CrossRef] [PubMed]

33. Zhang, W.; Zhang, Y.; Bakht, S.; Aleman, J.; Shin, S.; Yue, K.; Sica, M.; Ribas, J.; Duchamp, M.; Ju, J.; et al. Elastomeric free-form blood vessels for interconnecting organs on chip systems. *Lab Chip* **2016**, *16*, 1579–1586. [CrossRef] [PubMed]

34. Patan, S. Vasculogenesis and angiogenesis. *Cancer Treat. Res.* **2004**, *117*, 3–32. [CrossRef] [PubMed]

35. Potente, M.; Makinen, T. Vascular heterogeneity and specialization in development and disease. *Nat. Rev. Mol. Cell Biol.* **2017**, *16*, 477–494. [CrossRef] [PubMed]

36. Herbert, S.P.; Stainier, D.Y. Molecular control of endothelial cell behaviour during blood vessel morphogenesis. *Nat. Rev. Mol. Cell Biol.* **2011**, *12*, 551–564. [CrossRef] [PubMed]

37. Tepper, O.M.; Capla, J.M.; Galiano, R.D.; Ceradini, D.J.; Callaghan, M.J.; Kleinman, M.E.; Gurtner, G.C. Adult vasculogenesis occurs through in situ recruitment, proliferation, and tubulization of circulating bone marrow-derived cells. *Blood* **2005**, *105*, 1068–1077. [CrossRef] [PubMed]

38. McLoughlin, P.; Keane, M.P. Physiological and pathological angiogenesis in the adult pulmonary circulation. *Compr. Physiol.* **2011**, *1*, 1473–1508. [CrossRef] [PubMed]

39. Hsu, Y.; Moya, M.; Hughes, C.; George, S.; Lee, A. A microfluidic platform for generating large-scale nearly identical human microphysiological vascularized tissue arrays. *Lab Chip* **2013**, *13*, 2990–2998. [CrossRef] [PubMed]

40. Hsu, Y.; Moya, M.; Abiri, P.; Hughes, C.; George, S.; Lee, A. Full range physiological mass transport control in 3D tissue cultures. *Lab Chip* **2013**, *13*, 81–89. [CrossRef] [PubMed]

41. Moya, M.; Hsu, Y.; Lee, A.; Hughes, C.; George, S. In vitro perfused human capillary networks. *Tissue Eng. Part C* **2013**, *19*, 730–737. [CrossRef] [PubMed]

42. Newman, A.C.; Nakatsu, M.N.; Chou, W.; Gershon, P.D.; Hughes, C.C. The requirement for fibroblasts in angiogenesis: Fibroblast-derived matrix proteins are essential for endothelial cell lumen formation. *Mol. Biol. Cell* **2011**, *22*, 3791–3800. [CrossRef] [PubMed]

43. Raghavan, S.; Nelson, C.; Baranski, J.; Lim, E.; Chen, C. Geometrically controlled endothelial tubulogenesis in micropatterned gels. *Tissue Eng. Part A* **2010**, *16*, 2255–2263. [CrossRef] [PubMed]

44. Ghajar, C.M.; Chen, X.; Harris, J.W.; Suresh, V.; Hughes, C.C.; Jeon, N.L.; Putnam, A.J.; George, S.C. The effect of matrix density on the regulation of 3-D capillary morphogenesis. *Biophys. J.* **2008**, *94*, 1930–1941. [CrossRef] [PubMed]

45. Whisler, J.; Chen, M.; Kamm, R. Control of perfusable microvascular network morphology using a multiculture microfluidic system. *Tissue Eng. Part C* **2014**, *20*, 543–552. [CrossRef] [PubMed]

46. Mendes, L.F.; Pirraco, R.P.; Szymczyk, W.; Frias, A.M.; Santos, T.C.; Reis, R.L.; Marques, A.P. Perivascular-like cells contribute to the stability of the vascular network of osteogenic tissue formed from cell sheet-based constructs. *PLoS ONE* **2012**, *7*, e41051. [CrossRef] [PubMed]

47. Logsdon, E.; Finley, S.; Popel, A.; Gabhann, F. A systems biology view of blood vessel growth and remodelling. *J. Cell. Mol. Med.* **2013**, *18*, 1491–1508. [CrossRef] [PubMed]

48. Vickerman, V.; Kamm, R. Mechanism of a flow-gated angiogenesis switch: Early signaling events at cell–matrix and cell–cell junctions. *Integr. Biol.* **2012**, *4*, 863. [CrossRef] [PubMed]

49. Shin, Y.; Jeon, J.; Han, S.; Jung, G.; Shin, S.; Lee, S.; Sudo, R.; Kamm, R.; Chung, S. In vitro 3D collective sprouting angiogenesis under orchestrated ANG-1 and VEGF gradients. *Lab Chip* **2011**, *11*, 2175–2181. [CrossRef] [PubMed]

50. Jeong, G.; Han, S.; Shin, Y.; Kwon, G.; Kamm, R.; Lee, S.; Chung, S. Sprouting angiogenesis under a chemical gradient regulated by interactions with an endothelial monolayer in a microfluidic platform. *Anal. Chem.* **2011**, *83*, 8454–8459. [CrossRef] [PubMed]

51. Kim, S.; Chung, M.; Ahn, J.; Lee, S.; Jeon, N. Interstitial flow regulates the angiogenic response and phenotype of endothelial cells in a 3D culture model. *Lab Chip* **2016**, *16*, 4189–4199. [CrossRef] [PubMed]

52. Yeon, J.; Ryu, H.; Chung, M.; Hu, Q.; Jeon, N. In vitro formation and characterization of a perfusable three-dimensional tubular capillary network in microfluidic devices. *Lab Chip* **2012**, *12*, 2815–2822. [CrossRef] [PubMed]

53. Song, J.; Bazou, D.; Munn, L. Anastomosis of endothelial sprouts forms new vessels in a tissue analogue of angiogenesis. *Integr. Biol.* **2012**, *4*, 857–862. [CrossRef] [PubMed]

54. Kim, S.; Lee, H.; Chung, M.; Jeon, N. Engineering of functional, perfusable 3D microvascular networks on a chip. *Lab Chip* **2013**, *13*, 1489–1500. [CrossRef] [PubMed]

55. Tzima, E.; Irani-Tehrani, M.; Kiosses, W.; Dejana, E.; Schultz, D.; Engelhardt, B.; Cao, G.; DeLisser, H.; Schwartz, M. A mechanosensory complex that mediates the endothelial cell response to fluid shear stress. *Nature* **2005**, *437*, 426–431. [CrossRef] [PubMed]

56. Khan, O.; Sefton, M. Endothelial cell behaviour within a microfluidic mimic of the flow channels of a modular tissue engineered construct. *Biomed. Microdevices* **2010**, *13*, 69–87. [CrossRef] [PubMed]

57. Abaci, H.; Shen, Y.; Tan, S.; Gerecht, S. Recapitulating physiological and pathological shear stress and oxygen to model vasculature in health and disease. *Sci. Rep.* **2014**, *4*, 4951. [CrossRef] [PubMed]

58. Nguyen, D.; Stapleton, S.; Yang, M.; Cha, S.; Choi, C.; Galie, P.; Chen, C. Biomimetic model to reconstitute angiogenic sprouting morphogenesis in vitro. *Proc. Nat. Acad. Sci. USA* **2013**, *110*, 6712–6717. [CrossRef] [PubMed]

59. Song, J.; Munn, L. Fluid forces control endothelial sprouting. *Proc. Nat. Acad. Sci. USA* **2011**, *108*, 15342–15347. [CrossRef] [PubMed]

60. Galie, P.; Nguyen, D.; Choi, C.; Cohen, D.; Janmey, P.; Chen, C. Fluid shear stress threshold regulates angiogenic sprouting. *Proc. Nat. Acad. Sci. USA* **2014**, *111*, 7968–7973. [CrossRef] [PubMed]

61. Swartz, M.; Fleury, M. Interstitial flow and its effects in soft tissues. *Annu. Rev. Biomed. Eng.* **2007**, *9*, 229–256. [CrossRef] [PubMed]

62. Ng, C.; Helm, C.; Swartz, M. Interstitial flow differentially stimulates blood and lymphatic endothelial cell morphogenesis in vitro. *Microvasc. Res.* **2004**, *68*, 258–264. [CrossRef] [PubMed]

63. Chung, S.; Sudo, R.; Mack, P.; Wan, C.; Vickerman, V.; Kamm, R. Cell migration into scaffolds under co-culture conditions in a microfluidic platform. *Lab Chip* **2009**, *9*, 269–275. [CrossRef] [PubMed]

64. Seo, H.; Jeong, H.; Joo, H.; Choi, S.; Park, C.; Kim, J.; Choi, J.; Cui, L.; Hong, S.; Chung, S.; et al. Intrinsic FGF2 and FGF5 promotes angiogenesis of human aortic endothelial cells in 3D microfluidic angiogenesis system. *Sci. Rep.* **2016**, *6*, 28832. [CrossRef] [PubMed]

65. Laco, F.; Grant, M.; Black, R. Collagen-nanofiber hydrogel composites promote contact guidance of human lymphatic microvascular endothelial cells and directed capillary tube formation. *J. Biomed. Mater. Res. A* **2012**, *101A*, 1787–1799. [CrossRef] [PubMed]

66. Rosenfeld, D.; Landau, S.; Shandalov, Y.; Raindel, N.; Freiman, A.; Shor, E.; Blinder, Y.; Vandenburgh, H.H.; Mooney, D.J.; Levenberg, S. Morphogenesis of 3D vascular networks is regulated by tensile forces. *Proc. Natl. Acad. Sci. USA* **2016**, *113*, 3215–3220. [CrossRef] [PubMed]

67. Ferrara, N. VEGF-A: A critical regulator of blood vessel growth. *Eur. Cytokine Netw.* **2009**, *20*, 158–163. [CrossRef] [PubMed]

68. Nakatsu, M.; Sainson, R.; Aoto, J.; Taylor, K.; Aitkenhead, M.; Pérez-del-Pulgar, S.; Carpenter, P.; Hughes, C. Angiogenic sprouting and capillary lumen formation modeled by human umbilical vein endothelial cells (HUVEC) in fibrin gels: The role of fibroblasts and Angiopoietin-1. *Microvasc. Res.* **2003**, *66*, 102–112. [CrossRef]

69. Gerhardt, H.; Golding, M.; Fruttiger, M.; Ruhrberg, C.; Lundkvist, A.; Abramsson, A.; Jeltsch, M.; Mitchell, C.; Alitalo, K.; Shima, D.; et al. VEGF guides angiogenic sprouting utilizing endothelial tip cell filopodia. *J. Cell Biol.* **2003**, *161*, 1163–1177. [CrossRef] [PubMed]

70. Del Amo, C.; Borau, C.; Gutiérrez, R.; Asín, J.; García-Aznar, J. Quantification of angiogenic sprouting under different growth factors in a microfluidic platform. *J. Biomech.* **2016**, *49*, 1340–1346. [CrossRef] [PubMed]

71. Perets, A.; Baruch, Y.; Weisbuch, F.; Shoshany, G.; Neufeld, G.; Cohen, S. Enhancing the vascularization of three-dimensional porous alginate scaffolds by incorporating controlled release basic fibroblast growth factor microspheres. *J. Biomed. Mater. Res. A* **2003**, *65A*, 489–497. [CrossRef] [PubMed]

72. Lim, S.; Kim, C.; Aref, A.; Kamm, R.; Raghunath, M. Complementary effects of ciclopirox olamine, a prolyl hydroxylase inhibitor and sphingosine 1-phosphate on fibroblasts and endothelial cells in driving capillary sprouting. *Integr. Biol.* **2013**, *5*, 1474–1484. [CrossRef] [PubMed]

73. Hao, X.; Silva, E.; Manssonbroberg, A.; Grinnemo, K.; Siddiqui, A.; Dellgren, G.; Wardell, E.; Brodin, L.; Mooney, D.; Sylven, C. Angiogenic effects of sequential release of VEGF-A165 and PDGF-BB with alginate hydrogels after myocardial infarction. *Cardiovasc. Res.* **2007**, *75*, 178–185. [CrossRef] [PubMed]

74. Nillesen, S.; Geutjes, P.; Wismans, R.; Schalkwijk, J.; Daamen, W.; van Kuppevelt, T. Increased angiogenesis and blood vessel maturation in acellular collagen–heparin scaffolds containing both FGF2 and VEGF. *Biomaterials* **2007**, *28*, 1123–1131. [CrossRef] [PubMed]

75. Baker, B.; Trappmann, B.; Stapleton, S.; Toro, E.; Chen, C. Microfluidics embedded within extracellular matrix to define vascular architectures and pattern diffusive gradients. *Lab Chip* **2013**, *13*, 3246–3252. [CrossRef] [PubMed]

76. Jeon, J.; Bersini, S.; Whisler, J.; Chen, M.; Dubini, G.; Charest, J.; Moretti, M.; Kamm, R. Generation of 3D functional microvascular networks with human mesenchymal stem cells in microfluidic systems. *Integr. Biol.* **2014**, *6*, 555–563. [CrossRef] [PubMed]

77. Silva, E.; Mooney, D. Effects of VEGF temporal and spatial presentation on angiogenesis. *Biomaterials* **2010**, *31*, 1235–1241. [CrossRef] [PubMed]

78. Krock, B.; Skuli, N.; Simon, M. Hypoxia-induced angiogenesis good and evil. *Genes Cancer* **2011**, *2*, 1117–1133. [CrossRef] [PubMed]

79. Brennan, M.; Rexius-Hall, M.; Elgass, L.; Eddington, D. Oxygen control with microfluidics. *Lab Chip* **2014**, *14*, 4305–4318. [CrossRef] [PubMed]

80. Oomen, P.; Skolimowski, M.; Verpoorte, E. Implementing oxygen control in chip-based cell and tissue culture systems. *Lab Chip* **2016**, *16*, 3394–3414. [CrossRef] [PubMed]

81. Mehta, G.; Lee, J.; Cha, W.; Tung, Y.; Linderman, J.; Takayama, S. Hard top soft bottom microfluidic devices for cell culture and chemical analysis. *Anal. Chem.* **2009**, *81*, 3714–3722. [CrossRef] [PubMed]

82. Chang, C.; Cheng, Y.; Tu, M.; Chen, Y.; Peng, C.; Liao, W.; Tung, Y. A polydimethylsiloxane–polycarbonate hybrid microfluidic device capable of generating perpendicular chemical and oxygen gradients for cell culture studies. *Lab Chip* **2014**, *14*, 3762–3772. [CrossRef] [PubMed]

83. Kim, J.; Yang, K.; Park, H.; Cho, S.; Han, S.; Shin, Y.; Chung, S.; Lee, J. Implantable microfluidic device for the formation of three-dimensional vasculature by human endothelial progenitor cells. *Biotechnol. Bioprocess Eng.* **2014**, *19*, 379–385. [CrossRef]

84. Dai, X.; Cai, S.; Ye, Q.; Jiang, J.; Yan, X.; Xiong, X.; Jiang, Q.; Wang, A.; Tan, Y. A novel in vitro angiogenesis model based on a microfluidic device. *Chin. Sci. Bull.* **2011**, *56*, 3301–3309. [CrossRef] [PubMed]

85. Samuel, R.; Daheron, L.; Liao, S.; Vardam, T.; Kamoun, W.; Batista, A.; Buecker, C.; Schafer, R.; Han, X.; Au, P.; et al. Generation of functionally competent and durable engineered blood vessels from human induced pluripotent stem cells. *Proc. Nat. Acad. Sci. USA* **2013**, *110*, 12774–12779. [CrossRef] [PubMed]

86. Reed, D.; Foldes, G.; Harding, S.; Mitchell, J. Stem cell-derived endothelial cells for cardiovascular disease: A therapeutic perspective. *Br. J. Clin. Pharmacol.* **2013**, *75*, 897–906. [CrossRef] [PubMed]

87. Wang, Z.; Au, P.; Chen, T.; Shao, Y.; Daheron, L.; Bai, H.; Arzigian, M.; Fukumura, D.; Jain, R.; Scadden, D. Endothelial cells derived from human embryonic stem cells form durable blood vessels in vivo. *Nat. Biotechnol.* **2007**, *25*, 317–318. [CrossRef] [PubMed]

88. Bergers, G. The role of pericytes in blood-vessel formation and maintenance. *Neuro Oncol.* **2005**, *7*, 452–464. [CrossRef] [PubMed]

89. Van der Meer, A.; Orlova, V.; ten Dijke, P.; van den Berg, A.; Mummery, C. Three-dimensional co-cultures of human endothelial cells and embryonic stem cell-derived pericytes inside a microfluidic device. *Lab Chip* **2013**, *13*, 3562–3568. [CrossRef] [PubMed]

90. Liu, Y.; Rayatpisheh, S.; Chew, S.; Chan-Park, M. Impact of endothelial cells on 3D cultured smooth muscle cells in a biomimetic hydrogel. *ACS Appl. Mater. Interfaces* **2012**, *4*, 1378–1387. [CrossRef] [PubMed]

91. Kim, J.; Chung, M.; Kim, S.; Jo, D.; Kim, J.; Jeon, N. Engineering of a biomimetic pericyte-covered 3D microvascular network. *PLoS ONE* **2015**, *10*, e0133880. [CrossRef] [PubMed]

92. Boyd, N.; Nunes, S.; Jokinen, J.; Krishnan, L.; Chen, Y.; Smith, K.; Stice, S.; Hoying, J. Microvascular mural cell functionality of human embryonic stem cell derived mesenchymal cells. *Tissue Eng. Part A* **2011**, *17*, 1537–1548. [CrossRef] [PubMed]

93. Booth, R.; Kim, H. Characterization of a microfluidic in vitro model of the blood-brain barrier (μBBB). *Lab Chip* **2012**, *12*, 1784–1792. [CrossRef] [PubMed]

94. Brown, J.; Pensabene, V.; Markov, D.; Allwardt, V.; Neely, M.; Shi, M.; Britt, C.; Hoilett, O.; Yang, Q.; Brewer, B.; et al. Recreating blood-brain barrier physiology and structure on chip: A novel neurovascular microfluidic bioreactor. *Biomicrofluidics* **2015**, *9*, 054124. [CrossRef] [PubMed]

95. Herland, A.; van der Meer, A.; FitzGerald, E.; Park, T.; Sleeboom, J.; Ingber, D. Distinct contributions of astrocytes and pericytes to neuroinflammation identified in a 3D human blood-brain barrier on a chip. *PLoS ONE* **2016**, *11*, e0150360. [CrossRef] [PubMed]

96. Jang, K.; Mehr, A.; Hamilton, G.; McPartlin, L.; Chung, S.; Suh, K.; Ingber, D. Human kidney proximal tubule-on-a-chip for drug transport and nephrotoxicity assessment. *Integr. Biol.* **2013**, *5*, 1119–1129. [CrossRef] [PubMed]

97. Kelly, E.; Wang, Z.; Voellinger, J.; Yeung, C.; Shen, D.; Thummel, K.; Zheng, Y.; Ligresti, G.; Eaton, D.; Muczynski, K.; et al. Innovations in preclinical biology: Ex vivo engineering of a human kidney tissue microperfusion system. *Stem Cell Res. Ther.* **2013**, *4*, S17. [CrossRef] [PubMed]

98. Prodanov, L.; Jindal, R.; Bale, S.; Hegde, M.; McCarty, W.; Golberg, I.; Bhushan, A.; Yarmush, M.; Usta, O. Long-term maintenance of a microfluidic 3D human liver sinusoid. *Biotechnol. Bioeng.* **2015**, *113*, 241–246. [CrossRef] [PubMed]

99. Lee, P.; Hung, P.; Lee, L. An artificial liver sinusoid with a microfluidic endothelial-like barrier for primary hepatocyte culture. *Biotechnol. Bioeng.* **2007**, *97*, 1340–1346. [CrossRef] [PubMed]

100. Birukova, A.; Chatchavalvanich, S.; Rios, A.; Kawkitinarong, K.; Garcia, J.; Birukov, K. Differential regulation of pulmonary endothelial monolayer integrity by varying degrees of cyclic stretch. *Am. J. Pathol.* **2006**, *168*, 1749–1761. [CrossRef] [PubMed]

101. Chung, H.; Mireles, M.; Kwarta, B.; Gaborski, T. Use of porous membranes in tissue barrier and co-culture models. *Lab Chip* **2018**, *18*, 1671–1689. [CrossRef] [PubMed]

102. Pasman, T.; Grijpma, D.; Stamatialis, D.; Poot, A. Flat and microstructured polymeric membranes in organs-on-chips. *J. R. Soc. Interface* **2018**, *15*, 20180351. [CrossRef] [PubMed]

103. Huh, D.; Kim, H.J.; Fraser, J.P.; Shea, D.E.; Khan, M.; Bahinski, A.; Hamilton, G.A.; Ingber, D.E. Microfabrication of human organs-on-chips. *Nat. Protoc.* **2013**, *8*, 2135–2157. [CrossRef] [PubMed]

104. Abhyankar, V.V.; Wu, M.; Koh, C.Y.; Hatch, A.V.A. reversibly sealed, easy access, modular (SEAM) microfluidic architecture to establish in vitro tissue interfaces. *PLoS ONE* **2016**, *11*, e0156341. [CrossRef] [PubMed]

105. Huh, D.; Matthews, B.; Mammoto, A.; Montoya-Zavala, M.; Hsin, H.; Ingber, D. Reconstituting organ-level lung functions on a chip. *Science* **2010**, *328*, 1662–1668. [CrossRef] [PubMed]

106. Kim, H.; Huh, D.; Hamilton, G.; Ingber, D. Human gut-on-a-chip inhabited by microbial flora that experiences intestinal peristalsis-like motions and flow. *Lab Chip* **2012**, *12*, 2165–2174. [CrossRef] [PubMed]

107. Yuan, W.; Lv, Y.; Zeng, M.; Fu, B. Non-invasive measurement of solute permeability in cerebral microvessels of the rat. *Microvasc. Res.* **2009**, *77*, 166–173. [CrossRef] [PubMed]

108. Shi, L.; Zeng, M.; Sun, Y.; Fu, B. Quantification of blood-brain barrier solute permeability and brain transport by multiphoton microscopy. *J. Biomech. Eng.* **2014**, *136*, 031005. [CrossRef] [PubMed]

109. Srinivasan, B.; Kolli, A.; Esch, M.; Abaci, H.; Shuler, M.; Hickman, J. TEER measurement techniques for in vitro barrier model systems. *J. Lab. Autom.* **2015**, *20*, 107–126. [CrossRef] [PubMed]

110. Barakat, A. Blood flow and arterial endothelial dysfunction: Mechanisms and implications. *C. R. Phys.* **2013**, *14*, 479–496. [CrossRef]

111. Hulkower, K.; Herber, R. Cell migration and invasion assays as tools for drug discovery. *Pharmaceutics* **2011**, *3*, 107–124. [CrossRef] [PubMed]

112. Friedrich, J.; Seidel, C.; Ebner, R.; Kunz-Schughart, L. Spheroid-based drug screen: Considerations and practical approach. *Nat. Protoc.* **2009**, *4*, 309–324. [CrossRef] [PubMed]

113. Yabushita, H.; Ohnishi, M.; Komiyama, M.; Mori, T.; Noguchi, M.; Kishida, T.; Noguchi, Y.; Sawaguchi, K.; Noguchi, M. Usefulness of collagen gel droplet embedded culture drug sensitivity testing in ovarian cancer. *Oncol. Rep.* **2004**, *12*, 307–311. [CrossRef] [PubMed]

114. Butler, J.; Kobayashi, H.; Rafii, S. Instructive role of the vascular niche in promoting tumour growth and tissue repair by angiocrine factors. *Nat. Rev. Cancer* **2010**, *10*, 138–146. [CrossRef] [PubMed]

115. Chaffer, C.; Weinberg, R. A perspective on cancer cell metastasis. *Science* **2011**, *331*, 1559–1564. [CrossRef] [PubMed]

116. Weis, S.M.; Cheresh, D.A. Tumor angiogenesis: Molecular pathways and therapeutic targets. *Nat. Med.* **2011**, *17*, 1359–1370. [CrossRef] [PubMed]

117. Chiang, S.P.; Cabrera, R.M.; Segall, J.E. Tumor cell intravasation. *Am. J. Physiol. Cell Physiol.* **2016**, *311*, C1–C14. [CrossRef] [PubMed]

118. Stoletov, K.; Kato, H.; Zardouzian, E.; Kelber, J.; Yang, J.; Shattil, S.; Klemke, R. Visualizing extravasation dynamics of metastatic tumor cells. *J. Cell Sci.* **2010**, *123*, 2332–2341. [CrossRef] [PubMed]

119. Fukumura, D.; Duda, D.; Munn, L.; Jain, R. Tumor microvasculature and microenvironment: Novel insights through intravital imaging in pre-clinical models. *Microcirculation* **2010**, *17*, 206–225. [CrossRef] [PubMed]

120. Lee, H.; Park, W.; Ryu, H.; Jeon, N. A microfluidic platform for quantitative analysis of cancer angiogenesis and intravasation. *Biomicrofluidics* **2014**, *8*, 054102. [CrossRef] [PubMed]

121. Buchanan, C.; Verbridge, S.; Vlachos, P.; Rylander, M. Flow shear stress regulates endothelial barrier function and expression of angiogenic factors in a 3D microfluidic tumor vascular model. *Cell Adh. Migr.* **2014**, *8*, 517–524. [CrossRef] [PubMed]

122. Ehsan, S.M.; Welch-Reardon, K.M.; Waterman, M.L.; Hughes, C.C.; George, S.C. A three-dimensional in vitro model of tumor cell intravasation. *Integr. Biol* **2014**, *6*, 603–610. [CrossRef] [PubMed]

123. DelNero, P.; Lane, M.; Verbridge, S.S.; Kwee, B.; Kermani, P.; Hempstead, B.; Stroock, A.; Fischbach, C. 3D culture broadly regulates tumor cell hypoxia response and angiogenesis via pro-inflammatory pathways. *Biomaterials* **2015**, *55*, 110–118. [CrossRef] [PubMed]

124. Zervantonakis, I.; Hughes-Alford, S.; Charest, J.; Condeelis, J.; Gertler, F.; Kamm, R. Three-dimensional microfluidic model for tumor cell intravasation and endothelial barrier function. *Proc. Nat. Acad. Sci. USA* **2012**, *109*, 13515–13520. [CrossRef] [PubMed]

125. Wong, A.; Searson, P. Live-cell imaging of invasion and intravasation in an artificial microvessel platform. *Cancer Res.* **2014**, *74*, 4937–4945. [CrossRef] [PubMed]

126. Chen, M.; Whisler, J.; Jeon, J.; Kamm, R. Mechanisms of tumor cell extravasation in an in vitro microvascular network platform. *Integr. Biol* **2013**, *5*, 1262–1271. [CrossRef] [PubMed]

127. Jeon, J.; Bersini, S.; Gilardi, M.; Dubini, G.; Charest, J.; Moretti, M.; Kamm, R. Human 3D vascularized organotypic microfluidic assays to study breast cancer cell extravasation. *Proc. Nat. Acad. Sci. USA* **2015**, *112*, 214–219. [CrossRef] [PubMed]

128. Guo, P.; Cai, B.; Lei, M.; Liu, Y.; Fu, B.M. Differential arrest and adhesion of tumor cells and microbeads in the microvasculature. *Biomech. Model. Mechanobiol.* **2014**, *13*, 537–550. [CrossRef] [PubMed]

129. Polacheck, W.J.; Charest, J.L.; Kamm, R.D. Interstitial flow influences direction of tumor cell migration through competing mechanisms. *Proc. Natl. Acad. Sci. USA* **2011**, *108*, 11115–11120. [CrossRef] [PubMed]

130. Carey, S.P.; Kraning-Rush, C.M.; Williams, R.M.; Reinhart-King, C.A. Biophysical control of invasive tumor cell behavior by extracellular matrix microarchitecture. *Biomaterials* **2012**, *33*, 4157–4165. [CrossRef] [PubMed]

131. Foda, H.D.; Zucker, S. Matrix metalloproteinases in cancer invasion, metastasis and angiogenesis. *Drug Discov. Today* **2001**, *6*, 478–482. [CrossRef]

132. Raman, D.; Baugher, P.J.; Thu, Y.M.; Richmond, A. Role of chemokines in tumor growth. *Cancer Lett.* **2007**, *256*, 137–165. [CrossRef] [PubMed]

133. Esch, E.W.; Bahinski, A.; Huh, D. Organs-on-chips at the frontiers of drug discovery. *Nat. Rev. Drug Discov.* **2015**, *14*, 248–260. [CrossRef] [PubMed]

134. Xu, Z.; Gao, Y.; Hao, Y.; Li, E.; Wang, Y.; Zhang, J.; Wang, W.; Gao, Z.; Wang, Q. Application of a microfluidic chip-based 3d co-culture to test drug sensitivity for individualized treatment of lung cancer. *Biomaterials* **2013**, *34*, 4109–4117. [CrossRef] [PubMed]

135. Sobrino, A.; Phan, D.; Datta, R.; Wang, X.; Hachey, S.; Romero-López, M.; Gratton, E.; Lee, A.; George, S.; Hughes, C. 3D microtumors in vitro supported by perfused vascular networks. *Sci. Rep.* **2016**, *6*, 31589. [CrossRef] [PubMed]

136. Phan, D.; Wang, X.; Craver, B.; Sobrino, A.; Zhao, D.; Chen, J.; Lee, L.; George, S.; Lee, A.; Hughes, C. A vascularized and perfused organ-on-a-chip platform for large-scale drug screening applications. *Lab Chip* **2017**, *17*, 511–520. [CrossRef] [PubMed]

137. Jain, R. Normalization of tumor vasculature: An emerging concept in antiangiogenic therapy. *Science* **2005**, *307*, 58–62. [CrossRef] [PubMed]

138. Boussommier-Calleja, A.; Li, R.; Chen, M.B.; Wong, S.C.; Kamm, R.D. Microfluidics: A new tool for modeling cancer-immune interactions. *Trends Cancer* **2016**, *2*, 6–19. [CrossRef] [PubMed]

139. Mattei, F.; Schiavoni, G.; De Ninno, A.; Lucarini, V.; Sestili, P.; Sistigu, A.; Fragale, A.; Sanchez, M.; Spada, M.; Gerardino, A.; et al. A multidisciplinary study using in vivo tumor models and microfluidic cell-on-chip approach to explore the cross-talk between cancer and immune cells. *J. Immunotoxicol.* **2014**, *11*, 337–346. [CrossRef] [PubMed]

140. Zhang, B.; Montgomery, M.; Chamberlain, M.; Ogawa, S.; Korolj, A.; Pahnke, A.; Wells, L.; Massé, S.; Kim, J.; Reis, L.; et al. Biodegradable scaffold with built-in vasculature for organ-on-a-chip engineering and direct surgical anastomosis. *Nat. Mater.* **2016**, *15*, 669–678. [CrossRef] [PubMed]

141. Van Midwoud, P.M.; Janse, A.; Merema, M.T.; Groothuis, G.M.; Verpoorte, E. Comparison of biocompatibility and adsorption properties of different plastics for advanced microfluidic cell and tissue culture models. *Anal. Chem.* **2012**, *84*, 3938–3944. [CrossRef] [PubMed]

 *micromachines*

Article

Nanogroove-Enhanced Hydrogel Scaffolds for 3D Neuronal Cell Culture: An Easy Access Brain-on-Chip Model

Alex Bastiaens [1,†], **Sijia Xie** [2,†,‡] **and Regina Luttge** [1,*]

[1] Neuro-Nanoscale Engineering Group, Department of Mechanical Engineering and Institute of Complex Molecular Systems (ICMS), Eindhoven University of Technology, 5600 MB Eindhoven, The Netherlands; a.j.bastiaens@tue.nl

[2] MESA+ Institute for Nanotechnology, University of Twente, 7500AE Enschede, The Netherlands; sijiaxie1987@gmail.com

* Correspondence: r.luttge@tue.nl; Tel.: +31-40-247-5235

† These authors contributed equally to the manuscript.

‡ Current affiliation: Laboratory for Micro- and Nanotechnology, Paul Scherrer Institute (PSI), 5232 Villigen PSI, Switzerland.

Received: 30 August 2019; Accepted: 20 September 2019; Published: 23 September 2019

Abstract: In order to better understand the brain and brain diseases, in vitro human brain models need to include not only a chemically and physically relevant microenvironment, but also structural network complexity. This complexity reflects the hierarchical architecture in brain tissue. Here, a method has been developed that adds complexity to a 3D cell culture by means of nanogrooved substrates. SH-SY5Y cells were grown on these nanogrooved substrates and covered with Matrigel, a hydrogel. To quantitatively analyze network behavior in 2D neuronal cell cultures, we previously developed an automated image-based screening method. We first investigated if this method was applicable to 3D primary rat brain cortical (CTX) cell cultures. Since the method was successfully applied to these pilot data, a proof of principle in a reductionist human brain cell model was attempted, using the SH-SY5Y cell line. The results showed that these cells also create an aligned network in the 3D microenvironment by maintaining a certain degree of guidance by the nanogrooved topography in the z-direction. These results indicate that nanogrooves enhance the structural complexity of 3D neuronal cell cultures for both CTX and human SH-SY5Y cultures, providing a basis for further development of an easy access brain-on-chip model.

Keywords: 3D cell culture; neuronal cells; SH-SY5Y cells; image-based screening; nanogrooves; neuronal cell networks; neuronal guidance

---

1. Introduction

Current models to study the brain and brain diseases are limited in their capabilities to translate findings toward the discovery of drugs that help treat these diseases [1]. In particular, the failure rate in drug development for brain diseases is disproportionately high compared to other drug discovery areas [2] and has yet to provide drugs that can slow, halt or reverse neurodegenerative diseases such as Alzheimer's disease (AD) [3] or Parkinson's disease (PD) [4]. While improvements can be made toward animal model studies, another approach is to study in vitro models. The so-called organ-on-chips (OOC) technology provides an opportunity to study human cells or organoids in a physiologically relevant microenvironment, potentially bridging the gap between current pre-clinical studies and human-based clinical trials [5,6].

To study the brain and brain diseases in an OOC platform, coined a brain-on-chip (BOC), we require a well-designed microsystem that can incorporate an environment for brain cells in a culture which mimics structural complexity in 3D [7]. In the natural cerebral cortex, a layered construction is formed during development. Through the cortical layers, so called "cortical columns" consist of aggregated cell bodies and neuronal outgrowths that are alternatively organized in laminar layers, and perpendicularly distributed in each layer as columns [8,9]. It is reported that neurons locating at the same radial column show similar response properties recorded by microelectrodes [10]. While brain organoids can be cultured and exhibit such 3D structural complexity [11], there is little control over the location of regions, where specific brain cell types or structures can be generated or the actual control of the distal arrangement of cells to each other. The use of micro- and nanotechnology can aid the design of BOC platforms that offer more control of these parameters and potentially more reproducible experiments.

Current research shows that nanotopography can guide neuronal cells' outgrowth and hence neuronal cell network organization, thereby creating more in vivo-like structures in in vitro models [12–19]. Previously, our group has investigated a range of nanogrooved patterns in different substrate materials and their impact on primary rat brain cortical (CTX) cells and the neuroblastoma cell line SH-SY5Y in 2D cultures. We have shown, in these studies, that slight dimensional changes in these nanogroove patterns elicit different responses with regards to the extent of the guidance effect, or alignment, for both CTX cultures and SH-SY5Y neuronal outgrowths [20,21]. Specifically, a pattern with a ridge width of 230 nm and pattern periodicity of 600 nm provided good alignment results for CTX cultures, as compared to other nanogroove dimensions. Higher alignment was also seen for SH-SY5Y cells when cultured on patterns with a ridge width of 230 nm and pattern periodicity of 1000 nm, compared to other nanogroove dimensions. Based on an automated image-based screening method we developed, it was shown that for SH-SY5Y cells, a smaller ridge width compared to the pattern periodicity of a nanogrooved pattern resulted in an increased alignment of neuronal outgrowths. Also, an increase in alignment of such outgrowths correlated to increased differentiation of SH-SY5Y cells and increased outgrowth length [22]. The material properties of the nanogrooved substrate material, here the stiffness of either silicon, glass or polydimethylsiloxane (PDMS), has shown to influence neurite length, too [23]. We first investigated if our previously developed automated imaged-based neuronal network screening method also applied to 3D CTX cell cultures. Since the method was successfully adopted in the data set of the pilot study, a proof of principle in a reductionist human brain cell model was attempted using also the SH-SY5Y cell line in 3D. In this paper, it is our aim to demonstrate that it is possible to control outgrowth direction in 3D by simply applying soft scaffolds, such as Matrigel, atop of the cells initially seeded on nanogrooved substrates. The benefits of a 3D microenvironment using a hydrogel, and the interfacing with a nanogrooved substrate, are complementary, and enable us to generate structural complexity in BOC platforms.

Applying this new culture concept, the cells will form a cellular network on top of the nanogrooves and inside the 3D volume of the hydrogel. By performing immunofluorescence staining and subsequent confocal microscopy, the cell cultures can be visualized. However, to analyze the generated data quantitatively, our previously developed automated image-based screening method was used to assess outgrowth alignment and length throughout the 3D volume of the cell cultures. In this paper, we describe how the screening method was tested using z-stack data of astrocytes in a 3D CTX culture on nanogrooves and subsequently applied to a human brain cell model culture experiment using SH-SY5Y. This type of data analysis confirmed that these neuronal cell cultures both contain a degree of outgrowth alignment, both at the nanogrooved substrate surface, as well as up to several micrometers away from that surface.

The results further show that the concept of nanogrooved-enhanced 3D scaffolds provides structural complexity in a reductionist human brain model. This type of BOC culture format hence offers a viable and accessible tool for creating more in vivo-like brain models using nanotechnology, thereby bringing advances in our understanding of the brain and brain disease.

2. Materials and Methods

2.1. Nanogroove Substrate Fabrication

The fabrication details for the nanoresist scaffolds were previously published by Xie and Luttge [20]. In brief, nanoresist was patterned using jet and flash imprint lithography (J-FIL) on a standard double-sided polished 100 mm diameter silicon wafer (Figure 1A). The wafer was first coated with a bottom anti-reflective coating (BARC; DUV30J, Brewer Science, Rolla, MO, USA) layer using a quartz master, kindly provided by the Bijkerk group at the University of Twente. The nanogrooved patterns had dimensions in the range of 200–2000 nm pattern periodicity, a ridge width of 100–1340 nm, and a height of 118 nm. Subsequently, the nanoresist patterns were used directly as a template in thermal nanoimprint lithography, creating a negative copy in cyclic olefin copolymer (COC; optical grade TOPAS 8007S-04, Topas Advanced Polymers, Frankfurt am Main, Germany) using a thermal nanoimprint lithography system (EITRE 6, Obducat, Lund, Sweden) at 108 °C and applying a pressure of 4 MPa (Figure 1B). After cooling the COC to room temperature, it was peeled off the nanoresist and was ready to be used as a negative mold for further replication steps.

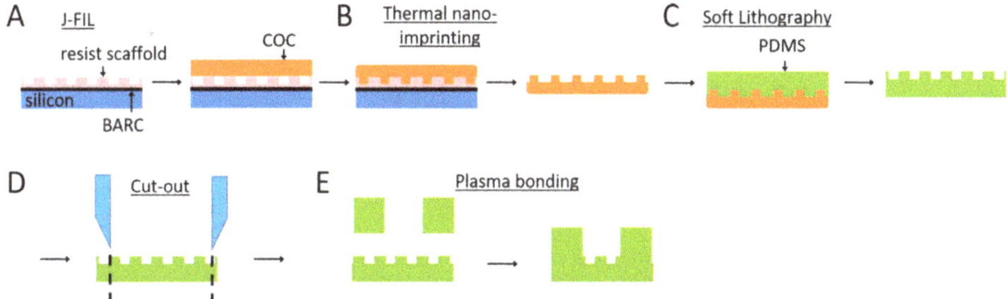

Figure 1. Fabrication process of polydimethylsiloxane (PDMS) constructs for 3D neuronal cell culture on nanogrooved substrates. (**A**) Nanoresist patterning through jet and flash imprint lithography on a silicon wafer coated with a bottom anti-reflective coating using a quartz master. (**B**) Thermal nano-imprinting of the nanogrooved patterns in the nanoresist to a negative mold in cyclic olefin copolymer. (**C**) Soft lithography from the negative mold into a layer of polydimethylsiloxane (PDMS). (**D**) A 10 mm puncher was used to cut out the required nanogrooved pattern from the PDMS. (**E**) A PDMS ring was cut out using punchers. After plasma oxidation of both the nanogrooved PDMS from (**D**) and the PDMS ring, these parts were bonded to create a PDMS construct with a well for 3D neuronal cell culture containing a nanogrooved substrate on the bottom.

Replication into culture substrates was performed using soft lithography of PDMS (Sylgard 184, Dow Corning, Midland, MI, USA) on either the nanoresist or the COC molds (Figure 1C). As the original nanoresist master contained complementary nanogrooves dimensions, e.g., complementary ridge widths of 230 nm and 370 nm with a 600 nm pattern periodicity, PDMS copies with similar dimensions could be obtained from both the nanoresist and the COC molds. PDMS elastomer and curing agent were mixed at a 10:1 ratio and degassed for 10 min using a vacuum desiccator. PDMS was spin-coated onto the mold at 500 rpm for 60 s to achieve a 100 μm layer of PDMS. Subsequently, the PDMS was cured in an oven at 65 °C for 4 h and peeled off for the nanogrooved PDMS cell culture substrates in these experiments.

For the SH-SY5Y cells, the nanogrooved patterns with a 1000 nm pattern periodicity and 230 nm ridge width, copied from the COC mold, were used in the experiments. A 10 mm diameter punch was used to punch out this specific nanogrooved pattern from the range of patterns (Figure 1D). A 100 μm layer of PDMS was spin-coated onto a 100 mm diameter silicon wafer and cured in the same way as described for the COC mold. The layer of PDMS was peeled off and placed in a polystyrene Petri dish. A 10 mm and a 3 mm punch were used to punch out rings of the same size as the nanogrooved PDMS, with a hole in the center. Plasma oxidation (EMITECH K1050X, Quorum, Lewes, UK) at 50 W for 60 s using an air plasma was performed on the nanogrooved PDMS and the PDMS ring prior to bonding the ring onto the nanogrooved PDMS to create a PDMS construct with a nanogrooved well of defined dimensions (Figure 1E). To strengthen the bond, the PDMS constructs were placed in an oven at 65 °C for 1 h after plasma oxidation.

For CTX cell culture experiments, patterns with a 600 nm pattern periodicity and 230 nm ridge width were copied directly from the nanoresist mold. The same preparation method of the PDMS constructs were applied for the CTX cell culture experiments as well.

Atomic force microscopy (AFM) was used to verify the presence and fidelity of the nanogrooved patterns in the PDMS constructs. The XE-100 (Park Systems, Santa Clara, CA, USA) with non-contact cantilever tips (PPP-NCHR, Park Systems) was used to assess the nanogrooved patterns used for SH-SY5Y cell culture. An AFM (Bruker, Santa Barbara, CA, USA) with FastScan cantilever tips (Bruker) was used to assess the nanogrooved patterns used for the CTX cell culture.

2.2. 3D Cell Culture on Nanogrooved Substrates

2.2.1. 3D SH-SY5Y Cells

After fabrication, the PDMS constructs were placed into the wells of a 24-well plate and subsequently sterilized by submersion in 70% ethanol for 5 min. Next, each construct was washed three times using sterilized, deionized water. The well of the PDMS constructs was coated with 10 µg/cm fibronectin (FC010, Sigma Aldrich, St. Louis, MI, USA) in phosphate buffered saline (PBS; LO BE02-017F, Westburg, Leusden, the Netherlands) for 30 min. Following that, the fibronectin coating was aspirated from the PDMS substrate, and the cell suspension was immediately dispensed on the surface.

The cell suspension consisted of the human neuroblastoma SH-SY5Y cell line (94030304, Sigma Aldrich), a reductionist model of brain cells. Cells were taken from cryovials in liquid nitrogen and thawed, after which the cells were plated in a T75 flask. Cells were kept in growth medium, composed of Dulbecco's modified Eagle's medium and Ham's F12 medium at a 1:1 ratio (L0093-500, VWR, Amsterdam, the Netherlands) supplemented with 10% fetal bovine serum (SFBS lot 11113, Bovogen, East Keilor, Australia) and 1% penicillin/streptomycin (LODE17-602E, Westburg). The cells were kept in an incubator at 37 °C and 5% CO_2 until the cell culture reached 70–80% confluency. At the start of the experiments with the PDMS constructs, cells were trypsinized, counted and then seeded onto the fibronectin-coated substrates in growth medium at 20,000 cells/cm^2.

As cells were seeded in a 2D fashion onto the nanogrooved surface of the PDMS constructs, a hydrogel layer covering the cells was subsequently added. For the hydrogel layer, growth factor-reduced Matrigel (734-0269, VWR) was used. Cell cultures were left for at least 3 h, during which cells could adhere to the PDMS cell culture substrate. An hour prior to adding Matrigel to the cell culture, the Matrigel was thawed on ice. When cells had been inspected for substrate adhesion and the Matrigel was fully thawed, the medium was taken off the cell culture and Matrigel was added into the well of the PDMS construct. The cultures where then placed in an incubator at 37 °C and 5% CO_2 for 15 min for the Matrigel to gel, after which 1 mL of growth medium was added to the wells of the 24-well plate hosting the PDMS constructs.

At 1 days in vitro (DIV), the growth medium was replaced with growth medium supplemented with 10 µM retinoic acid (RA; R2625, Sigma Aldrich), so as to initiate neuronal differentiation [24,25]. The medium supplemented with RA was maintained in the cell cultures for 72 h and refreshed halfway. Subsequently, cell differentiation was further enhanced by adding growth medium with 50 ng/mL brain-derived neurotrophic factor (B2795, Sigma Aldrich) for 72 h [26], with the medium being refreshed halfway. After differentiation, cells were kept in growth medium, with the growth medium being refreshed every other day. At 21 DIV, cell cultures were fixed using 3.7% formaldehyde in PBS for 1 h prior to immunofluorescence staining. During cell culture, cells were observed using an EVOS FL microscope (ThermoFisher Scientific, Eindhoven, The Netherlands).

2.2.2. 3D CTX Cells

The PDMS constructs for culturing CTX cells were coated with a non-protein-based polymer to enhance the cell-substrate adhesion. Branched polyethylenimine (PEI; approximate molecular weight 60,000, 50 wt % aq. solution; Acros Organics, Morris Plains, NJ, USA; CAS: 9002-98-6) was used as the coating material [27]. The PEI coating solution was prepared with a concentration of 50 µg/mL in sterile MilliQ water. The PDMS constructs were first sterilized by submersion in 70% ethanol for 5 min and air-dried, then treated with oxygen plasma and subsequently immersed in the coating solution at 37 °C for a minimum duration of 2.5 h or overnight. Before culturing, the residual coating solution was aspirated in a biological safety cabinet. The PDMS constructs were then air-dried and ready for cell culturing.

CTX cells were isolated from new-born rat's brain (Mother rat: Wistar Crl:WU) and dissociated in R12H medium [28], then seeded on the coated PDMS constructs with an approximate number of 3×10^5 cells/well (6000 cells/mm^2). After seeding, the 2D seeded CTX culture was kept in a cell incubator at 37 °C, 5% CO_2 and 95% humidity for the first 24 h. At 1 DIV, Matrigel was diluted with R12H medium to 75% of its original concentration and subsequently added to the 2D CTX culture, in the same way as it was for the SH-SY5Y cells. Without any differentiation treatment, the CTX cells were cultured with R12H medium containing 100 units penicillin/streptomycin, at 37 °C, 5% CO_2 and 95% humidity. The medium was refreshed every 2 days until the culture was terminated at 14 DIV. Cell cultures were fixed using 4% formaldehyde solution diluted from 37% formaldehyde aqueous solution (Sigma Aldrich) to 1/8 (v/v) with PBS (Sigma Aldrich, D8537) for 1 h prior to immunofluorescence staining.

2.3. Immunofluorescence Staining and Imaging

2.3.1. SH-SY5Y Cells

Immunofluorescence staining was performed on the cell cultures. The SH-SY5Y cells were stained using anti-β-Tubulin III (T8578, Sigma Aldrich) and anti-mouse IgG Alexa 555 (A21424, Molecular Probes, Eugene, OR, USA) as primary and secondary antibodies to selectively stain SH-SY5Y that differentiated into a neuron-like phenotype [29]. Cells were permeabilized for one hour with 0.1% Triton X-100 (Merck Millipore, Burlington, MA, USA), then washed twice for 30 min with PBS. Then, for four hours a blocking buffer of 10% FBS in PBS was applied, followed by incubation overnight with 1:100 primary antibody and 10% FBS in PBS. Cells were washed four times for 2 h with 10% FBS in PBS, then incubated overnight with 1:100 secondary antibody and 10% FBS in PBS. Additionally, 2 drops/mL of NucBlue™ (Thermofisher Scientific) and 2 drops/mL of Actingreen™ (Thermofisher Scientific) were added to the secondary antibody solution to stain the cell nuclei and the cytoskeletal protein F-actin, respectively. Finally, samples were rinsed four times for 1 h with PBS prior to imaging. To ensure cell cultures would not dry out during confocal imaging, PBS was replaced with the embedding medium Mowiol (Sigma Aldrich), which is typically used to mount 2D samples on cover slips prior to imaging, but can also be used for 3D samples. Z-stack images of the stained cells were obtained using a confocal laser scanning microscope (Leica TCS SP5X, Leica, Milton Keynes, UK).

2.3.2. CTX Cells

Fluorescent immunostaining was performed on the CTX cell culture to identify the major cell types within the neuronal networks. Glial fibrillary acidic protein (GFAP) and microtubule-associated protein 2 (MAP2) were marked as specific indications for astrocytes and neurons, respectively. Anti-GFAP antibody (goat; Sigma Aldrich, SAB2500462; 1:200) and anti-MAP2 antibody (mouse; Sigma Aldrich, M9942; 1:200) were used as the primary antibodies. Anti-goat IgG (H + L), CF™ 488A (donkey; Sigma Aldrich, SAB4600032; 1:500), anti-mouse IgG (H + L), both highly cross-adsorbed, and CF™ 640R antibody (donkey; Sigma Aldrich, SAB4600154; 1:500) were used as secondary antibodies. The nuclei of the cortical cells were stained with DAPI (FluoroShield™, ImmunoBioScience Corp., Mukilteo,

WA, USA). The stained cell cultures were imaged with a confocal laser scanning fluorescence microscope (Nikon A1 confocal, Nikon, Tokyo, Japan). Imaging of 3D cultures was realized by the z-stage stack scanning in the confocal scanning mode. The images were analyzed with the NIS Elements Analysis software provided by Nikon, and Fiji (Image J) software provided by NIH (Bethesda, MA, USA).

2.4. Cell Culture Analysis: CTX Set and SH-SY5Y

An automated image-based screening method developed by our group [22] was tested toward its use for 3D z-stacks, instead of the originally intended use of 2D cell culture images (Figure 2). In brief, the screening method takes slices at an approximate interval of 1.8 μm from a z-stack (Figure 2A) and uses these images in neuronal cell image analysis software (HCA-Vision, CSIRO, Canberra, Australia) to identify cell bodies and cell outgrowths (Figure 2B). From this automated process, numerous parameters can be extracted, but here the focus lies on determining the number of cells and the length of the outgrowths. The cell bodies are then subtracted from the original image, which results in images with only outgrowths present. A Frangi vesselness algorithm determines the orientation of all of the outgrowths (Figure 2C). Outgrowths are considered aligned when the orientation is <30° relative to the underlying nanogrooved pattern, implying that isotropic orientation of outgrowths is at 33% alignment. Together, the number of cells, outgrowth length and outgrowth alignment describe how the 3D neuronal cell network is shaped by the cells and outgrowths, and how these are influenced by the nanogrooved substrate.

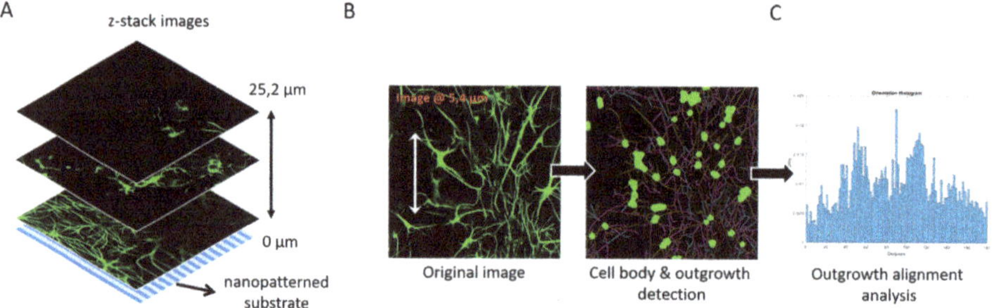

Figure 2. Overview of image-based screening analysis on z-stack images from a 3D neuronal cell culture. (**A**) Schematic example using several image slices from a z-stack showing astrocytes (green) in a primary rat brain CTX culture on top of a nanogrooved substrate. Astrocytes and their outgrowths can be seen across the multiple slices, showing their presence farther away from the nanogrooved surface. (**B**) Cell image analysis software (HCA-Vision, CSIRO) was used to identify cell bodies and their respective outgrowth for each of the slices in the z-stack. (**C**) The identification of outgrowths in (**B**) was analyzed using a Frangi vesselness algorithm to determine the percentage of outgrowths aligned to the underlying nanogrooved substrate for each slice of the z-stack to determine the degree of alignment as function of the distance away from the nanogrooved substrate.

Testing of the image-based screening method was done by first using the method on the data from a 3D CTX culture experiment. Previously, this data was analyzed manually and only generated results with regard to the alignment of the astrocyte outgrowths [7]. Here, the screening method was used on the same dataset to generate results on the outgrowth alignment, as well as the number of cells and the total length of all outgrowths for each of the analyzed slices of the z-stack. Based on the assessment of these results, the screening method was also used to analyze these parameters for 3D SH-SY5Y cell cultures (n = 3).

3. Results

3.1. Nanogrooved Substrate Fidelity

The nanogrooved wells of the fabricated PDMS constructs for 3D neuronal cell cultures were measured to assess the nanogrooved pattern fidelity. An AFM measurement was performed for the pattern with a ridge width of 230 nm and pattern periodicity of 1000 nm, as used in the SH-SY5Y cell culture (Figure 3A), resulting in a pattern with an actual ridge width of 246 ± 22 nm and a pattern periodicity of 990 ± 22 nm (n = 5). The same was done for the pattern with a ridge width of 230 nm and a pattern periodicity of 600 nm, as used in the CTX cell culture (Figure 3B), resulting in a pattern with an actual ridge width of 219 ± 5 nm and a pattern periodicity of 587 ± 0 nm (n = 5). Qualitatively, both patterns showed good fidelity with regard to overall pattern dimensions.

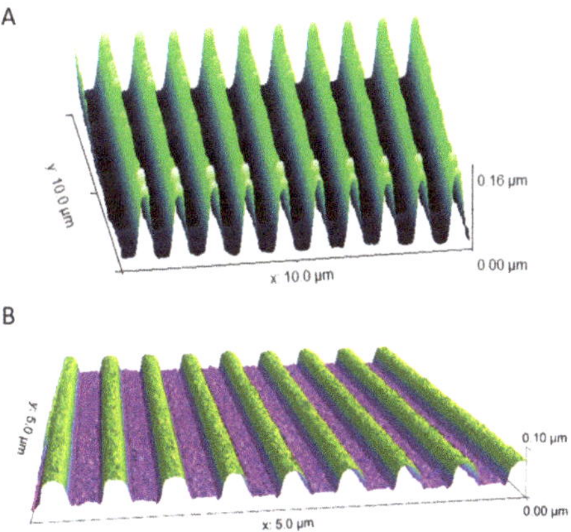

Figure 3. AFM measurements on nanogrooved PDMS substrates. (**A**) Nanogrooved pattern with a ridge width of 246 ± 22 nm and a pattern periodicity of 990 ± 22 nm (n = 5). This pattern was used for SH-SY5Y cell culture experiments. (**B**) Nanogrooved pattern with a ridge width of 219 ± 5 nm and a pattern periodicity of 587 ± 0 nm (n = 5). This pattern was used for primary rat brain cortical cell culture experiments.

3.2. 3D SH-SY5Y Cell Culture

The SH SY5Y cell culture experiments were observed for cell survival during the duration of the experiment (Figure 4A). Cells exhibited neuronal differentiation, as seen from the neuronal outgrowths. Also, cells would migrate into the 3D microenvironment, seen from the cells that were out of focus during bright field microscopy of the running experiments when focusing on the cells on the nanogrooved substrate. After immunofluorescence staining, confocal microscopy of the cells (Figure 4B–D) showed that indeed part of the cell population had migrated into the hydrogel, with cells visible up to ~80 µm distance from the nanogrooved substrate. Qualitatively, outgrowths could be observed to preferentially orient along the nanogrooved pattern (Figure 4B) when close to the substrate's surface. This alignment effect decreased the farther away as cells were from the surface (Figure 4C,D).

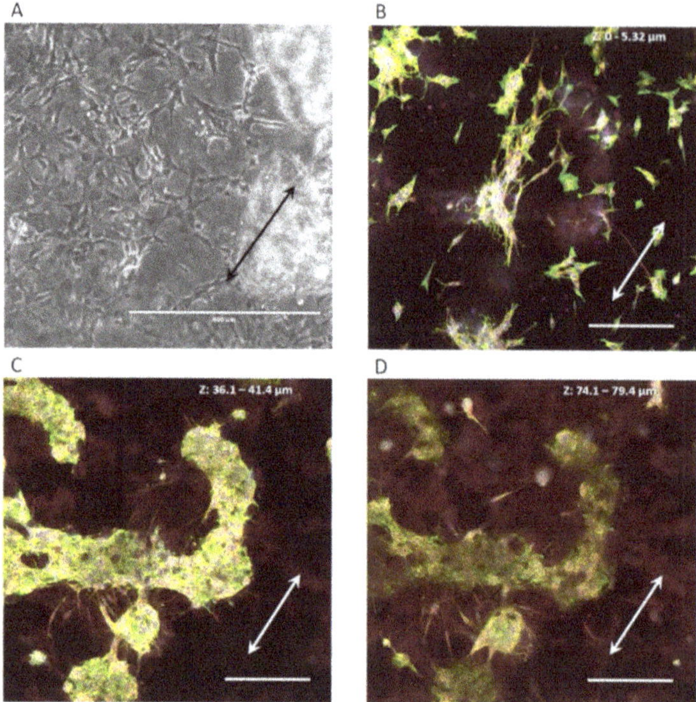

Figure 4. Aligned 3D SH-SY5Y cell cultures on nanogrooved substrates. (**A**) SH-SY5Y cells at 11 days in vitro (DIV) showing differentiation into the neuronal lineage during the experiment, as seen from the neuronal outgrowths. Cells show a qualitative preference to outgrowth direction along the nanogrooves (direction of the double-headed arrows). Blurry areas in the image indicate "clusters" of cells that have migrated into the 3D hydrogel away from their original position on the nanogrooved substrate. The scale bar denotes 400 μm. (**B–D**) Maximum intensity projections of subsections of slices from a z-stack showing the extent of SH-SY5Y cells and their outgrowths throughout the 3D cell culture. Qualitatively, cell bodies and outgrowths can be observed to align to the nanogrooved substrate at close range (**B**), with the effect decreasing as the distance to the substrate increases (**C,D**). The white, double-sided arrows denote the orientation of the nanogrooves. Scale bars denote 200 μm. Immunofluorescence staining shows neuron-specific staining for β-Tubulin III in red, F-actin in green and cell nuclei in blue.

3.3. 3D CTX Cell Culture

3.3.1. Neuron-Astrocytes Alignment in 2D Culture

Our initial study on the CTX culture has shown associated neuronal outgrowth, also referred to as neurites, with directional growth of the astrocytes on a soft linear nanoscaffold, i.e., PDMS nanogrooves. It has been demonstrated in our previous work that a softer scaffold material such as PDMS seems to affect the growth of the astrocyte and its network formation in a 2D culture [23], owing to the lower stiffness of the PDMS (Young's modulus 0.01–1 MPa level) than the rigid silicon (Young's modulus at GPa level). Interestingly, further morphology evaluation on the neurons in the mixed co-culture has shown that the neurons growing on the soft scaffolds have a tendency to form neurites in parallel to the directional outgrowth of the astrocytes., Meanwhile, the neurons on a rigid silicon scaffold do not show similar behavior, despite the outgrowths of the astrocytes showing similar alignment behavior on these two materials (Figure 5A,C). As reported before, there was no distinguishable difference in the astrocyte network formation on the 10:1 PDMS and the silicon [23]. In addition, the organization of the neuronal cell bodies appears to be associated with the "clustering" of the astrocyte's cell bodies. The cell "clusters" and the neurite bundles interconnect with the adjacent ones, resulting in a neuronal network with a few hundred micrometers of interspacing (Figure 5B), which displays similar organization as the neuron bundles in the brain cortical layers. Again the unique phenomenon was not observed in the culture on silicon scaffolds (Figure 5C,D).

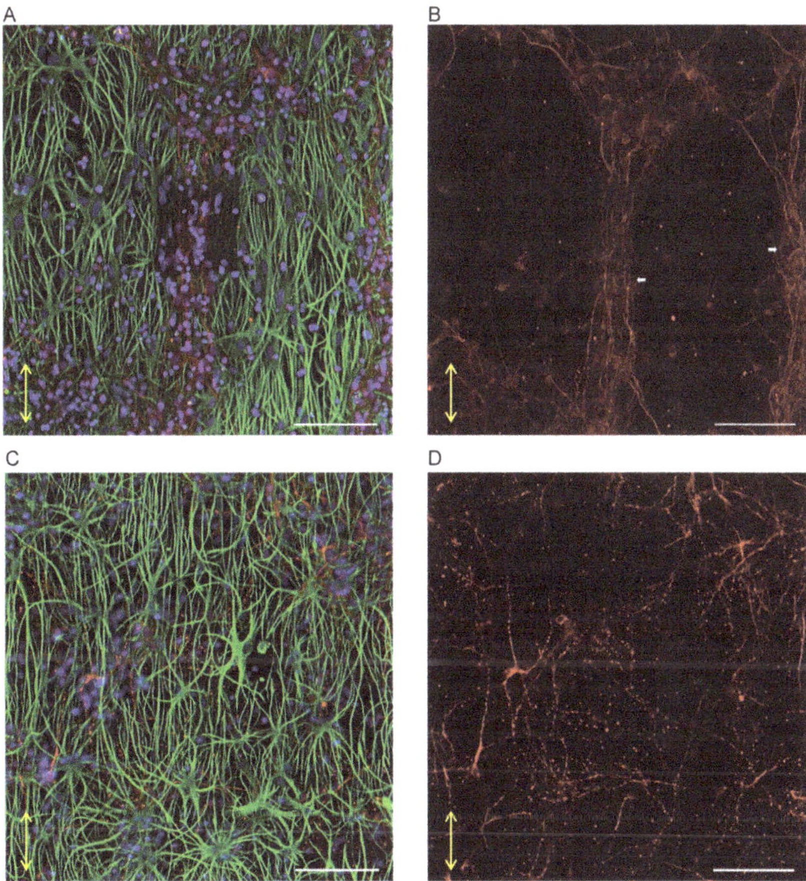

Figure 5. Different neuronal network formation in CTX culture on nanoscaffolds. (**A,B**) Outgrowth alignment of the astrocytes and neurite bundles formation on the soft PDMS nanoscaffold. The white arrows in (**B**) indicate the neurite bundles that aligned with the directional outgrowth of the astrocyte. (**C,D**) Outgrowth of astrocytes and neurites on the rigid silicon scaffold. Red staining indicates the microtubule-associated protein 2 (MAP2) of neurons, green staining indicates the glial fibrillary acidic protein (GFAP) of astrocytes, and blue staining by DAPI indicates the cell nuclei. Scale bars denote 100 μm. The double-sided yellow arrows indicate the direction of the nanogrooves. Adapted from Figure 4.7 in [30].

3.3.2. Neuron-Astrocyte Alignment in 3D Culture

We have previously demonstrated our concept of nanogroove-enhanced hydrogel scaffolds for 3D neuronal cell culture using the combination of nanogrooves and Matrigel to emulate the environment of the extracellular microenvironment, in vivo, applying a manual analysis [7]. As reported in this previous publication, approximately 50% of the outgrowth alignment was still observed in the astrocytes, which had migrated into the Matrigel and were ~6 µm away from the surface of the nanogrooves acting as a scaffold. In comparison, the anisotropic distribution of outgrowths, as seen for the cells on nanogrooved substrates, is not present in controls where the used substrate is flat (Figure S1A), even when cells migrate away from the substrate into the Matrigel (Figure S1B). As for the neurons, the similar aligning behavior of the neurites remained in the 3D culture: the associated neurite bundles in Figure 5 can also be observed in the cells that are growing in the gel. The results suggest that, within the gel scaffold, the aligned neurite bundles are able to maintain ~20 µm height from the bottom into the 3D hydrogel (Figure 6), which corresponds with 2~3 layers of cells. This indicates that the neurite organization formed at the cell-nanogroove interface is stronger than the adhesion between cells and anchoring points in the gel, wherein cells are capable of migrating into the gel space as far as ~80 µm away from the nanogrooved substrate surface (Figure S2A).

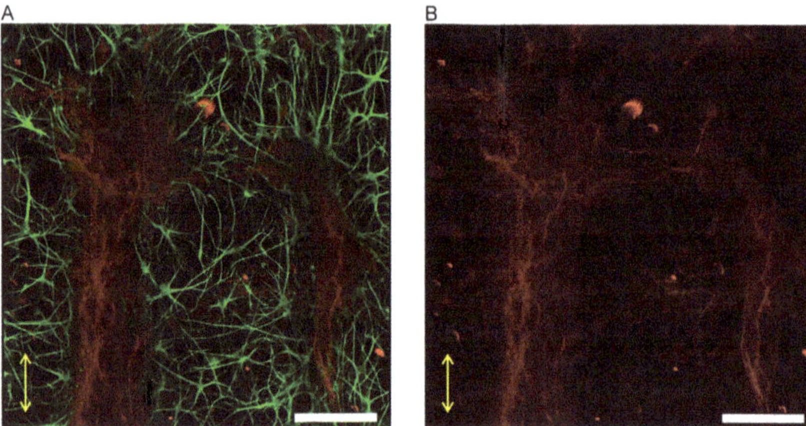

Figure 6. Neurite bundles in 3D-like culture. (**A**) Green staining indicates the GFAP for astrocytes, and red staining (shown alone in (**B**)) indicates the MAP2 for neurons. The double-sided yellow arrows indicate the direction of the nanogrooves. Scale bar: 100 µm. Adapted from Figure 5.11 in [30].

3.4. Image-Based Screening Method Analysis Using 3D CTX Data Set

As a means to test the image-based screening method against the manually performed measurements published elsewhere already [7], the screening method was used to analyze the 3D CTX data set as a pilot in this work. Quantitative results were obtained for the following parameters: the number of astrocytes, total outgrowth length and the outgrowth alignment relative to their distance from the substrate surface (Figure 7). The number of astrocytes at the nanogrooved substrate was around the 48 cells, whereas after this initial layer of cells near the surface, the number of cells dropped to approximately 13 cells around 10–15 µm from the surface, up to roughly 3 cells at 20 µm and farther away from the surface (Figure 7A). The total outgrowth length is around 14,097 µm at the substrate surface, but then peaks at approximately 24,643 µm, after which there is a sharp decline to 8930 µm around 8 µm and a gradual decline thereafter to 1054 µm at 25.2 µm (Figure 7B). The outgrowth alignment to the underlying nanogrooved pattern is 58.5% at the substrate surface and declines to 35.5% at around 8 µm distance from the surface and a gradual increase to 47.1% at around 18 µm before receding to 35.5% at 25.2 µm (Figure 7C). Compared to the manual analysis previously performed [7], additional information about the cell culture was obtained in the form of the number of cells and the total outgrowth length present in the z-stack. Since the bottom surface of a sample cannot perfectly coincide with the first slice image in a z-stack, the initial values for the number of cells and total outgrowth length are lower compared to the following values found for the next slice image. The decline of all three parameters indicates the decrease of cells when investigating areas farther away from the substrate surface, which coincides with the experimental setup of culturing cells first in a 2D format, prior to adding the hydrogel for the 3D microenvironment.

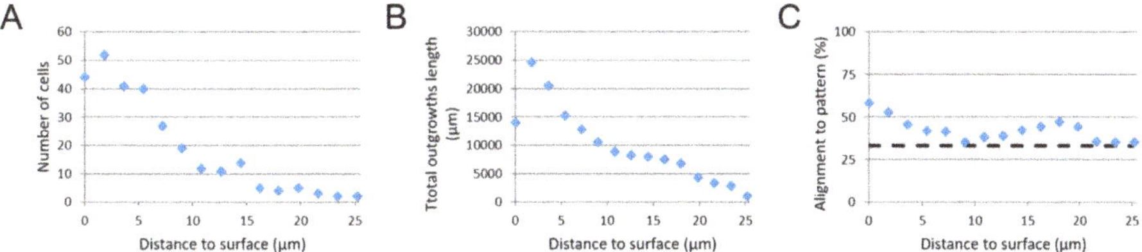

Figure 7. Results from analysis of 3D CTX culture on a nanogrooved substrate. (**A**) The number of cells, (**B**) the total outgrowth length (μm) and (**C**) outgrowth alignment (%) relative to the distance (μm] away from the nanogrooved substrate. The dashed line at 33% alignment in (**C**) represents an isotropic distribution of outgrowths without preferred direction.

Considering the lack of detectable outgrowths in particular, z-stacks were not analyzed beyond the 25.2 μm range from the substrate surface upwards. Although cells were detected up to ~80 μm away from the substrate, these were not considered for the investigation of the structural complexity that nanogrooved patterns could induce in the neuronal cell cultures. The results for alignment, shown in Figure 7C, indicate that up to a height of approximately 6–8 μm the screening method also finds outgrowths to be preferentially aligned with the nanogrooved pattern.

3.5. Image-Based Screening of 3D SH-SY5Y Cell Cultures

Based on the results as shown in Section 3.4, the image-based screening method was considered an appropriate way of generating quantitative data on z-stacks from 3D SH-SY5Y cell cultures on nanogrooved PDMS substrates. Again, results were acquired for the parameters of the number of cells, total outgrowth length and outgrowth alignment relative to their distance from the substrate surface (Figure 8). Three z-stacks were analyzed, numbered 1 through 3.

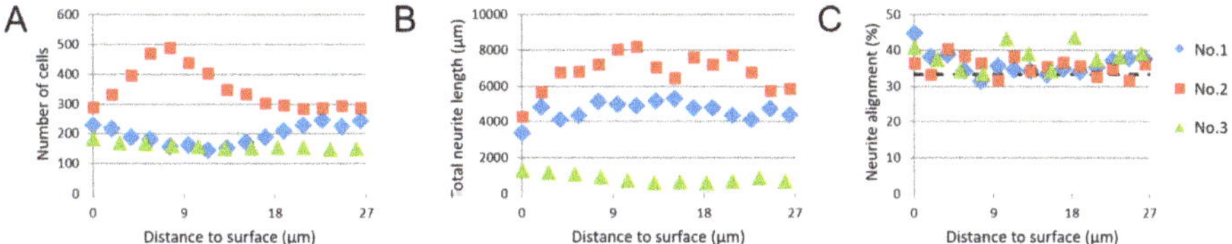

Figure 8. Results from analysis of 3D SH-SY5Y cell cultures on nanogrooved substrates. (**A**) The number of cells, (**B**) the total outgrowth length (μm) and (**C**) outgrowth alignment (%) relative to the distance (μm) away from the nanogrooved substrate. The dashed line at 33% alignment in (**C**) represents an isotropic distribution of outgrowths without preferred direction. For visibility of the data points, the alignment axis was capped at 50% instead of 100% alignment. The legend to the right of (**C**) describes the samples as shown in (**A–C**).

A range of 151–490 cells can be observed, regardless of distance from the substrate surface for the three z-stacks that were analyzed. For both z-stack 1 and 3, little change was observed in the number of cells relative to the distance to surface. A peak could be observed for z-stack 2 around 7.6 μm, with approximately 190 cells more relative to cell numbers at the surface or farthest from the surface. This is likely caused by the 3D SH-SY5Y cell culture sample being inserted slightly skewed into the confocal microscopy. As a result, cells adhering to the substrate surface seemed to be spread out over several slices within the z-stack when imaged, which most likely has resulted in a higher number of cells being farther away from the surface than anticipated.

As with the number of cells, the total outgrowth length, as seen within the slices of the z-stacks, changes only slightly for z-stacks 1 and 3, averaging at 4636 µm for z-stack 1 and 855 µm for z-stack 3. Z-stack 2 shows a total outgrowth length starting at 4307 µm, increasing to 8214 µm at 11.3 µm distance from the surface, before declining to 5871 µm at 26.4 µm away from the surface. These values can be compared against our previous experiments, which were on 2D SH-SY5Y cell cultures on nanogrooved PDMS substrates, and performed with a similar experimental protocol, but without the addition of the Matrigel to create 3D cell cultures [22]. In those experiments, we considered parameters such as the percentage of differentiated cells and the average outgrowth length of differentiated cells, as opposed to the total number of cells and the total outgrowth length. When taking the values for the total number of cells and the total outgrowth length from the previous dataset (Table S1), which have not been assessed in a context such as here, the number of cells and the total outgrowth length within the 3D SH-SY5Y cell culture are comparable to the 2D SH-SY5Y cell cultures.

The outgrowth alignment to the nanogrooved pattern was, for all z-stacks, in the range of 33.5–44.8%, with a higher percentage of alignment mostly occurring up to 11.3 µm away from the surface. While comparatively low in alignment percentage, as compared to the results of the CTX cells, the alignment percentages do show that some preference for outgrowth direction along the nanogrooves is retained for these reductionist type of neuronal cell cultures.

4. Discussion

In this work, we have aimed to demonstrate that it is possible to control outgrowth direction in 3D by simply applying soft scaffolds, such as Matrigel, atop of neuronal cell cultures initially seeded on nanogrooved substrates. In the case of the reductionist human neuronal cell line SH-SY5Y, results show that a layer of cells cultured in a 2D conformation will migrate into their 3D environment and also retain a degree of alignment to the underlying nanogrooved PDMS substrate. To our knowledge, results for neuronal cell cultures in 2D find similar topographical guidance effects of nano- and micro-scale features on neuronal cell outgrowths and morphology [14,31–33], however these do not investigate the emergence of structural complexity in 3D. The approach of a reductionist cell model allows for a more straightforward investigation of the neuronal cell morphology and neuronal network formation as opposed to the complex nature of primary cell cultures or co-cultures. As seen from the confocal imaging z-stack shown in Figure 4, cells can form dense cell clusters, and background interference from the hydrogel's autofluorescence can interfere with the quantitative analysis of such images. The cell clusters are even more challenging for assessment when dealing with multiple cell types, such as the CTX cultures. In this particular case, the autofluorescence of the hydrogel is due to the formaldehyde fixation of the Matrigel. Recent work into clearance techniques to enhance visualization of 3D organoids [34,35] may prove beneficial toward better visualization, thereby alleviating these challenges. In turn, improved visualization may lead to more capabilities in the quantitative analysis of neuronal cell culture images, such as investigating the branching of outgrowths, neuronal polarity and for co-cultures the number of cells per cell type, e.g., neurons, astrocytes. The current image-based screening method works by analyzing the slices from each z-stack separately. Typically, the resolution of data across slices in the z-direction is limited and relatively noisy compared to the data within each slice in the x- and y-direction. This effect currently limits both the usefulness and implementation of whole 3D analysis algorithms.

Compared to the CTX cell culture, the SH-SY5Y cell line provides a reductionist human brain model, but without the interplay complexity of co-cultured glia cells and neurons. This reduction allows for a thorough analysis of neuronal network formation parameters dependent on differentiation-stimuli at the cellular scale. Although we have used different dimensions of the nanogrooved patterns for CTX cells and SH-SY5Y cell culture experiments [21–30], these different patterns elicited similarly high responses with regard to outgrowth alignment, which was essential for analyzing the feasibility of our approach. Also, this shows that the application of a reductionist neuronal cell model, here the SH-SY5Y cell line, can still express structural complexity in nanogroove-enhanced hydrogel scaffolds.

On the other hand, these simple cell line models do not show the emergence of "meta-structures" in 3D conformation by applying our 3D culture format. Although neurite bundles, as seen in Figures 5 and 6, did not emerge from the SH-SY5Y cells in our experiments, such reductionist human brain models still open up possibilities to investigate the changes in neuronal network behavior, due to pharmacological interventions and set-up experiments at a scale of relevance for the industry, with a predefined set of clear parameters. In particular, our quantitative findings that compare the results for the number of cells and the total outgrowth length between 2D and 3D SH-SY5Y cell culture, as described in Table S1, can be used for setting up a baseline for optimization procedures toward robust system design and to establish a baseline expectation from cell culture results.

With regard to the CTX cultures, it has been reported that astrocytes can act as a layer of structural "scaffold" for directional neurite growth on 2D surface [36,37], where the environment lacks biochemical cues for neuron adhesion. In those peer studies, the neuron-astrocyte co-culture was realized by seeding the astrocytes first, followed by the neurons on top, separately. Instead of directly contacting the culturing substrate, the neurons face the "natural" surface of the pre-allocated astrocytes, and aligned neurites were thus induced by the morphology of the astrocytes. A similar phenomenon was observed in our study as well. In Figure S3, immunostaining of GFAP and MAP2 in the gel indicates the interconnection between astrocyte outgrowths and neurites. The neurite grows along with astrocyte outgrowth, extended in parallel direction within the gel in different layers of height. The step of the z-stage stack scanning was 300 nm, which is thinner than the average diameter of these branches [38]. The parallel trajectory of neurites and astrocyte outgrowths suggests that the astrocyte outgrowth may still assist or regulate the neuritogenesis or neurite orientation [39–43] with the presence of the extracellular environment provided by the Matrigel.

The unique associated network formation observed in our experiments using a combination of hard and soft scaffolds suggests the possibility of realizing brain-like architecture in vitro. This happens through the assistance of the directional guidance from the culturing environment: The hard nanogrooved "scaffold" at the bottom of the culture provides topographical information of the directional growth for the cells in a strongly determined manner, however remains effective in guiding the neuronal network formation in a few layers of the cells in the soft gel "scaffold". This phenomenon is visible for both nanogrooved pattern types, as well as both neuronal cell cultures. It suggests that the nanogrooves provide mechanical boundary conditions that limit the degree of freedom by strong cell–cell interactions also in the gel. The gel layer, on the other hand, offers spatially distributed anchor points for the cells in 3D with additional biochemical cues, facilitating the formation of an extended neuronal network throughout the culture volume. The bottom layer of cells in the culture, contacting the nanogrooved substrate, plays a crucial role as an interface connecting the two types of scaffolds.

A potential limitation of this 3D construction, regardless of the chosen neuronal cell model, would be the lack of sufficient circulation of the nutrition supply and the waste clearance during the culturing. In the experiments discussed in this work, the culture medium was only refreshed from the top of the Matrigel layer. Although the Matrigel allows the diffusion of the nutrients efficiently, the physical confinement of the gel layer may hinder the removal of the debris of the dead cells during the refreshment of the medium. It would, at the least, be less efficient than the same process in the conventional 2D cultures. This may result in accumulation of the waste biochemicals, particularly on the bottom of the culture, where the majority of the cells are located, as has been observed for primary rat CTX cells (Figure S2B). Conversely, the data shown in Table S1 suggests that the volume of Matrigel used for the 3D SH-SY5Y cell cultures was not detrimental with regard to nutrition supply, as the overall number of cells or outgrowths were within a similar range compared to 2D SH-SY5Y cell cultures. Nevertheless, an integrated microfluidic culturing chamber, additional to the 3D constructions, may provide improvement by creating a robust platform for the circulation of medium in the culture environment, and therefore may result in a higher consistency of the cell viability [44,45].

Furthermore, to seed a cell suspension within the gel onto the 2D cell layer could achieve an even more realistic 3D culture, also then offering multiple cell types. Alternatively, to prepare 3D scaffolds with nano- to micro-scale topographical guidance cues such as fiber-based scaffolds and microtunnels [46] can also help to realize a more advanced 3D culturing model. Other approaches have also been used to realize a more brain-like 3D neural cell culture, such as mixing neuronal cells in hydrogel and seeding the 3D construct on a petri-dish in a conventional way [47], as well as bioprinting neuronal cells or tissue within a medium or hydrogel [48,49]. Also, the layered 3D construction in cerebral cortex can be realized by stacking layers of cells confined in gel [50], however, the reproduction of the "cortical column" has not yet been discussed. Pluripotent stem cells are a promising cell type for realizing brain organoids, which have shown ability to reproduce both structural and functional properties of the brain cortex [51], while the formation of distinct cortical neuronal layers, as well as complex neuronal circuitry, still remain a challenge [52]. Despite these alternatives, nanogrooves can provide a valuable alternative from a system's design point of view, where the platform itself has integrated features, here nanogrooved patterns, to add structural complexity to 3D neuronal cell cultures. Such purely passive features, which can be an integral part of a standard culture wells plate or microelectrode array, are especially useful for high-throughput biological assays in a commercial setting. As stated in the previous paragraph, the use of an integrated microfluidic culturing chamber could extend the capabilities of such brain models, not only by providing enhanced nutrient and waste circulation, but also through the possibility of implementing multiple nano- and micro-scale features to enhance both formation of the brain cell model and the functional readout thereof. In turn, our results aid these models to potentially achieve such a level of control over the brain cell culture that specific brain regions, or structures such as cortical columns, can be mimicked in vitro.

5. Conclusions

In this work, we developed a method of adding structural network complexity in 3D in vitro neuronal cell models. This was performed by culturing neuronal cells on nanogrooved substrates covered with Matrigel. Two neuronal cell models, primary rat brain cortical cells and SH-SY5Y cells, were analyzed using image-based screening, for which the quantified results show that cells create a network in the 3D microenvironment and retain a degree of approximately 38–58% alignment, with regard to the underlying nanogrooved substrate in the z-direction for several micrometers. Also, the indication that 3D neuronal cell cultures that maintain a range of approximately 150–350 cells per mm^2 with an approximate total length of 4500–7500 μm of outgrowths, allows for accessible parameters to investigate during further optimization of a robust platform. In conclusion, these results indicate that nanogrooves enhance the structural complexity of 3D neuronal cell cultures for both primary rat brain cortical cell cultures and human SH-SY5Y cultures, providing a basis for advances in brain-on-chip technology.

Supplementary Materials: The following are available online at http://www.mdpi.com/2072-666X/10/10/638/s1, Table S1: Parameter values derived from 2D and 3D SH-SY5Y cell cultures on nanogrooved PDMS substrates, Figure S1: Neurite and astrocyte outgrowth of the 3D CTX cell culture on a flat PDMS substrate as control, Figure S2: Cell migration and viability in gel in the 3D CTX cultures, Figure S3: Parallel growth of neurite and astrocyte outgrowth in gel.

Author Contributions: Conceptualization: A.B., S.X. and R.L.; Experiments and data analysis: A.B. and S.X.; Writing—original draft preparation: A.B. and S.X.; Writing—review and editing: A.B., S.X. and R.L.; Visualization: A.B. and S.X.; Supervision: R.L.; Project administration: R.L.; Funding acquisition: R.L.; All authors have given their approval on the manuscript.

Funding: This research was funded by the Eindhoven University of Technology, the FET Proactive CONNECT project (grant no. 824070), ERC-STG project grant no. 280281 and ERC-PoC project grant no. 713732.

Acknowledgments: The authors thank the members of the Microfab/lab at the Eindhoven University of Technology and the members of the MESA+ Institute at the University of Twente for their experimental support.

Conflicts of Interest: The authors declare no conflict of interest.

References

1. Gribkoff, V.K.; Kaczmarek, L.K. The need for new approaches in CNS drug discovery: Why drugs have failed, and what can be done to improve outcomes. *Neuropharmacology* **2017**, *120*, 11–19. [CrossRef] [PubMed]
2. Finkbeiner, S. Bridging the Valley of Death of therapeutics for neurodegeneration. *Nat. Med.* **2010**, *16*, 1227–1232. [CrossRef] [PubMed]
3. Cummings, J.; Reiber, C.; Kumar, P. The price of progress: Funding and financing Alzheimer's disease drug development. *Alzheimer's Dement. Transl. Res. Clin. Interv.* **2018**, *4*, 330–343. [CrossRef] [PubMed]
4. Olanow, C.W.; Kieburtz, K.; Schapira, A.H.V. Why have we failed to achieve neuroprotection in Parkinson's disease. *Ann. Neurol.* **2008**, *64* (Suppl. 2), S101–S110. [CrossRef]
5. Sosa-Hernández, J.E.; Villalba-Rodríguez, A.M.; Romero-Castillo, K.D.; Aguilar-Aguila-Isaías, M.A.; García-Reyes, I.E.; Hernández-Antonio, A.; Ahmed, I.; Sharma, A.; Parra-Saldívar, R.; Iqbal, H.M.N. Organs-on-a-Chip module: A review from the development and applications perspective. *Micromachines* **2018**, *9*, 536. [CrossRef] [PubMed]
6. Moraes, C.; Mehta, G.; Lesher-Perez, S.C.; Takayama, S. Organs-on-a-Chip: A focus on compartmentalized microdevices. *Ann. Biomed. Eng.* **2012**, *40*, 1211–1227. [CrossRef] [PubMed]
7. Frimat, J.-P.; Xie, S.; Bastiaens, A.; Schurink, B.; Wolbers, F.; den Toonder, J.; Luttge, R. Advances in 3D neuronal cell culture. *J. Vac. Sci. Technol. B Nanotechnol. Microelectron. Mater. Process. Meas. Phenom.* **2015**, *33*, 06F902. [CrossRef]
8. Mountcastle, V. The columnar organization of the neocortex. *Brain* **1997**, *120*, 701–722. [CrossRef]
9. Oberlaender, M.; De Kock, C.P.J.; Bruno, R.M.; Ramirez, A.; Meyer, H.S.; Dercksen, V.J.; Helmstaedter, M.; Sakmann, B. Cell type-specific three-dimensional structure of thalamocortical circuits in a column of rat vibrissal cortex. *Cereb. Cortex* **2012**, *22*, 2375–2391. [CrossRef]
10. Shipp, S. Structure and function of the cerebral cortex. *Curr. Biol.* **2007**, *17*, R443–R449. [CrossRef]
11. Lancaster, M.A.; Renner, M.; Martin, C.A.; Wenzel, D.; Bicknell, L.S.; Hurles, M.E.; Homfray, T.; Penninger, J.M.; Jackson, A.P.; Knoblich, J.A. Cerebral organoids model human brain development and microcephaly. *Nature* **2013**, *501*, 373–379. [CrossRef]
12. Truskett, V.N.; Watts, M.P.C. Trends in imprint lithography for biological applications. *Trends Biotechnol.* **2006**, *24*, 312–317. [CrossRef]
13. Bremus-Koebberling, E.A.; Beckemper, S.; Koch, B.; Gillner, A. Nano structures via laser interference patterning for guided cell growth of neuronal cells. *J. Laser Appl.* **2012**, *24*, 042013. [CrossRef]
14. Johansson, F.; Carlberg, P.; Danielsen, N.; Montelius, L.; Kanje, M. Axonal outgrowth on nano-imprinted patterns. *Biomaterials* **2006**, *27*, 1251–1258. [CrossRef]
15. Tonazzini, I.; Cecchini, A.; Elgersma, Y.; Cecchini, M. Interaction of SH-SY5Y cells with nanogratings during neuronal differentiation: Comparison with primary neurons. *Adv. Healthc. Mater.* **2014**, *3*, 581–587. [CrossRef]
16. Miller, C.; Jeftinija, S.; Mallapragada, S. Synergistic effects of physical and chemical guidance cues on neurite alignment and outgrowth on biodegradable polymer substrates. *Tissue Eng.* **2002**, *8*, 367–378. [CrossRef]
17. Kim, Y.; Meade, S.M.; Chen, K.; Feng, H.; Rayyan, J.; Hess-Dunning, A.; Ereifej, E.S. Nano-architectural approaches for improved intracortical interface technologies. *Front. Neurosci.* **2018**, *12*, 1–20. [CrossRef]
18. Hoffman-Kim, D.; Mitchel, J.A.; Bellamkonda, R.V. Topography, cell response, and nerve regeneration. *Annu. Rev. Biomed. Eng.* **2010**, *12*, 203–231. [CrossRef]
19. Kim, H.N.; Jiao, A.; Hwang, N.S.; Kim, M.S.; Kang, D.H.; Kim, D.-H.; Suh, K.-Y. Nanotopography-guided tissue engineering and regenerative medicine. *Adv. Drug Deliv. Rev.* **2013**, *65*, 536–558. [CrossRef]
20. Xie, S.; Luttge, R. Imprint lithography provides topographical nanocues to guide cell growth in primary cortical cell culture. *Microelectron. Eng.* **2014**, *124*, 30–36. [CrossRef]
21. Bastiaens, A.J.; Xie, S.; Luttge, R. Investigating the interplay of lateral and height dimensions influencing neuronal processes on nanogrooves. *J. Vac. Sci. Technol. B Nanotechnol. Microelectron. Mater. Process. Meas. Phenom.* **2018**, *36*, 06J801. [CrossRef]
22. Bastiaens, A.J.; Xie, S.; Mustafa, D.A.M.; Frimat, J.-P.; den Toonder, J.M.J.; Luttge, R. Validation and optimization of an image-based screening method applied to the study of neuronal processes on nanogrooves. *Front. Cell. Neurosci.* **2018**, *12*, 1–14. [CrossRef]

23. Xie, S.; Schurink, B.; Wolbers, F.; Luttge, R.; Hassink, G. Nanoscaffold's stiffness affects primary cortical cell network formation. *J. Vac. Sci. Technol. B Nanotechnol. Microelectron. Mater. Process. Meas. Phenom.* **2014**, *32*, 06FD03. [CrossRef]

24. Dwane, S.; Durack, E.; Kiely, P.A. Optimising parameters for the differentiation of SH-SY5Y cells to study cell adhesion and cell migration. *BMC Res. Notes* **2013**, *6*, 366. [CrossRef]

25. Teppola, H.; Sarkanen, J.R.; Jalonen, T.O.; Linne, M.L. Morphological differentiation towards neuronal phenotype of SH-SY5Y neuroblastoma cells by estradiol, retinoic acid and cholesterol. *Neurochem. Res.* **2016**, *41*, 731–747. [CrossRef]

26. Encinas, M.; Iglesias, M.; Liu, Y.; Wang, H.; Muhaisen, A.; Ceña, V.; Gallego, C.; Comella, J.X. Sequential treatment of SH-SY5Y cells with retinoic acid and brain-derived neurotrophic factor gives rise to fully differentiated, neurotrophic factor-dependent, human neuron-like cells. *J. Neurochem.* **2000**, *75*, 991–1003. [CrossRef]

27. Wiertz, R. Regulation of In Vitro Cell-Cell and Cell-Substrate Adhesion. Ph.D. Thesis, University of Twente, Enschede, The Netherlands, 2010.

28. Romijn, H.J.; van Huizen, F.; Wolters, P.S. Towards an improved serum-free, chemically defined medium for long-term culturing of cerebral cortex tissue. *Neurosci. Biobehav. Rev.* **1984**, *8*, 301–334. [CrossRef]

29. Agholme, L.; Lindström, T.; Kågedal, K.; Marcusson, J.; Hallbeck, M. An In Vitro model for neuroscience: differentiation of SH-SY5Y cells into cells with morphological and biochemical characteristics of mature neurons. *J. Alzheimer's Dis.* **2010**, *20*, 1069–1082. [CrossRef]

30. Xie, S. Brain-On-A-Chip Integrated Neuronal Networks. Ph.D. Thesis, University of Twente, Enschede, The Netherlands, 2016.

31. Chua, J.S.; Chng, C.P.; Moe, A.A.K.; Tann, J.Y.; Goh, E.L.K.; Chiam, K.H.; Yim, E.K.F. Extending neurites sense the depth of the underlying topography during neuronal differentiation and contact guidance. *Biomaterials* **2014**, *35*, 7750–7761. [CrossRef]

32. Kang, K.; Park, Y.S.; Park, M.; Jang, M.J.; Kim, S.M.; Lee, J.; Choi, J.Y.; Jung, D.H.; Chang, Y.T.; Yoon, M.H.; et al. Axon-first neuritogenesis on vertical nanowires. *Nano Lett.* **2016**, *16*, 675–680. [CrossRef]

33. Ferrari, A.; Cecchini, M.; Dhawan, A.; Micera, S.; Tonazzini, I.; Stabile, R.; Pisignano, D.; Beltram, F. Nanotopographic control of neuronal polarity. *Nano Lett.* **2011**, *11*, 505–511. [CrossRef]

34. Boutin, M.E.; Hoffman-Kim, D. Application and assessment of optical clearing methods for imaging of tissue-engineered neural stem cell spheres. *Tissue Eng. Part C Methods* **2014**, *21*, 292–302. [CrossRef]

35. Grist, S.M.; Nasseri, S.S.; Poon, T.; Roskelley, C.; Cheung, K.C. On-chip clearing of arrays of 3-D cell cultures and micro-tissues. *Biomicrofluidics* **2016**, *10*, 044107. [CrossRef]

36. Biran, R.; Noble, M.D.; Tresco, P.A. Directed nerve outgrowth is enhanced by engineered glial substrates. *Exp. Neurol.* **2003**, *184*, 141–152. [CrossRef]

37. Alexander, J.K.; Fuss, B.; Colello, R.J. Electric field-induced astrocyte alignment directs neurite outgrowth. *Neuron Glia Biol.* **2006**, *2*, 93–103. [CrossRef]

38. Routh, B.N.; Johnston, D.; Harris, K.; Chitwood, R.A. Anatomical and electrophysiological comparison of CA1 pyramidal neurons of the rat and mouse. *J. Neurophysiol.* **2009**, *102*, 2288–2302. [CrossRef]

39. Price, J.; Hynes, R.O. Astrocytes in culture synthesize and secrete a variant form of fibronectin. *J. Neurosci.* **1985**, *5*, 2205–2211. [CrossRef]

40. Liesi, P.; Kirkwood, T.; Vaheri, A. Fibronectin is expressed by astrocytes cultured from embryonic and early postnatal rat brain. *Exp. Cell Res.* **1986**, *163*, 175–185. [CrossRef]

41. Neugebauer, K.M.; Tomaselli, K.J.; Lilien, J.; Reichardt, L.F. N-cadherin, NCAM, and integrins promote retinal neurite outgrowth on astrocytes In Vitro. *J. Cell Biol.* **1988**, *107*, 1177–1187. [CrossRef]

42. Lois, C.; García-Verdugo, J.M.; Alvarez-Buylla, A. Chain migration of neuronal precursors. *Science* **1996**, *271*, 978–981. [CrossRef]

43. Snow, D.M.; Lemmon, V.; Carrino, D.A.; Caplan, A.I.; Silver, J. Sulfated proteoglycans in astroglial barriers inhibit neurite outgrowth in vitro. *Exp. Neurol.* **1990**, *109*, 111–130. [CrossRef]

44. Schurink, B.; Luttge, R. Hydrogel/poly-dimethylsiloxane hybrid bioreactor facilitating 3D cell culturing. *J. Vac. Sci. Technol. B Nanotechnol. Microelectron. Mater. Process. Meas. Phenom.* **2013**, *31*, 06F903. [CrossRef]

45. Bastiaens, A.J.; Frimat, J.-P.; van Nunen, T.; Schurink, B.; Homburg, E.F.G.A.; Luttge, R. Advancing a MEMS-based 3D cell culture system for in vitro neuro-electrophysiological recordings. *Front. Mech. Eng.* **2018**, *4*, 1–10. [CrossRef]

46. Pirlo, R.K.; Sweeney, A.J.; Ringeisen, B.R.; Kindy, M.; Gao, B.Z. Biochip/laser cell deposition system to assess polarized axonal growth from single neurons and neuron/glia pairs in microchannels with novel asymmetrical geometries. *Biomicrofluidics* **2011**, *5*, 13408. [CrossRef]

47. Choi, S.H.; Kim, Y.H.; Hebisch, M.; Sliwinski, C.; Lee, S.; D'Avanzo, C.; Chen, H.; Hooli, B.; Asselin, C.; Muffat, J.; et al. A three-dimensional human neural cell culture model of Alzheimer's disease. *Nature* **2014**, *515*, 274–278. [CrossRef]

48. Xu, T.; Gregory, C.A.; Molnar, P.; Cui, X.; Jalota, S.; Bhaduri, S.B.; Boland, T. Viability and electrophysiology of neural cell structures generated by the inkjet printing method. *Biomaterials* **2006**, *27*, 3580–3588. [CrossRef]

49. Suri, S.; Han, L.H.; Zhang, W.; Singh, A.; Chen, S.; Schmidt, C.E. Solid freeform fabrication of designer scaffolds of hyaluronic acid for nerve tissue engineering. *Biomed. Microdevices* **2011**, *13*, 983–993. [CrossRef]

50. Lozano, R.; Stevens, L.; Thompson, B.C.; Gilmore, K.J.; Gorkin, R.; Stewart, E.M.; in het Panhuis, M.; Romero-Ortega, M.; Wallace, G.G. 3D printing of layered brain-like structures using peptide modified gellan gum substrates. *Biomaterials* **2015**, *67*, 264–273. [CrossRef]

51. Eiraku, M.; Watanabe, K.; Matsuo-Takasaki, M.; Kawada, M.; Yonemura, S.; Matsumura, M.; Wataya, T.; Nishiyama, A.; Muguruma, K.; Sasai, Y. Self-organized Formation of polarized cortical tissues from ESCs and its active manipulation by extrinsic signals. *Cell Stem Cell* **2008**, *3*, 519–532. [CrossRef]

52. Qian, X.; Song, H.; Ming, G.L. Brain organoids: Advances, applications and challenges. *Development* **2019**. [CrossRef]

 *micromachines*

Article

The Applications of Lattice Light-Sheet Microscopy for Functional Volumetric Imaging of Hippocampal Neurons in a Three-Dimensional Culture System

Chin-Yi Chen, Yen-Ting Liu, Chieh-Han Lu, Po-Yi Lee, Yun-Chi Tsai, Jyun-Sian Wu, Peilin Chen and Bi-Chang Chen *

Research Center for Applied Sciences, Academia Sinica, Taipei 11529, Taiwan
* Correspondence: chenb10@gate.sinica.edu.tw; Tel.: +886-2-2787-3133

Received: 12 August 2019; Accepted: 9 September 2019; Published: 11 September 2019

Abstract: The characterization of individual cells in three-dimensions (3D) with very high spatiotemporal resolution is crucial for the development of organs-on-chips, in which 3D cell cultures are integrated with microfluidic systems. In this study, we report the applications of lattice light-sheet microscopy (LLSM) for monitoring neuronal activity in three-dimensional cell culture. We first established a 3D environment for culturing primary hippocampal neurons by applying a scaffold-based 3D tissue engineering technique. Fully differentiated and mature hippocampal neurons were observed in our system. With LLSM, we were able to monitor the behavior of individual cells in a 3D cell culture, which was very difficult under a conventional microscope due to strong light scattering from thick samples. We demonstrated that our system could study the membrane voltage and intracellular calcium dynamics at subcellular resolution in 3D under both chemical and electrical stimulation. From the volumetric images, it was found that the voltage indicators mainly resided in the cytosol instead of the membrane, which cannot be distinguished using conventional microscopy. Neuronal volumetric images were sheet scanned along the axial direction and recorded at a laser exposure of 6 ms, which covered an area up to 4800 μm^2, with an image pixel size of 0.102 μm. When we analyzed the time-lapse volumetric images, we could quantify the voltage responses in different neurites in 3D extensions.

Keywords: lattice light-sheet microscopy; 3D cell culture system; functional neuron imaging

1. Introduction

In recent years, the rapid development of microfabrication techniques has enabled us to study cellular behavior in a well-controlled microenvironment, mimicking the native environment of the disease states [1–4]. With the integration of microfluidic systems and three-dimensional (3D) cell cultures, organs-on-chips have shown great potential in high throughput drug screening applications [5–8]. In order to faithfully reconstruct the in vivo microenvironment, 3D culture systems are often used for the organs-on-chips, which raises a challenging issue in detecting the response of individual cells with high spatiotemporal resolution while minimizing the damage of cells during the observation process. In the conventional approach, microfluidic devices are placed on an inverted microscope, where the scattering from multiple layers of cells hampers light penetration, thus leading to low imaging quality. Therefore, one of the key issues in the development of the organs-on-chips system is to monitor the spatiotemporal behavior of individual cells in the 3D cell culture.

One of the major advantages of the organs-on-chips is the capability to mimic diseases at the organ level on chips. Among various diseases, neurodegenerative diseases are of great research importance because these diseases are currently considered incurable [9]. Two-dimensional (2D) neuron cultures are the most studied systems for understanding neurodegenerative diseases, in which neuronal behavior can be manipulated and measured through various techniques [10–14]. The function of neurons can be monitored through membrane potential or calcium dynamics using optical microscopic tools [15–18]. In general, the application of calcium indicators has yielded satisfactory results for primary 2D cultures, slices of brain tissue, and in vivo brain imaging [19]. However, it remains difficult to measure neuronal activity in 3D due to the limited light penetration and imaging speed. At present, high-speed calcium imaging from 2D or 3D cultured neurons can be recorded at a fixed focal plane [20]. Alternatively, multi-point scanning spinning disc microscopy [21] also offers high speed imaging with low phototoxicity. However, the penetration depth of light in 3D culture is always problematic, which may be partially resolved by the use of two-photon microscopy [8,22,23]. However, recording neuronal activity in 3D cultures requires the development of high-speed volumetric imaging techniques. Recently, selective plane illumination-based techniques have been demonstrated to be capable of high-speed volumetric imaging, [24,25]. Among them, lattice light-sheet microscopy (LLSM) has been shown to offer several advantages over other volumetric imaging tools, including less photobleaching and phototoxicity and better subcellular imaging resolution [26,27]. In LLSM, a pair of microscope objectives with perpendicular orientation shares the same focal point, where the sample is placed. The specimen and both tapered ends of the objective are immersed in the medium filled, temperature-controlled chamber. A 2D optical lattice composed of hundreds of Bessel beams scans across the whole sample and generates a 3D fluorescent image with resolutions of 250 nm laterally and 500 nm axially, at a speed of several milliseconds per excitation plane. Therefore, in this study, we utilized LLSM to monitor neuronal activity via subcellular imaging of voltage and calcium dynamics.

2. Materials and Methods

2.1. Preparation of Rat Hippocampal Neurons for 3D Culture

Animal experiments were conducted under the guidelines of the Academia Sinica Institutional Animal Care. Postnatal day 0 (P0) rats were sacrificed, and hippocampal tissues were dissected as described in previous experiments [28,29]. Hippocampal neurons, isolated by papain-mediated isolation, were kept in ice-cold medium, mixed with Matrigel (Corning) in a 1:1 ratio [30], and seeded on 35 mm glass-bottom dishes (MatTek Corporation, Massachusetts, MA, USA) (Figure 1). The mixture of Matrigel and neurons was added in the center of a MatTek dish. An 18 mm coverslip was then placed on top of the mixture droplet, resulting in a spread round sheet with a thickness of ~2 mm. A consistent size of 3D culture gel could be realized by this approach. The mixture of Matrigel and neurons would rapidly become solidified at 22 °C to 35 °C. Those procedures were performed on ice to prevent gel polymerization. The neurons were allowed to form a round sheet of 3D gel at a density of $10^4/mm^3$ at 37 °C for 2 h. After gelation, 2 mL of neuronal medium were added into each dish. Hippocampal neurons were maintained in neuron culture medium (Neurobasal-A medium, Thermo Fisher Scientific, Massachusetts, MA, USA), supplemented with GlutaMAX (Thermo Fisher Scientific, Massachusetts, MA, USA), B-27 (Thermo Fisher Scientific, Massachusetts, MA, USA), and 20% glial conditional medium, and incubated at 37 °C in a 5% CO_2 environment. Glial cells were cultured only for collecting their medium, which have secreted growth factors and some unidentified factors for providing neuron health during culture.

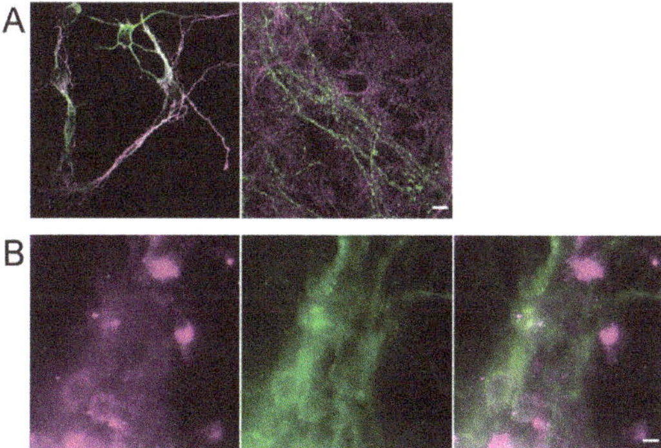

Figure 1. The characterization and comparison between 2D and 3D cultured hippocampal neurons. (**A**) Images of neurons cultured on 2D coverslips. (Left) early stage markers, Tau (magenta) and microtubule-associated protein 2 (MAP2) (green) are used to labeled axon and dendrites, respectively, at 4 days in vitro (DIV). (Right) mature neurons at 15 DIV were labeled with markers for synapse formation, Synaptotagmin 1 (magenta) for presynaptic terminals and postsynaptic density protein 95 (PSD-95) (green) for postsynaptic. Scale bar: 10 µm. (**B**) The differentiation of 3D cultured neurons was visualized by staining with the neuronal markers, Tau (magenta) and MAP2 (green) at 15 DIV. Scale bar: 10 µm.

2.2. Immunofluorescence Staining

In order to verify the condition of cultured neurons, we conducted neuronal immunostaining using various markers. The neurons were rinsed with ice-cold PBS-MC (phosphate-buffered saline, Sigma Aldrich, Missouri, MO, USA) with 1 mM $MgCl_2$ and 0.1 mM $CaCl_2$, and fixed with 4% paraformaldehyde (PFA) and 4% formaldehyde (paraformaldehyde aqueous solution; EM Grade, Electron Microscopy Sciences/4% sucrose/PBS-MC) for 20 min at room temperature (RT). Neurons were rinsed three times with PBS-MC and incubated with blocking buffer (4% goat serum (Gibco)/2% bovine serum albumin (BSA; Sigma Aldrich, Missouri, MO, USA)/PBS-MC) for 30 min at RT. Before antibody staining, neurons were permeabilized with blocking solution containing 0.2% TritonX-100 (Sigma Aldrich, Missouri, MO, USA) for 30 min at RT. The neurons were rinsed three times with PBS-MC and then stained with primary antibodies, such as anti-Tau-1 (1:500; Millipore, Burlington, Massachusetts, MA, USA), postsynaptic density protein-95 (PSD-95) (1:200; Millipore, Burlington, MA, USA), synaptotagmin-1 (1:200; Synaptic Systems, Beijing, China), and microtubule-associated protein-2 (1:200; MAP2; Chemicon, Osaki, Tokyo, Japan), which were diluted in blocking buffer and incubated at 4 °C overnight in a humid chamber. Neurons were rinsed three times with PBS-MC at RT. In the final stage, the secondary antibodies, goat anti-mouse-conjugated Alexa Fluor 488 and goat anti-rabbit-conjugated Alexa Fluor 546 (1:1000 for both antibodies; Jackson ImmunoResearch, Shanghai, China), were diluted in blocking buffer and incubated with neurons at 4 °C overnight in a humid chamber. After three rinses with PBS-MC at RT, neuron gels were kept within PBS-MC until imaging was acquired by an inverted research microscope system (Leica DMi8, Illinois, IL, USA) (Figure 1B). The neurons which were cultured on coverslips were mounted in Prolonged Gold antifade reagent (Invitrogen, Carlsbad, CA, USA) according to the manufacturer's instruction with overnight curing in the dark at RT, and images were recorded with a confocal microscopy LSM880 (ZEISS, New York, NY, USA) (Figure 1A).

For Figure 1A, immunostained images were acquired in z-stacks with an upright laser scanning confocal microscopy LSM880 (ZEISS, New York, NY, USA) with a 63× (NA = 1.4) oil objective. Alexa Fluor 488 and Alexa Fluor 546 fluorescent dyes were excited with the 488 and 561 nm laser lines, and collected within 510–550 and 570–620 band filters, respectively. Each channel was acquired separately to minimize bleed-through. Images were processed by ZEN software (ZEISS, New York, NY, USA).

The two-color Nneuronal 3D gel images (Figure 1B) were recorded with a confocal microscopy LSM880 (ZEISS, New York, NY, USA) (Figure 1A) and an inverted research microscope system (Leica DMi8), Leica DMi8 microscope equipped with a 40×/1.3 NA oil-immersion objective, Colibri LED Illumination system, 450–490 nm and 540–580 nm band pass filters. Neuronal 3D gel images were captured by an electron multiplying CCD (Evolve 512, Photometrics, Arizona, AZ, USA) (Figure 1B) in 16-bit scale and were processed by ImageJ (NIH).

2.3. Setup of Lattice Light-Sheet Microscopy

The schematic diagram of our microscopic system is depicted in Figure 2. In order to record 3D neuronal activity, we constructed LLSM as described previously [26] with a slightly modifications of sample holder and stages, which were built on an inverted microscope (Olympus IX71, Olympus Corporation, Tokyo, Japan). Two excitation lasers (λ = 488 and 561 nm) were used in this experiment. The light-sheet was generated via either a low numerical aperture (NA) configuration (excitation objective (Nikon, 40× CFI APO NIR, 0.8 NA, Tokyo, Japan) and detection objective (Nikon, 40× CFI APO NIR, 0.8 NA)) or a high NA configuration (excitation objective (special optics, NA = 0.66) and detection objective (Nikon, CFI Apo LWD 25×, NA = 1.1)). The emission signals were imaged by an sCMOS (Hamamatsu, Orca Flash 4.0 v2 sCMOS) detector, (Hamamatsu, Iwata, Japan). The high NA microscope features near-isotropic resolution in all directions by very thin plane scanning to get the lateral and axial resolution, with 200 nm and 400 nm. By using a piezo scanner (Physik Instrumente, P-726 PIFOC, Karlsruhe, Germany), the movement of the detection objective can be synchronized with the excitation objective. For image acquisition, a 3D gel sheet sample was mounted on a regular microscope glass slide attached on a 3D translational stage. The gel sheet was then immersed in the imaging buffer (Hank's balanced salt solution (HBSS) buffer) (Sigma Aldrich, Missouri, MO, USA) meniscus formed between the objectives and the glass slide.

2.4. Voltage and Calcium Dye Labeling for Functional Imaging

In order to record the neuronal activity, 4 μM Fluo-4AM (Thermo Fisher Scientific, Massachusetts, MA, USA) was added to the 14–19 days in vitro (DIV) neuronal culture for calcium imaging, whereas 0.1 μM di-4-ANEPPS (Thermo Fisher Scientific Massachusetts, MA, USA) was used for voltage imaging. During the experiment, neurons were supplied with Hank's balanced salt solution (HBSS) buffer (Sigma Aldrich, Missouri, MO, USA) (137 mM NaCl, 5.4 mM KCl, 0.25 mM Na_2HPO_4, 0.44 mM KH_2PO_4, 1.3 mM $CaCl_2$, 1.0 mM $MgSO_4$, 1.0 mM $MgCl_2$, 10 mM glucose, and 10 mM HEPES; pH 7.4) containing 0.01% Pluronic F127 (Thermo Fisher Scientific, Massachusetts, MA, USA) and 0.1% bovine serum albumin (Sigma Aldrich, Missouri, MO, USA) [31,32]. Calcium and voltage dyes were loaded at 37 °C under 5% CO_2 environment for 50 min.

The acquisition time for one 3D volume image was between 3 and 5 s All of the experiments were performed at room temperature. In order to obtain images, KCl-induced membrane potential and changes in calcium influx, images of 3D-cultured neurons were acquired for 1 min at baseline, in which no significant photobleaching was observed, and signals were averaged as F0.

2.5. Setup of Electrical Stimulation of Neurons

In order to electrically stimulate the neurons, two parallel electrodes were attached to the glass slide, with a separation distance of 1.5 cm. Electric pulses were generated through a data acquisition device (DAQ, USB-6341, National Instruments, Texas, TX, USA), in which the analog output was connected to one of the scaling amplifiers (SIM-983 in SIM-900 mainframe, Stanford Research Systems, California, CA, USA). The output of the amplifier was connected to the electrodes on the glass slide using a 50 cm off-the-shelf BNC-to-alligator cable (RG-58/U, Jun-Mao, Taipei, Taiwan). Electrodes ran parallel to the long side of the slide in order to avoid mechanical interference during acquisition. In order to calibrate the strength of the electrical field with respect to the provided voltage level, a secondary set of paraxial electrodes was placed in between the first set in order to measure the sensed voltage. Temporal parameters of the stimuli, including amplitude, pulse interval, and duration, were controlled through a customized LabVIEW program. During electrical stimulation, electric pulses with a magnitude of 10 V/cm at 10 Hz were applied to the samples for a total of 900 pulses through an amplifier (Stanford Research Systems, SR830, California, CA, USA).

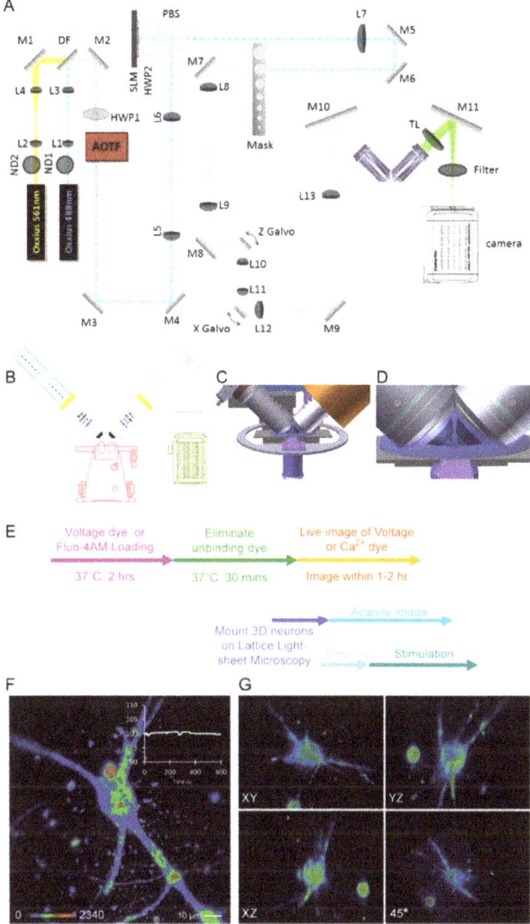

Figure 2. (**A**) The detailed optical schematic for the lattice light-sheet microscopy (LLSM) system. Abbreviations: L—lens, M—mirror, DF—dichroic filter, ND—neutral density filter, HWP—half-wave plate, AOTF—acousto–optical tunable filter, PBS—polarization beam cube, TL—tube lens. The system was built on an inverted microscope with two excitation lasers. (**B**) The excitation and detection objectives were mounted on an inverted microscope perpendicular to each other. (**C**) The design of the 3D culture sample holder and relative position of the orthogonal excitation (left) and detection (right) objectives. Samples were mounted on a slide, which was connected to a customized sample holder on the sample stage. (**D**) The sample was located in between two objectives and the space between objectives was filled with 1 mL imaging buffer. (**E**) The flowchart of voltage and calcium dye labeling procedures of 3D cultured neurons. (**F**) The maximum intensity projection of a live image of 3D hippocampal neuron labeled by voltage dye at T0. Inset shows the fluorescent intensity change of the 3D volumes over time. Scale bar: 10 µm. (**G**) Different angles of views of the 3D culture neuron shown in (F).

2.6. Image Processing and Analysis

In order to quantify the voltage responses, the maximum intensity at each time point was projected onto one 3D image and used for the selection of region of interest (ROI) in 3D. This 3D projection image of the time-lapse image was derived from registered images with maximum intensity projection over time, which served as the entire neuronal contour. The ROIs were depicted by hand drawn profiles for each z-stack. The label analysis function of Amira was used to calculate the values in 3D ROI.

Segmentations of neurites and soma were automatically traced using the XTracing pack of Amira (Thermo Fisher Scientific, Massachusetts, MA, USA) (Weber et al., 2012; Rigort et al., 2012). We followed the protocol of the Amira User Guide [33] using the parameters of the cylinder correlation module as listed below (μm): (1) cylinder length of 63.7; (2) angular sampling of 5; (3) mask cylinder radius of 14; (4) outer cylinder radius of 12; and (5) inner cylinder radius of 0. The color codes of each neurite and soma were selected independently.

The functional images of 3D neurons were processed by ImageJ (NIH) and Amira 3D Software (Thermo Fisher Scientific, Massachusetts, MA, USA). Prism (Graph Pad) and Excel (Microsoft Office) were used for data analysis. For quantitative analysis, the initial fluorescence Ca^{2+} and membrane potential intensities within 1 min were averaged and used as the baseline fluorescence intensity (F0). The fluorescence intensity of each time point was defined as Ft. For the fluorescence change (ΔF) before and after stimulation, we calculated the change by the equation $\Delta F = ([Ft-F0]/F0)$ (%).

3. Results

3.1. Characterization of Hippocampal Neurons in a 3D Culture System

There are various types of 3D culture systems for the study of neuronal development and differentiation [34–36]. Based on previous reports [30], we developed a modified protocol for a 3D culture system to observe the differentiation and maturation of primary hippocampal neurons. For primary hippocampal neurons, it takes approximately 15 days to develop complete neuronal networks and synapses. To establish the 3D environment for neurons, the neurons were cultured within 50% Matrigel [30] under neuronal growth and differentiation medium, as described in a regular culture system [28]. The thickness of the Matrigel was controlled by laminating a coverslip with a Matrigel droplet on a glass-bottom dish, which was around 2 mm. From the bright field z-stack images (approximately 250–300 μm in total depth) of 3D-cultured neurons (Movie S1 in Supplementary Materials with field of view at 400 μm × 400 μm × 160 μm, z step size of 1.6 μm), it was found that the neurons were evenly distributed within the 3D environment. This result suggests that our modified protocol could successfully create a 3D environment for neurons.

After 10 days in vitro (DIV) culture, synaptogenesis started in neurons and formed synapses between neurons for transmission at the neuronal network. A mature network was formed after 15 DIV. In order to compare hippocampal neurons cultured in 2D and 3D environments, we examined axon and dendrite formation by staining Tau and MAP2 proteins at an early stage (4 DIV), and the formation of synapses using synaptotagmin to label presynaptic terminals and PSD-95 for postsynaptic terminal staining at a later stage (15 DIV) (Figure 1). Our experimental results indicate that the 3D culture neurons could successfully differentiate and form a neuronal network, suggesting that this modified protocol can provide an environment for neuronal differentiation and maturation in a 3D culture system similar to those neurons in organoid or tissue cultures. Unlike the neurons grown in the 2D culture system, the neurons cultured in the 3D system could extend their neurites over 150 μm in the z direction, which is ~six-fold longer than those found in 2D cultured neurons (20–25 μm). Therefore, we used this modified 3D culture system to mimic the neuronal spheres or brain organoids grown on chips [36,37].

3.2. Lattice Light-Sheet Microscopy (LLSM) for 3D-Cultured Neuron

To visualize the neurons in the 3D culture system by conventional fluorescence microscopy, strong scattering from thick samples leads to blurred fluorescence signals (Movie S2 in SI). The limitation of scattering and light penetration can be improved by the use of light-sheet excitation, which offers several advantages for monitoring the neuronal activity in a 3D culture system, including a reduction in photobleaching, phototoxicity, and scattering [26,27,38]. In the original LLSM setup, two orthogonally aligned water dipping objectives were immersed in an imaging medium bath with a volume of ~10 mL [26]. For the drug treatment or chemical stimulation experiments, this large liquid-filled chamber will become problematic, especially for pricing chemicals. In order to study the 3D-cultured hippocampal neurons, we modified the original design of the LLSM by replacing the liquid-filled chamber with a microscope glass slide (Figure 2A) mounted on an inverted microscope (Figure 2B). The combination of LLSM with an inverted microscope allows us to monitor the locations of samples through eyepieces (Figure 2B). For refraction index matching, approximately 1 mL of imaging buffer (HBSS) was added to the center of the microscope glass slide in order to create a water bridge between the two objectives (Figure 2C,D) [39]. The 3D-cultured hippocampal neurons can be imaged in a small volume of liquid compared with the original design, which requires about 10 to 12 mL of imaging buffer [26]. The reduced volume of imaging buffer is especially beneficial for pharmaceutical studies, in which reagents are limited and sometimes very expensive.

In order to monitor neuronal activity, voltage probes for imaging the membrane potential as a record of neuronal activity of the 3D-cultured hippocampal neuronal activity are required. However, there are only limited probes options in the market. In this study, we utilized a commercially available voltage sensitive aminonaphthylethenylpyridinium dye, di-4-ANEPPS, to image the membrane potential, because of its high signal-to-noise ratio, positive correlation with voltage dynamic, and a red-shift emission [40,41]. In order to perform the live labeling of 3D-cultured neurons with voltage, we adapted previously published procedures [31,32]. The 3D spatial distribution of voltage response on the neuron membrane can be mapped out by the customized LLSM (Figure 2F and Movie S3 in SI), while different angles of views of a neuron can be seen in Figure 2G. These results revealed that the neurite outgrowths of the primary hippocampal neurons cultured on the Matrigel-based system were extended in 3D. In addition, we demonstrated that the LLSM is capable of recording three-dimensional voltage responses of the primary neuron culture. Indeed, image shown in Figure 2F exhibits significantly improved imaging quality when compared with those images taken by conventional microscope.

3.3. Quantification of Voltage Responses at Different Subcellular Areas

Quantification and analysis of neuronal function and morphology is an active research area that has been ongoing for decades because of its importance in our understanding of neuronal degenerative diseases [42–44]. In order to monitor neuronal activity, we utilized LLSM to record the three-dimensional voltage response of hippocampal neurons on a 3D culture system stimulated by 25 mM of KCl solution. The selected images at different time points are displayed in Figure 3A, and the entire time-lapse image can be found in the supporting information (Movie S4 in SI). The color-coded intensity indicates the change in fluorescent intensity throughout the acquisition period before and after the KCl treatment. Since LLSM provides the 3D voltage response information in neuronal cells, it is very important to develop a 3D quantitative analysis method in order to quantify the voltage response in 3D. We first conducted neuronal segmentation using the z-stack images, in which five neurites and soma were identified. In order to calculate intensity changes in each neurite over time, we selected a region of interest (ROI) in each neurite and soma. The segmented neuron is shown in Figure 3B(a) and the ROIs are displayed in Figure 3B(b), in which the overlay image is shown in Figure 3B(c) (Movie S5 in SI). Using the integrated intensity in the ROIs, we compared the change in fluorescence intensity at different locations and time points. In the initial stage, the absolute fluorescence intensity of soma was higher than those observed in the neurites (Figure 3C). After KCl stimulation, the fluorescence intensity of soma increased at a faster rate compared with that of neurites (Figure 3A). In order to quantify the voltage response at different times, we calculated the changes in integrated intensity ($\Delta F/F0$ (%)) at different ROIs using the integrated fluorescence intensity of each ROI at time zero as a reference (F0) (Figure 3D). Changes in fluorescence intensity were observed for each neurite after KCl stimulation, in which fluorescence intensity was found to increase in three out of five neurites.

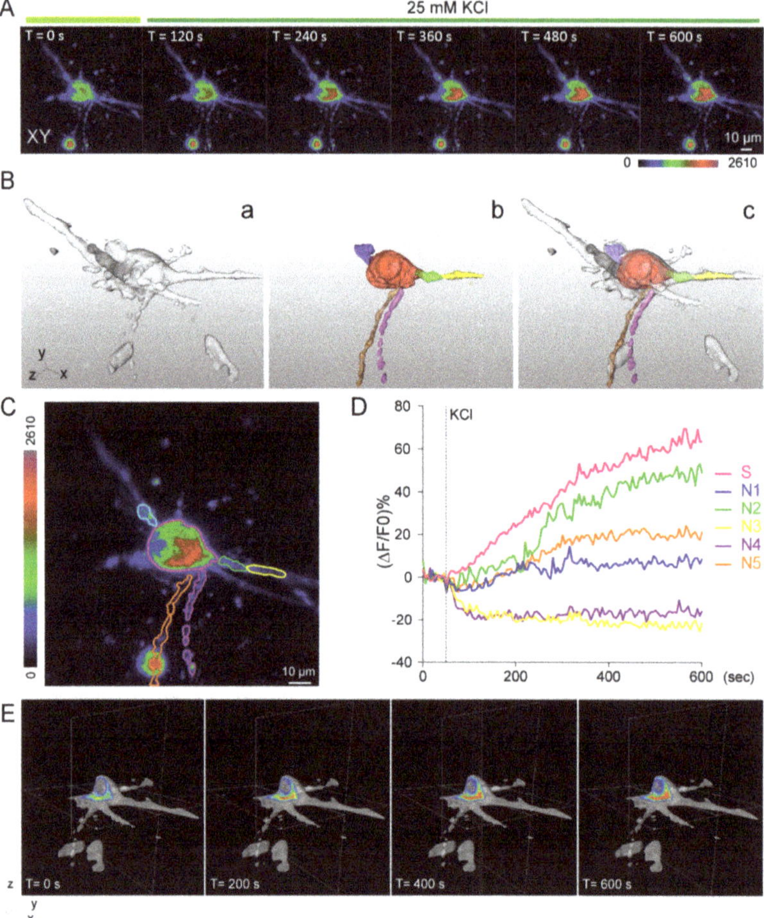

Figure 3. Voltage response of neuron to chemical stimulation (KCl). (**A**) Images of voltage dye-labeled neuron before and after 25 mM KCl treatment at different time points. Scale bar: 10 μm. (**B**) The 3D segmentation and ROI images of voltage dye labeling neurons. The T0 image of (A) was used for 3D reconstruction and segmentation. Selected 3D ROIs are shown in different colors. The relative positions between ROIs (a) and (b) are overlaid, as shown in (c). (**C**) The maximum intensity projection of the T0 image and the outlines of ROIs. (**D**) The changes in the integrated intensity in different ROIs at different times before and after 25 mM KCl stimulation. The dashed line indicates the time point of stimulation. The intensity is normalized with ROI volume. (**E**) Cross-sectional view of voltage dye at different time points.

During image analysis, we found that di-4-ANEPPS dyes labeled not only the plasma membrane but also the cytoplasm of a soma. We further used a 3D graphic software to visualize the intensity change in soma after KCl treatment (Figure 3E). The orthoslice images of the voltage response show that the area with maximum intensity was adjacent to the nucleus instead of the plasma membrane. This result suggests that the voltage dye, di-4-ANEPPS, may not only anchor on the membrane but also accumulate in the cytosol, which could not be resolved by conventional microscopic tools, indicating the superior performance of LLSM for 3D cell cultures.

3.4. Calcium Imaging of Neuronal Activity under External Stimulations

Calcium ions are another common indicator for measuring neuron activity. In a typical cell, the intracellular free calcium concentration is about 50–100 nM, which is 104 times lower than the extracellular concentration [45]. In order to monitor changes in calcium concentration, fluorescence dyes, such as Fluo-4AM, are often used [19,46]. In neuronal cultures, the neuronal network would release neuronal transmitters, which results in spontaneous neuronal firing. In this experiment, we employed LLSM to record the spontaneous activity of 3D-cultured hippocampal neurons labeled with Fluo-4AM [47]. We observed low-frequency spontaneous spiking in our 3D culture system (Figure 4A). In addition to spontaneous firing, neuronal activity can also be stimulated by chemicals or electric pulses. We first stimulated the 3D-cultured hippocampal neurons with 25 mM KCl (Figure 4A, Movies S6 and S7 in Supplementary Materials). A significant increase in the fluorescence intensity of Fluo-4-AM was observed, indicating functional activity and calcium influx in 3D-cultured hippocampal neurons (Movies S8 and S9 in Supplementary Materials). In order to test the behavior of the 3D-cultured neurons under electric pulse stimulation, two electrodes were placed in parallel on the microscope glass slide, and an electric field of 10 V/cm was applied to the electrodes at 10 Hz. We observed wave-like calcium responses [48] (Figure 4B and Movies S10 and S11 in Mupplementary Materials), suggesting that the 3D-cultured neurons could form a functional neuronal network and the neuronal activity could be stimulated by both chemicals and electric pulses.

In order to quantify the spontaneous neuronal firing, we analyzed the fluorescence intensity change of neurons in the selected ROIs (Figure S1A in SI). Figure S1A shows the maximum intensity projection image of calcium response in neurons, in which volumetric images were acquired at a rate of 1 second per volume. Figure 4C shows the changes in mean fluorescence intensity. In this experiment, spontaneous firing behavior was observed in 27.3% of neurons in 3D culture (3 out of 11 neurons from two independent experiments). The calcium response can also be found in neurons under chemical stimulation (KCl). Different temporal behaviors were observed in different neurons (Figure 4D). The fluorescence intensities of four neurons (Figure S1B of Supplementary Materials) were found to increase immediately after stimulation and decay with different behavior. Whereas neuron 3 exhibited a single decay, bimodal responses were found in neurons 1 and 2.

In order to further investigate neuronal activity under electrical stimulation, we applied an electrical field of 10 V/cm between two paralleled electrodes on the slide [49], which were generated by scaling amplifiers using customized software. After acquiring the baseline images, the neurons were stimulated with a sequence of 900 electric pulses with 1 msec pulse width at 10 Hz. The electrical pulse stimulation can induce the release of neuronal transmitters, resulting in neuronal activation. By applying the electric field to the 3D-cultured neurons, we could monitor the synaptic activity of neurons via analysis of calcium influx of spines and boutons (Figure 4E). The intensities of calcium were quantified through the selected ROIs (Figure S1C in Supplementary Materials). Some (but not all) of the boutons also presented a similar behavior after several hundred rounds of electric pulse stimulation. These results indicate that this 3D-cultured neuron system could provide a functional network, and the synaptic plasticity of spine and boutons in the 3D-cultured system could be observed and analyzed by LLSM.

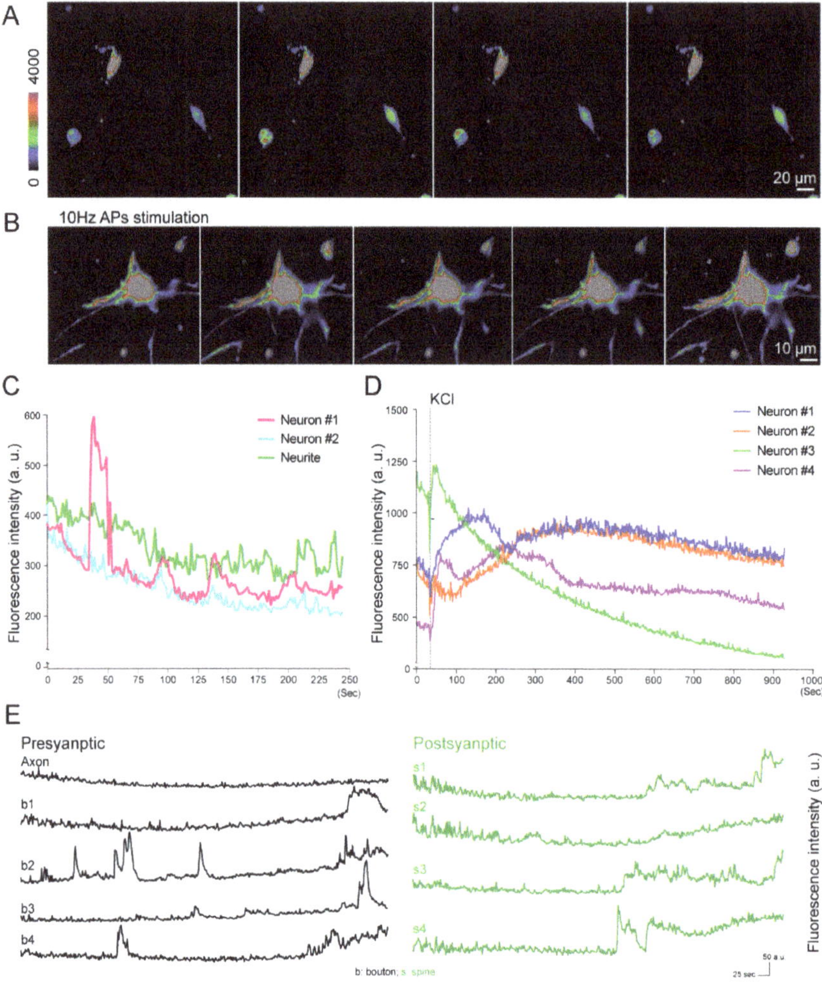

Figure 4. The calcium images of neuronal activity of 3D-cultured neurons. (**A**) Maximum projection intensity images (XY plane) of the spontaneous calcium activity of 3D-cultured neurons at different time points. Scale bar: 20 μm. (**B**) Maximum projection intensity images (XY plane) of 3D-cultured neurons under 10 Hz electric stimulation. Scale bar: 10 μm. (**C**) The fluorescence intensity of the calcium influx of neurite and somas from selected neurons in image S1 of supporting information. (**D**) The fluorescence intensity of calcium influx of selected neurons in image S2 of supporting information. (**E**) The fluorescent intensity dynamic of calcium responses from selected spines and boutons of 3D cultured neuron in images S3 of supporting information. Scale bars: 25 s and 50 a.u. (arbitrary units); b: bouton; s: spine.

4. Discussion

Electrophysiological techniques are currently regarded as the most accurate approach for providing membrane potential information in neurons, since this approach directly measures and records the conductance of a single neuron [50]. However, this approach provides limited information on the directional and spatiotemporal dynamics of the voltage change. In addition, such manual operation is difficult to scale up, especially for on-chip screening applications. In this experiment, we demonstrated that LLSM could be used as an alternative tool for measuring neuronal activity with very high spatial resolution in 3D culture systems. Overcoming the limitations of conventional fluorescence microscopy, LLSM has presented great potential in the study of functional aspects of hippocampal neurons in a 3D environment. It provides excellent optical properties for time-lapse imaging, allows parallelization for large-scale studies, and reveals detailed subcellular information with low photobleaching and phototoxicity effects [38]. In this study, the 3D Matrigel-embedded hippocampal neurons could differentiate and mature as regular 2D culture (Figure 1). Figure 1A shows the confocal images of 2D fixed cultured neurons with good spatial resolution because of point-scanning with a placed pinhole, while Figure 1B shows the images of 3D cultured neurons taken by wide-field microscopy with worse 3D image quality due to the fluorescent background from the thick sample. In order to record the chemical or electrical response of 3D cultured neurons at high spatiotemporal resolution, we built LLSM to perform such experiments for its good 3D imaging capability, such as optical sectioning and live imaging. The 3D images acquired by LLSM demonstrate that the hippocampal neurons could elongate the neurite outgrowth toward different orientations (Figure 2F,G). Moreover, the inset of Figure 2F shows the photobleaching curve of the observed samples, there was no obvious bleaching happening over our experimental time. LLSM was used to observe the functional dynamics of 3D-cultured hippocampal neurons (Figures 3 and 4), which is the most fundamental requirement for visualized neuron activity. The chemically-induced increase in voltage was observed with a subcellular resolution by monitoring the intensity dynamic of di-4-ANEPPS in 3D (Figure 3) or with Fluo-4AM (Figure 4), while the spontaneous firing of hippocampal neurons monitored by a calcium indicator was captured in a large imaging field (Figure 4A,C). The calcium influx spikes of presynaptic terminal and postsynaptic spines could be observed by LLSM after the neurons were stimulated with electric pulses (Figure 4B,E). These results demonstrate the possibility of live 3D culture imaging, in which LLSM can serve as a simple and powerful approach for the study of neuronal activity in a 3D culture environment.

Although LLSM serves as excellent approach for functional imaging in the neuroscience field, there are still some limitations in the selection of fluorescent indicators. The major technical challenges of voltage imaging are the limited selection of voltage sensitive dyes, genetically encoded voltage indicators, and rapid firing events of action potential, which take sub-milliseconds to occur [16]. Even though we are unable to increase the temporal resolution of the voltage image as fast as electrophysiological changes occur, we could still progressively improve the spatial resolution for observing the dynamics of the subcellular region, which allows for direct analysis of the neuronal response from the soma and axon to dendrites (Figures 3 and 4). Voltage-sensitive dyes were loaded into plasma membrane via the internalization by the neurons [51,52], which may have led to the observation that the di-4-ANEPPS dye accumulated in the cytosol and nuclear regions [53] (Figure 3E). Similar results were observed in a previous study, in which the voltage-sensitive dye, the fluorescent imaging plate reader membrane potential dye, FMP, behaved as a charged molecule and accumulated in the cytosol of soma [54].

According to our voltage and calcium neuronal images, these live imaging results revealed that not all neurons or subcellular structures respond in the same way during chemical stimulation (Figures 3D and 4D). In this study, we used di-4-ANEPPS as a voltage dye, which was excited by a 488 nm laser and had a red emission shift, while the same laser was used to excite the Fluo4-AM calcium dye. Emission of Fluo4-AM could be separated by different band-pass filters. Therefore, the combination of voltage (di-4-ANEPPS) and calcium (Fluo4-AM) dyes with the same excitation but disparate emission wavelengths within the same neuron could provide advanced information about the neuronal connection and neuronal activity within a 3D network [32,47].

Although a two-photon imaging system could monitor the calcium dynamic in 3D-cultured systems or tissues [55,56], the images are restricted to a single focal plane, which loses some spatial neuronal information. Here, we used the LLSM to image the whole live 3D z-stack image and then analyzed the synaptic plasticity in a 3D-cultured system. Our results might open the road toward live functional imaging of 3D-cultured systems and show the possibility of imaging specimen in a 3D controlled microenvironment at high spatial and temporal resolution by LLSM. However, the throughput of preset microscopes will be the main drawback. The microfluidics can contribute to developing a high-throughput LLSM microscope. If LLSM is built based on microfluidics, the 3D controlled microenvironment for biological research could be studied easily and extensively for live imaging. Therefore, with the powerful 3D culture ability, microfluidics can be incorporated with advanced LLSM to develop a new platform for biology and biomedical research fields. Instead of PDMS, used mostly in microfluids, integrating these two systems with a water refractive index matching environment will be of interest for us.

5. Conclusions

Cell populations in diseased tissues are very heterogeneous and extend over a 3D space, which is very different behavior than that seen in cells in 2D culture dishes. Therefore, in order to improve the efficiency in drug screening applications, organs-on-chips, which integrate 3D cultures with microfluidic systems, may provide more reliable results, because they better represent the native microenvironment within tissues. However, some cells in the 3D cultures are buried deep inside the cell mass, which cannot be easily accessed by conventional microscopic tools. In this study, we demonstrated that individual neurons in 3D culture can be visualized by LLSM. The volumetric images of neurons were done by scanning the light-sheet plane along the axial direction with frame exposure of 6 ms, which covered an area up to 60 μm × 80 μm with an image pixel size of 0.102 μm. With the voltage probe di-4-ANEPPS, 3D membrane voltage responses of neurons under chemical stimulation were monitored at different time points with subcellular resolution. It was noticed through the volumetric images of neurons that the voltage probes accumulated near the nucleus instead of the membrane, which cannot be resolved by conventional fluorescence microscopy. Calcium influx in neurons under external stimuli, such as chemical and electric stimulation, was investigated by LLSM, in which spontaneous spiking and wave-like behaviors were observed for neurons in 3D culture. In summary, LLSM can be used as a platform for the characterization of organs-on-chips with several advantages, including volumetric imaging, high spatiotemporal resolution, reduced photo-toxicity and photo-bleaching, deeper penetration, and the capability of recoding functional imaging in 3D.

Supplementary Materials: The following are available online at http://www.mdpi.com/2072-666X/10/9/599/s1, Figure S1: The calcium images of neurons in 3D culture. (A) The maximum intensity projection of 3D-cultured neurons. Scale bar: 5 μm. (B) The maximum intensity projection of 3D neurons under chemical stimulation. Scale bar: 5 μm. (C) The calcium image of neurons with neighboring boutons and spines. Scale bar: 5 μm. b: bouton; s: spine, Movie S1: Bright field image of z-stack in three-dimensional (3D)-cultured neurons by conventional microscopy, Movie S2: Immunofluorescence labeling of neuronal markers in 3D-cultured neurons by conventional microscopy, Movie S3: The entire z-stack image of voltage dye-labeled 3D neurons was observed by lattice light-sheet microscopy (LLSM), Movie S4: Time-lapse image of membrane potential dynamic from a voltage dye-labeled 3D neuron was observed by LLSM, Movie S5: The 3D visualization image of the ROIs selected from the voltage dye-labeled 3D neuron, Movie S6: Live responses of calcium influx within the calcium dye-labeled 3D neurons were observed by LLSM, Movie S7: The live responses of calcium influx within the calcium dye-labeled 3D neurons were observed by LLSM (larger imaging field), Movie S8: KCl treatment resulted in calcium influx within the calcium dye labeled 3D neuron were observed by LLSM (larger imaging field), Movie S9: The KCl treatment resulted in calcium influx within the calcium dye-labeled 3D neuron were observed by LLSM, Movie S10: Electrical pulse stimulation caused by action potential inspired the calcium influx within the calcium dye-labeled 3D neurons, observed by LLSM, Movie S11: The electrical pulse stimulation caused action potential inspired the calcium influx within the calcium dye-labeled 3D neuron, observed by LLSM (elevated background signals).

Author Contributions: Conceptualization, C.-Y.C. and B.-C.C.; data curation, C.-Y.C.; formal analysis, C.-Y.C., Y.-T.L., P.-Y.L. and Y.-C.T.; funding acquisition, P.C. and B.-C.C.; investigation, C.-Y.C.; methodology, C.-Y.C., C.-H.L., Y.-T.L. and J.-S.W.; project administration, C.-Y.C.; resources, P.C. and B.-C.C.; software, Y.-T.L. and P.-Y.L.; supervision, B.-C.C.; validation, C.-Y.C. and B.-C.C.; visualization, C.-Y.C., P.-Y.L. and Y.-C.T.; writing—original draft, C.-Y.C.; Writing—review & editing, P.C. and B.-C.C.

Funding: P.C. and B.-C.C. acknowledge the support from the Ministry of Science and Technology of Taiwan under contract numbers 107-2119-M-001-007, 105-2119-M-001-031-MY2, 106-2811-M-001-160, and 106-2811-M-001-145. P.C. is grateful for the support of the Thematic project of Academia Sinica (AS-TP-107-ML9).

Acknowledgments: Howard Hughes Medical Institute, Janelia Farm Research Campus, licenses the microscope control software.

Conflicts of Interest: The authors declare no conflict of interest. The funders had no role in the design of the study; in the collection, analyses, or interpretation of data; in the writing of the manuscript, or in the decision to publish the results.

References

1. Shin, Y.; Han, S.; Jeon, J.S.; Yamamoto, K.; Zervantonakis, I.K.; Sudo, R.; Kamm, R.D.; Chung, S. Microfluidic assay for simultaneous culture of multiple cell types on surfaces or within hydrogels. *Nat. Protoc.* **2012**, *7*, 1247–1259. [CrossRef] [PubMed]

2. Rothbauer, M.; Zirath, H.; Ertl, P. Recent advances in microfluidic technologies for cell-to-cell interaction studies. *Lab Chip* **2018**, *18*, 249–270. [CrossRef] [PubMed]

3. Kuo, C.W.; Chueh, D.Y.; Chen, P. Investigation of size-dependent cell adhesion on nanostructured interfaces. *J. Nanobiotechnol.* **2014**, *12*, 54. [CrossRef] [PubMed]

4. Chien, F.C.; Kuo, C.W.; Yang, Z.H.; Chueh, D.Y.; Chen, P. Exploring the formation of focal adhesions on patterned surfaces using super-resolution imaging. *Small* **2011**, *7*, 2906–2913. [CrossRef] [PubMed]

5. Huh, D.; Hamilton, G.A.; Ingber, D.E. From 3D cell culture to organs-on-chips. *Trends Cell Biol.* **2011**, *21*, 745–754. [CrossRef] [PubMed]

6. Bhatia, S.N.; Ingber, D.E. Microfluidic organs-on-chips. *Nat Biotechnol.* **2014**, *32*, 760–772. [CrossRef] [PubMed]

7. Zhang, B.; Korolj, A.; Lai, B.F.L.; Radisic, M. Advances in organ-on-a-chip engineering. *Nat. Rev. Mater.* **2018**, *3*, 257–278. [CrossRef]

8. Sarkar, S.; Peng, C.-C.; Kuo, C.W.; Chueh, D.-Y.; Wu, H.-M.; Liu, Y.-H.; Chen, P.; Tung, Y.-C. Study of oxygen tension variation within live tumor spheroids using microfluidic devices and multi-photon laser scanning microscopy. *RSC Adv.* **2018**, *8*, 30320–30329. [CrossRef]

9. Gan, L.; Cookson, M.R.; Petrucelli, L.; La Spada, A.R. Converging pathways in neurodegeneration, from genetics to mechanisms. *Nat. Neurosci.* **2018**, *21*, 1300–1309. [CrossRef]

10. Kuhlenbeck, H. *The Central Nervous System of Vertebrates: A General Survey of Its Comparative Anatomy with an Introduction to the Pertinent Fundamental Biologic and Logical Concepts*; Karger Medical and Scientific Publishers: Basel, Switzerland, 1967.

11. Hsiao, Y.-S.; Lin, C.-C.; Hsieh, H.-J.; Tsai, S.-M.; Kuo, C.-W.; Chu, C.-W.; Chen, P. Manipulating location, polarity, and outgrowth length of neuron-like pheochromocytoma (PC-12) cells on patterned organic electrode arrays. *Lab Chip* **2011**, *11*, 3674–3680. [CrossRef]

12. Lu, C.H.; Hsiao, Y.S.; Kuo, C.W.; Chen, P. Electrically tunable organic bioelectronics for spatial and temporal manipulation of neuron-like pheochromocytoma (PC-12) cells. *Biochim. Biophys. Acta.* **2013**, *1830*, 4321–4328. [CrossRef] [PubMed]

13. Hsiao, Y.S.; Liao, Y.H.; Chen, H.L.; Chen, P.; Chen, F.C. Organic Photovoltaics and Bioelectrodes Providing Electrical Stimulation for PC12 Cell Differentiation and Neurite Outgrowth. *ACS Appl. Mater. Interfaces* **2016**, *8*, 9275–9284. [CrossRef] [PubMed]

14. Tsai, N.-C.; She, J.-W.; Wu, J.-G.; Chen, P.; Hsiao, Y.-S.; Yu, J. Poly(3,4-ethylenedioxythiophene) Polymer Composite Bioelectrodes with Designed Chemical and Topographical Cues to Manipulate the Behavior of PC12 Neuronal Cells. *Appl. Mater. Interfaces* **2019**, *6*, 1801576. [CrossRef]

15. Hiruma, H.; Uemura, T.; Kimura, F. Neuronal Synchronization and Ionic Mechanisms for Propagation of Excitation in the Functional Network of Immortalized GT1-7 Neurons: Optical Imaging with a Voltage-Sensitive Dye. *J. Neuroendocrinol.* **1997**, *9*, 835–840. [CrossRef]

16. Peterka, D.S.; Takahashi, H.; Yuste, R. Imaging voltage in neurons. *Neuron* **2011**, *69*, 9–21. [CrossRef] [PubMed]

17. Jin, L.; Han, Z.; Platisa, J.; Wooltorton, J.R.; Cohen, L.B.; Pieribone, V.A. Single action potentials and subthreshold electrical events imaged in neurons with a fluorescent protein voltage probe. *Neuron* **2012**, *75*, 779–785. [CrossRef] [PubMed]

18. Shoham, D.; Glaser, D.E.; Arieli, A.; Kenet, T.; Wijnbergen, C.; Toledo, Y.; Hildesheim, R.; Grinvald, A. Imaging Cortical Dynamics at High Spatial and Temporal Resolution with Novel Blue Voltage-Sensitive Dyes. *Neuron* **1999**, *24*, 791–802. [CrossRef]

19. Grienberger, C.; Konnerth, A. Imaging calcium in neurons. *Neuron* **2012**, *73*, 862–885. [CrossRef]

20. Ulloa Severino, F.P.; Ban, J.; Song, Q.; Tang, M.; Bianconi, G.; Cheng, G.; Torre, V. The role of dimensionality in neuronal network dynamics. *Sci. Rep.* **2016**, *6*, 29640. [CrossRef]

21. Anderson, H.E.; Fontaine, A.K.; Caldwell, J.H.; Weir, R.F. Imaging of electrical activity in small diameter fibers of the murine peripheral nerve with virally-delivered GCaMP6f. *Sci. Rep.* **2018**, *8*, 3219. [CrossRef]

22. Johenning, F.W.; Theis, A.K.; Pannasch, U.; Ruckl, M.; Rudiger, S.; Schmitz, D. Ryanodine Receptor Activation Induces Long-Term Plasticity of Spine Calcium Dynamics. *PLoS Biol.* **2015**, *13*, e1002181. [CrossRef] [PubMed]

23. Chamberland, S.; Yang, H.H.; Pan, M.M.; Evans, S.W.; Guan, S.; Chavarha, M.; Yang, Y.; Salesse, C.; Wu, H.; Wu, J.C.; et al. Fast two-photon imaging of subcellular voltage dynamics in neuronal tissue with genetically encoded indicators. *eLife* **2017**, *6*, e25690. [CrossRef] [PubMed]

24. Huisken, J.; Swoger, J.; Del Bene, F.; Wittbrodt, J.; Stelzer, E.H. Optical sectioning deep inside live embryos by selective plane illumination microscopy. *Science* **2004**, *305*, 1007–1009. [CrossRef] [PubMed]

25. Keller, P.J.; Schmidt, A.D.; Wittbrodt, J.; Stelzer, E.H. Reconstruction of zebrafish early embryonic development by scanned light sheet microscopy. *Science* **2008**, *322*, 1065–1069. [CrossRef] [PubMed]

26. Chen, B.C.; Legant, W.R.; Wang, K.; Shao, L.; Milkie, D.E.; Davidson, M.W.; Janetopoulos, C.; Wu, X.S.; Hammer, J.A., 3rd; Liu, Z.; et al. Lattice light-sheet microscopy: imaging molecules to embryos at high spatiotemporal resolution. *Science* **2014**, *346*, 1257998. [CrossRef] [PubMed]

27. Chatterjee, K.; Pratiwi, F.W.; Wu, F.C.M.; Chen, P.; Chen, B.C. Recent Progress in Light Sheet Microscopy for Biological Applications. *Appl. Spectrosc.* **2018**, *72*, 1137–1169. [CrossRef] [PubMed]

28. Chen, C.Y.; Chen, Y.T.; Wang, J.Y.; Huang, Y.S.; Tai, C.Y. Postsynaptic Y654 dephosphorylation of beta-catenin modulates presynaptic vesicle turnover through increased n-cadherin-mediated transsynaptic signaling. *Dev. Neurobiol.* **2017**, *77*, 61–74. [CrossRef]

29. Fath, T.; Ke, Y.D.; Gunning, P.; Gotz, J.; Ittner, L.M. Primary support cultures of hippocampal and substantia nigra neurons. *Nat. Protoc.* **2009**, *4*, 78–85. [CrossRef] [PubMed]

30. Kim, Y.H.; Choi, S.H.; D'Avanzo, C.; Hebisch, M.; Sliwinski, C.; Bylykbashi, E.; Washicosky, K.J.; Klee, J.B.; Brustle, O.; Tanzi, R.E.; et al. A 3D human neural cell culture system for modeling Alzheimer's disease. *Nat. Protoc.* **2015**, *10*, 985–1006. [CrossRef] [PubMed]

31. Gee, K.R.; Brown, K.A.; Chen, W.N.U.; Bishop-Stewart, J.; Gray, D.; Johnson, I. Chemical and physiological characterization of fluo-4 Ca^{2+}-indicator dyes. *Cell Calcium* **2000**, *27*, 97–106. [CrossRef]

32. Johnson, P.L.; Smith, W.; Baynham, T.C.; Knisley, S.B. Errors Caused by Combination of Di-4 ANEPPS and Fluo3/4 for Simultaneous Measurements of Transmembrane Potentials and Intracellular Calcium. *Ann. Biomed. Eng.* **1999**, *27*, 563–571. [CrossRef] [PubMed]

33. Amira User Guide. Available online: https://www.fei.com/software/amira-user-guide/ (accessed on 11 September 2019).

34. Camp, J.G.; Badsha, F.; Florio, M.; Kanton, S.; Gerber, T.; Wilsch-Brauninger, M.; Lewitus, E.; Sykes, A.; Hevers, W.; Lancaster, M.; et al. Human cerebral organoids recapitulate gene expression programs of fetal neocortex development. *Proc. Nat. Acad. Sci. USA* **2015**, *112*, 15672–15677. [CrossRef] [PubMed]

35. Kelava, I.; Lancaster, M.A. Stem Cell Models of Human Brain Development. *Cell Stem Cell* **2016**, *18*, 736–748. [CrossRef] [PubMed]

36. Lancaster, M.A.; Renner, M.; Martin, C.A.; Wenzel, D.; Bicknell, L.S.; Hurles, M.E.; Homfray, T.; Penninger, J.M.; Jackson, A.P.; Knoblich, J.A. Cerebral organoids model human brain development and microcephaly. *Nature* **2013**, *501*, 373–379. [CrossRef] [PubMed]

37. Shamir, E.R.; Ewald, A.J. Three-dimensional organotypic culture: experimental models of mammalian biology and disease. *Nat. Rev. Mol. Cell Biol.* **2014**, *15*, 647–664. [CrossRef] [PubMed]

38. Lu, C.H.; Tang, W.C.; Liu, Y.T.; Chang, S.W.; Wu, F.C.M.; Chen, C.Y.; Tsai, Y.C.; Yang, S.M.; Kuo, C.W.; Okada, Y.; et al. Lightsheet localization microscopy enables fast, large-scale, and three-dimensional super-resolution imaging. *Commun. Biol.* **2019**, *2*, 177. [CrossRef] [PubMed]

39. Aerov, A.A. Why the water bridge does not collapse. *Phys. Rev. E Stat. Nonlin. Soft Matter. Phys.* **2011**, *84*, 036314. [CrossRef] [PubMed]

40. Fluhler, E.; Burnham, V.G.; Loew, L.M. Spectra, membrane binding, and potentiometric responses of new charge shift probes. *Biochemistry* **1985**, *24*, 5749–5755. [CrossRef] [PubMed]

41. Hassner, A.; Birnbaum, D.; Loew, L.M. Charge-shift probes of membrane potential. Synthesis. *J. Org. Chem.* **1984**, *49*, 2546–2551. [CrossRef]

42. Peng, H.; Hawrylycz, M.; Roskams, J.; Hill, S.; Spruston, N.; Meijering, E.; Ascoli, G.A. BigNeuron: Large-Scale 3D Neuron Reconstruction from Optical Microscopy Images. *Neuron* **2015**, *87*, 252–256. [CrossRef] [PubMed]

43. Peng, H.; Meijering, E.; Ascoli, G.A. From DIADEM to BigNeuron. *Neuroinformatics* **2015**, *13*, 259–260. [CrossRef] [PubMed]

44. Gillette, T.A.; Brown, K.M.; Ascoli, G.A. The DIADEM metric: comparing multiple reconstructions of the same neuron. *Neuroinformatics* **2011**, *9*, 233–245. [CrossRef] [PubMed]

45. Maravall, M.; Mainen, Z.F.; Sabatini, B.L.; Svoboda, K. Estimating Intracellular Calcium Concentrations and Buffering without Wavelength Ratioing. *Biophys. J.* **2000**, *78*, 2655–2667. [CrossRef]

46. Helmchen, F.; Imoto, K.; Sakmann, B. Ca^{2+} buffering and action potential-evoked Ca^{2+} signaling in dendrites of pyramidal neurons. *Biophys. J.* **1996**, *70*, 1069–1081. [CrossRef]

47. Canepari, M.; Djurisic, M.; Zecevic, D. Dendritic signals from rat hippocampal CA1 pyramidal neurons during coincident pre- and post-synaptic activity: a combined voltage- and calcium-imaging study. *J. Physiol.* **2007**, *580*, 463–484. [CrossRef] [PubMed]

48. Deneke, V.E.; Di Talia, S. Chemical waves in cell and developmental biology. *J. Cell Biol.* **2018**, *217*, 1193–1204. [CrossRef]

49. Kim, S.H.; Ryan, T.A. CDK5 serves as a major control point in neurotransmitter release. *Neuron* **2010**, *67*, 797–809. [CrossRef]

50. Wickenden, A.D. Overview of Electrophysiological Techniques. *Curr. Protoc. Pharmacol.* **2000**, *11*, 11.1.1–11.1.17. [CrossRef]

51. Obaid, A.L.; Loew, L.M.; Wuskell, J.P.; Salzberg, B.M. Novel naphthylstyryl-pyridium potentiometric dyes offer advantages for neural network analysis. *J. Neurosci. Methods* **2004**, *134*, 179–190. [CrossRef]

52. Rohr, S.; Salzberg, B.M. Multiple site optical recording of transmembrane voltage (MSORTV) in patterned growth heart cell cultures: assessing electrical behavior, with microsecond resolution, on a cellular and subcellular scale. *Biophys. J.* **1994**, *67*, 1301–1315. [CrossRef]

53. Preuss, S.; Stein, W. Comparison of two voltage-sensitive dyes and their suitability for long-term imaging of neuronal activity. *PLoS ONE* **2013**, *8*, e75678. [CrossRef] [PubMed]

54. Fairless, R.; Beck, A.; Kravchenko, M.; Williams, S.K.; Wissenbach, U.; Diem, R.; Cavalie, A. Membrane potential measurements of isolated neurons using a voltage-sensitive dye. *PLoS ONE* **2013**, *8*, e58260. [CrossRef] [PubMed]

55. Park, J.; Wetzel, I.; Marriott, I.; Dreau, D.; D'Avanzo, C.; Kim, D.Y.; Tanzi, R.E.; Cho, H. A 3D human triculture system modeling neurodegeneration and neuroinflammation in Alzheimer's disease. *Nat. Neurosci.* **2018**, *21*, 941–951. [CrossRef] [PubMed]

56. Mansour, A.A.; Goncalves, J.T.; Bloyd, C.W.; Li, H.; Fernandes, S.; Quang, D.; Johnston, S.; Parylak, S.L.; Jin, X.; Gage, F.H. An in vivo model of functional and vascularized human brain organoids. *Nat. Biotechnol.* **2018**, *36*, 432–441. [CrossRef] [PubMed]

Article

Micro Vacuum Chuck and Tensile Test System for Bio-Mechanical Evaluation of 3D Tissue Constructed of Human Induced Pluripotent Stem Cell-Derived Cardiomyocytes (hiPS-CM)

Kaoru Uesugi [1,2], Fumiaki Shima [3], Ken Fukumoto [3], Ayami Hiura [3], Yoshinari Tsukamoto [3], Shigeru Miyagawa [2,4], Yoshiki Sawa [2,4], Takami Akagi [3], Mitsuru Akashi [2,3] and Keisuke Morishima [1,2,*]

[1] Department of Mechanical Engineering, Osaka University, 2-1 Yamada-oka, Suita, Osaka 565-0871, Japan
[2] Global Center for Medical Engineering and Informatics, Osaka University, 2-1 Yamada-oka Suita, Osaka 565-0871, Japan
[3] Building Block Science Joint Research Chair, Graduate School of Frontier Biosciences, Osaka University, 1-3 Yamada-oka, Suita, Osaka 565-0871, Japan
[4] Department of Cardiovascular Surgery, Graduate School of Medicine, Osaka University, 2-2 Yamada-oka, Suita, Osaka 565-0871, Japan
*   Correspondence: morishima@mech.eng.osaka-u.ac.jp; Tel.: +81-6-6879-7343

Received: 19 June 2019; Accepted: 17 July 2019; Published: 19 July 2019

Abstract: In this report, we propose a micro vacuum chuck (MVC) which can connect three-dimensional (3D) tissues to a tensile test system by vacuum pressure. Because the MVC fixes the 3D tissue by vacuum pressure generated on multiple vacuum holes, it is expected that the MVC can fix 3D tissue to the system easily and mitigate the damage which can happen by handling during fixing. In order to decide optimum conditions for the size of the vacuum holes and the vacuum pressure, various sized vacuum holes and vacuum pressures were applied to a normal human cardiac fibroblast 3D tissue. From the results, we confirmed that a square shape with 100 µm sides was better for fixing the 3D tissue. Then we mounted our developed MVCs on a specially developed tensile test system and measured the bio-mechanical property (beating force) of cardiac 3D tissue which was constructed of human induced pluripotent stem cell-derived cardiomyocytes (hiPS-CM); the 3D tissue had been assembled by the layer-by-layer (LbL) method. We measured the beating force of the cardiac 3D tissue and confirmed the measured force followed the Frank-Starling relationship. This indicates that the beating property of cardiac 3D tissue obtained by the LbL method was close to that of native cardiac tissue.

Keywords: beating force; bio-mechanical property; cardiac 3D tissue; human induced pluripotent Stem cell-derived cardiomyocytes (hiPS-CM); tissue engineering; vacuum chuck

---

1. Introduction

Three-dimensional (3D) tissues which are constructed by cells in the environment *in vitro* have been applied in a wide range of fields such as regenerative medicine [1–26], drug development [2,9,11–15,26–42], disease modelling for pathology [11–15,36,43–46], bioactuators [47–52], food industry [42,52–56], and BioArt [52,55–62]. With the advancement of 3D tissue technologies, evaluation methods for them have been demanded. Conventionally, analytical approaches such as biochemical, immunological, morphological [24,63], electrophysiological, and motion image analysis [31] methods have been applied to evaluate artificial tissues. These methods, however, cannot evaluate bio-mechanical properties directly.

To construct the 3D tissues, it is necessary to make not only biochemical and electrophysiological evaluations, but also bio-mechanical evaluations. By evaluating bio-mechanical properties of 3D tissues, various conditions of 3D tissues can be understood. For example, in the field of regenerative medicine, bio-mechanical evaluations are important for constructing 3D tissues which replicate living tissues. The behaviors of cells in tissues are affected by the surrounding environment [64,65]. If the surrounding environment affects the bio-mechanical properties in a different manner from that *in vivo*, tissues might develop in an undesirable form and might act in ways which are different from their true functions. The fact that conditions of the extracellular matrix influence bio-mechanical properties of 3D tissues also indicates the importance of measurement of 3D tissues [4,5,18,23]. When constructing thick tissues, a luminal structure is needed for nutritional transport [29]; hence, the luminal structure is an important parameter for 3D tissue construction. The bio-mechanical properties of 3D tissues are affected by the luminal structure because the mechanical properties of the luminal structure and other components of the tissues are each different. Thus, measurement of bio-mechanical properties can be evaluated by using the luminal structure of 3D tissues. Additionally, it is important for transplantation of 3D tissues *in vitro* to ascertain whether or not the 3D tissues are broken by the *in vivo* pressure, whether or not the 3D tissues satisfy the desired mechanical properties, and whether or not the 3D tissues have the optimum stiffness or viscoelasticity. For the above reasons, various researchers have statically measured bio-mechanical properties of artificial tissues such as bone [16,17], cartilage [18,19], tendon [20–22], vein [23], and skin [24,25]. Additionally, if large-scale manufacturing of 3D tissues is to be done, automatic handling of 3D tissues may be demanded, and then, information on mechanical properties may be important also. In the field of drug development, bio-mechanical properties are also important for the reasons described above. In the food industry field, artificial meats must possess mechanical properties (texture) similar to that of real meats [54,55]. In the field of BioArt, to get the sense of touch for semi-living, that is, artificial, tissues, it is necessary to start creating some sort of bond [56]. Because the sense of touch mainly depends on mechanical information, the mechanical properties of artificial 3D tissues must be known.

As mentioned above, the bio-mechanical properties are important parameters for constructing 3D tissues. However, conventional methods, such as biochemical and electrophysiological methods and microscopic observations have not been able to evaluate bio-mechanical properties of 3D tissues. Therefore, we tried to measure bio-mechanical properties of a sheet shape tissue (a cell sheet). For example, we measured the adhesion force of a cell sheet by the ninety-degree peel test with our newly developed system [66–68]. We also measured stiffness of a cell sheet with a tensile test system which we developed [69,70]. Since the procedures for the ninety-degree peel test and tensile test are defined by the International Standard Organization (ISO), a proposed method using them should be standardized easily.

Next, we looked at the importance of mechanical properties of cardiac 3D tissue beating. The bio-mechanical properties such as adhesion force and stiffness, which we measured previously, were static properties. Nowadays, dynamical bio-mechanical property, such as beating force, is beginning to receive attention with the appearance of cardiac 3D tissues which have been realized by tissue engineering technologies, and embryo-stem (ES) cell and induced pluripotent stem cell (iPSC) technologies. For example, cardiac 3D tissues were constructed by a layer-by-layer (LbL) method [26,31]. Thus, various methods which can evaluate beating properties of cardiac 3D tissues which were constructed by various methods have been proposed and beating force of cardiac 3D tissues has been measured [3–13,15,32–37,43–46]. The cardiac 3D tissues assembled by the LbL method were evaluated by biochemical, immunological, electrophysiological, and motion image analysis methods [31]. However, these evaluation methods for cardiac 3D tissues which were constructed by LbL method cannot evaluate bio-mechanical properties directly. Therefore, our objective in this report was to overcome this deficiency.

However, fixing 3D tissues onto a measurement system is difficult because of their mechanical and chemical fragility. Many studies have tried to fix 3D tissues by various methods [3–13,15,32–37,43–46].

These methods have limits to the size and the shape of the 3D tissues because of how the fixing to the force measurement system is done. Therefore, in the measurement of mechanical properties for cardiac 3D tissues, most cardiac 3D tissues for which mechanical properties were measured did not consist of just cells. These 3D tissues consisted of cells and scaffold (e.g., fibrin gel and collagen gel). And even though cardiac 3D tissues consisted mainly of cells, the sizes of those 3D tissues were small (length, smaller than 2 mm; width, 200 µm) [37]. Additionally, these fixing methods needed a lot of preparation steps and complex operations.

Evaluation for dynamical bio-mechanical properties of cardiac 3D tissues which consist of just cells (for example, tissue obtained by the LbL method) and whose size is large has been needed. In order to mount the 3D tissue on the evaluation system, a fixture which enables fixing of various sizes and various shaped 3D tissues is necessary. Additionally, for reliable and efficient measurement, the fixture demands easy connection of the 3D tissues to the evaluation system. Therefore, the objective of this study was development of a system which can solve these issues.

In this study, we newly developed a special chucking tool and a tensile test system. The chucking tool can chuck the 3D tissue by vacuum pressure which is loaded through multi micro meter size holes. The tensile system can measure a small force (from sub-micro newton to milli newton level) and can drive the vacuum chucking tool. To confirm applicability of the system, we measured beating property of cardiac 3D tissue that consisted mostly of cells (with a tiny amount of extracellular matrix) and the tissue size was larger than that of conventional studies (diameter: 12 mm).

2. Materials and Methods

2.1. Design of the Micro Vacuum Chuck (MVC)

In order to apply a tensile test for 3D tissues, tissues have to be connected to the tensile test system by using a chucking tool. In general, the mechanical and chemical strengths of conventionally chucked samples (i.e., metal, latex, plastic, cloth, and ceramic, etc.) are larger than those of 3D tissues and the sample sizes (millimeters to centimeters) are also bigger than those of 3D tissues (micrometers to millimeters). Thus, when conventional samples are connected to the tensile test system, a clamp type fixture or chemical bonding is applied. On the other hand, because 3D tissues are very soft (mechanically fragile), the clamp can crush them and chemical bonding can cause chemical damage to them. Additionally, the small 3D tissue size makes a fixing operation which employs the above methods difficult. Some reports which measured contractile force of 3D tissues have used fixing by piercing or hooking with a fine hook [3–8,12,37], tying with nylon fiber [10,13,46], embedding a special fixture in an early preparation stage [11,15,32,35,43–45], and using a silicon post (a micro pillar) [9,33,34,36] in a manner similar to the embedding method. We have also employed hooking for fixing sheet shape tissues [66–70]. However, when applying these fixing methods to the tensile test of 3D tissue, there are some problems. For instance, by hooking or piercing the 3D tissue, it may tear away from the pierced holes. Hooking also restricts the shape of the 3D tissues somewhat; however, a ring and a fiber (rod) shape are possible examples. When using the ring shape 3D tissue, mechanical properties due to the shape and friction between the hook and tissue become a concern. Tying is a complicated operation and it does not allow evaluation of many 3D tissues. Additionally, tying causes variation of the measured force and load mechanical damage to 3D tissues. Operations such as hooking, piercing, and tying may also cause unsure measurements because of deformation of the 3D tissue by stress concentration. To use a special embedded fixture, the shape of the 3D tissue and preparation method are restricted. Force generated by the 3D tissue cannot be measured directly by the micro pillar. It is difficult to apply the tensile test for the micro pillar since the length of the 3D tissues cannot change due to the structure of the system. Additionally, for the micro pillar, shapes of 3D tissues are limited (i.e., a fiber (rod) shape).

Therefore, in this report, we have suggested an MVC which can fix 3D tissue by using vacuum pressure (Figure 1). The MVC has four advantages. First, the MVC can fix various shaped 3D tissues such as sheet, tube, and block shapes. Second, the MVC can fix 3D tissues easily without complex operations. Third, the MVC causes less damage to the 3D tissues and the uncertainty of the measurement is less by decreasing the effect of stress concentration because the MVC does not use fixation tools such as the fine hook, tying fiber, and embedding fixture. Finally, the MVC causes less chemical damage to the 3D tissue since the MVC does not need chemical bonding.

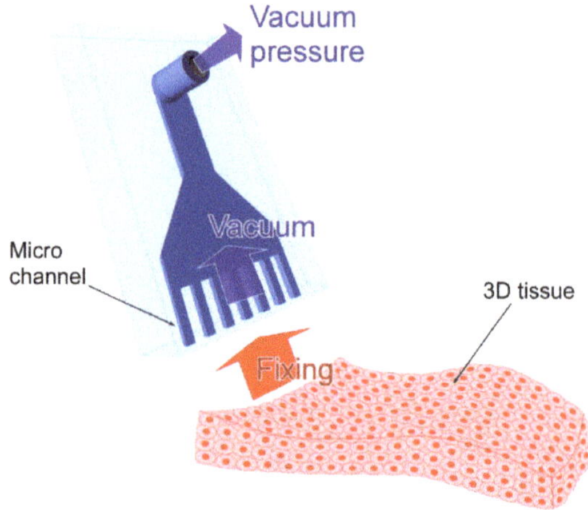

Figure 1. Conceptual illustration of the micro vacuum chuck (MVC). The 3D tissue is suctioned toward the micro size vacuum holes and fixed there by vacuum pressure which is generated at the holes.

The 3D tissue was fixed with a negative pressure generated at vacuum holes of the MVC. When the single vacuum hole was applied for tissue fixing, an excessive vacuum was confirmed to cause a load on the tissue [71]. The excessive vacuum could cause damage to the 3D tissue, including unexpected length changes and deformation. Additionally, when fixing the 3D tissue with the single vacuum hole, the 3D tissue has to be detached from the culture surface before fixing. Therefore, the initial length of the 3D tissue changes by pre-detachment of the 3D tissue because the 3D tissue length is not fixed by the fixtures in advance. If the 3D tissue is fixed by using the single vacuum hole, the 3D tissue becomes detached from the vacuum hole when the 3D tissue is detached from the culture surface because the vacuum pressure which is generated by only one hole is weak (that is, the vacuum pressure is not strong enough to pull the 3D tissue into the vacuum hole). In order to fix the 3D tissue to the tensile test system while maintaining the initial length of the 3D tissue, a fixture which has multi vacuum holes is needed.

Therefore, we used the MVC with several vacuum holes. By increasing the number of vacuum holes, the 3D tissue could be fixed with a small stress concentration. Additionally, we expect that the excessive vacuum on the 3D tissue would be decreased because multiple close-by areas of the tissue would be pulled into the micro vacuum holes at the same time. By increasing the number of the vacuum holes, we can also expect certainty of fixation. Additionally, unlike hooking and tying, the MVC is able to fix variously sized and shaped 3D tissues by adjusting size and shape of the MVC. While, the method which uses an embedding special fixture may fix variously sized and shaped 3D tissues, it has to use impurities (e.g., fibrin gel and collagen gel) for formation of 3D tissues.

Figure 2 shows a photo of the MVC. The MVC had 17 vacuum holes. The distance between the vacuum holes was 206 µm. Each vacuum hole was square and each side length was 100 µm. The vacuum hole size was decided by an experiment (see Section 2.3, Section 3.1, and Section 4). The MVC width was 5 mm. Because the single hole type vacuum chucks were made of fine tube, fabrication of the multi hole type vacuum chuck was difficult. We used a micro fluid channel fabricated by a photolithography technique (see Supplementary Materials and Figure S1).

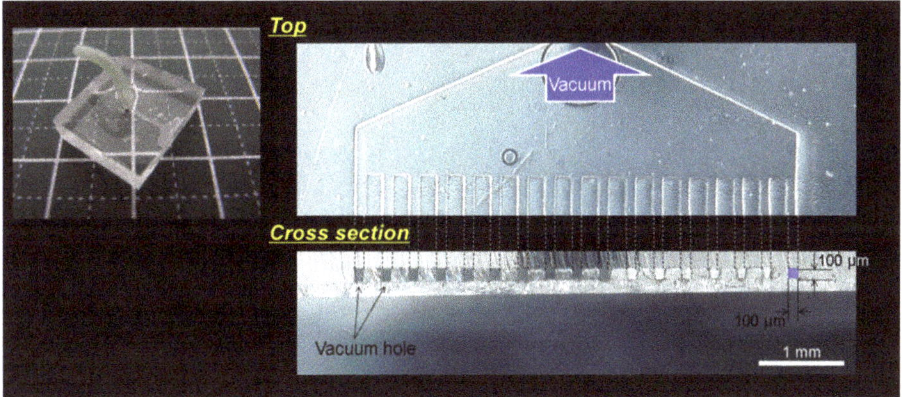

Figure 2. Photo of a micro vacuum check (MVC). The MVC had 17 micro vacuum holes and each hole was a square shape with a side length of 100 μm. The MVC was 5 mm wide.

2.2. Design of Tensile Test System

Since there were no tensile test systems for 3D tissue, we also developed a special tensile test system. We designed the system configuration and devices conditions and ordered the system fabrication from a maker (HS101, Tech Alpha, Tokyo, Japan) (Figure 3a). The tensile test system for 3D tissues consisted of the six parts indicated in Figure 3b. (1) The force transducer was used to measure contractile force of the 3D tissue. (2) The MVCs fixed the 3D tissue to the force transducer and motorized stage. (3) The vacuum pressure generating system adjusted the vacuum pressure of the MVCs. This vacuum pressure system was constructed with a vacuum pump, an electromagnetic valve and a pressure sensor. The vacuum pressure was feedback controlled. (4) The motorized stage loaded a tensile force onto the 3D tissue. (5) The culture dish protected the 3D tissue from drying out during the tensile test. (6) The charge-coupled device (CCD) camera and microscope were used in recording the test conditions. Measured force data were analyzed by commercial analysis software (Igor Pro, Wave Metrics, Lake Oswego, OR, USA). Video data and measurement data were synchronized by turning on an LED lamp and recording the LED voltage signal simultaneously.

Figure 3c shows a schematic illustration of the region around one MVC which was connected to the force transducer of the tensile test system. In order to get close contact between the micro vacuum holes and the 3D tissue which adheres on the culture surface, the MVC was connected to the tensile test system vertically. When measuring the tensile force of the 3D tissue, stiffness of the vacuum tube which was supplying the vacuum pump suction at the MVC may affect the force measurement. Thus, a small diameter (ID, 0.5 mm; OD, 1 mm) and flexible silicon tube (CP-N-0.5-1-10, Shin-Etsu Polymer Co., Ltd., Tokyo, Japan) was used and the tube was positioned vertically relative to the tensile force loaded direction. If the tube connects the force transducer and the vacuum pressure generating system directly, the force transducer can measure mechanical noise (i.e., vibration) which is generated by the vacuum pump. Thus, a tube fixture kept the tube in place mechanically near the force transducer, and the tube fixture mechanically separated the force transducer from the vacuum pressure generating system.

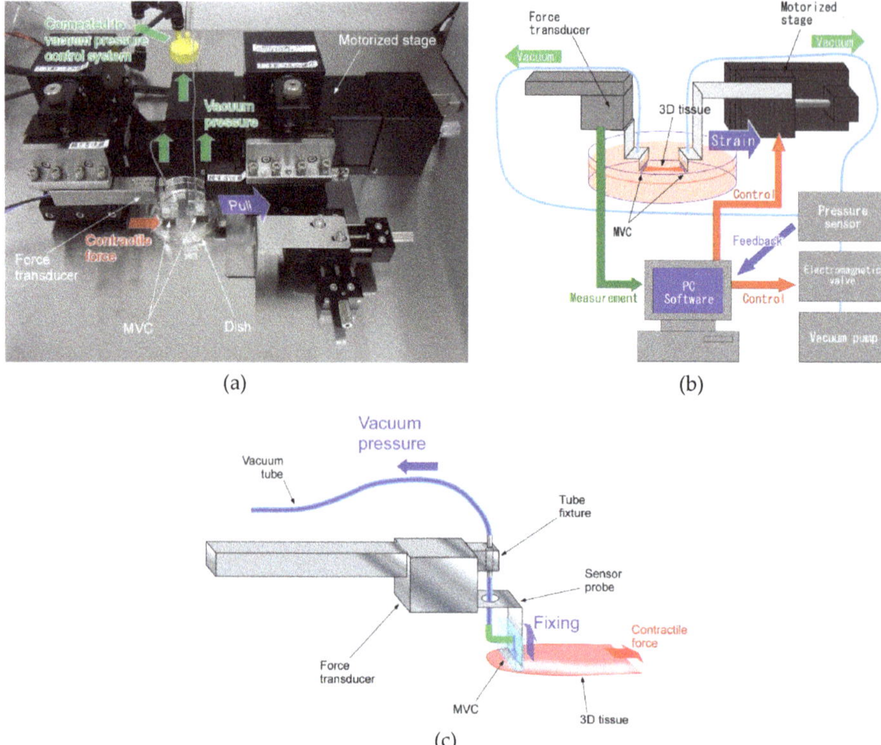

(a)

(b)

(c)

Figure 3. (**a**) Photo of the newly developed tensile test system with attached micro vacuum chuck (MVCs). (**b**) The tensile test system for 3D tissues consisted of six parts. (1) The force transducer was used to measure the bio-mechanical properties of the 3D tissue. (2) The MVCs fixed the 3D tissue to the force transducer and motorized stage. (3) The vacuum pressure generating system adjusted the vacuum pressure of the MVCs. This vacuum pressure system had a vacuum pump, an electromagnetic valve and a pressure sensor. (4) The motorized stage loaded tensile force onto the 3D tissue. (5) The culture dish protected the 3D tissue from drying out during the tensile test. (6) The charge-coupled device (CCD) camera and microscope were used in recording the test conditions. (**c**) Schematic illustration around one MVC connected to the force transducer of the tensile test system. The vacuum tube (a flexible silicon tube) was positioned vertically relative to the tensile force loaded direction. A tube fixture kept the tube in place mechanically.

2.3. Confirmation Experiment on the Effect of Vacuum Hole Size and Vacuum Pressure

To decide the optimum size of the vacuum hole of the MVCs, we confirmed amounts of 3D tissues which were pulled into variously sized vacuum holes at different vacuum pressures. We applied 3D tissues with pressures of 1 kPa, 3 kPa, 10 kPa, and 20 kPa. The square-shaped vacuum holes had side lengths of 50 μm, 100 μm and 200 μm. To simply evaluate these experimental results, we made MVCs with a single hole; the fabrication process was as described in the Supplementary Materials.

The reason for setting the end of the vacuum pressure range as 20 kPa was as follows. In this study, we measured beating properties of cardiac 3D tissues with a loaded strain stimulation by tensile force. When a large strain stimulation is loaded, the 3D tissues become detached from the MVCs by the large tensile force and beating properties cannot be measured. Therefore, vacuum hole size and vacuum pressure needed to be determined in order to prevent the detachment of the 3D tissues from the MVCs by large strain stimulation. The normal human cardiac fibroblast (NHCF, CC-2509, Lonza,

Basel, Switzerland) 3D tissue had a contractile force of about 400 μN (sample width, 3 mm; strain, 0.6) which was measured by using a prototype MVC and a prototype tensile test system [72]. If the contractile force increases linearly with increase of loaded strain, the contractile force could be about 667 μN when NHCF 3D tissue has a loaded strain of 1. In measurement of cardiac tissue contractile force for this study, we used 12-well cell culture inserts (diameter, 12 mm; this diameter would mean our tissue sample width was four times that of the reported sample width with presumed contractile force of about 667 μN). Thus, we hypothesized that the contractile force of cardiac tissue was about four times larger (about 2668 μN) than that of NHCF tissue for the loaded strain of 1 (667 μN). Since the number of vacuum holes of the multi-hole MVCs was 17 and the necessary fixing force was at least over 2668 μN, we hypothesized that the needed force generated by one vacuum hole was about 157 μN. Then, because the area of one vacuum hole was 0.01 μm^2, the needed vacuum pressure was more than 15.7 kPa. Thus, we applied the maximum pressure of 20 kPa to the single-hole MVC. In this confirmation experiment, 3D tissues constructed from NHCF was used (see Supplementary Materials).

The confirmation experiment was done in the culture medium (Dulbecco's Modified Eagle Medium (D-MEM, High glucose, Nacalai Tesque, Kyoto, Japan) containing 10% fetal bovine serum (FBS, Life Technologies Co., Grand Island, NY, USA) at about 25 °C. About one minute after vacuuming NHCF 3D tissues using the single-hole MVCs, the amounts of pulled-in 3D tissue were observed through a microscope (Leica DMi1, Leica Microsystems, Tokyo, Japan) and recorded by the attached CCD camera. Then, recorded images were analyzed by the image processing software (ImageJ, 1.48v). After every observation was finished, the NHCF 3D tissue was removed from the single-hole MVCs, and then, another vacuuming and observation was made for the rest of the same 3D tissue. The amount of NHCF 3D tissue which was pulled into the single-hole MVC was defined as the distance between the tip of the pulled-in 3D tissue and the vacuum hole opening. Experiments were carried out once at each pressure and each size vacuum hole.

2.4. Tensile Test of 3D Tissue Constructed of Human Induced Pluripotent Stem Cell-Derived Cardiomyocytes (hiPS-CM)

In this test, 3D tissues constructed of human induced pluripotent stem cell-derived cardiomyocytes (hiPS-CM) were used (see Supplementary Materials). When cardiac 3D tissues are detached from the culturing surface, the 3D tissues shrink and their length changes. So first, the MVCs were used to hold the 3D tissue which was on the culturing surface that consisted of a culturing insert membrane by vacuum pressure (about 30 kPa) (Figure 4a). After confirming that the MVCs were holding the tissue, the culturing surface was detached from the adhered cardiac 3D tissue with a knife and tweezers (Figure 4b). By following these steps, we were able to measure the beating properties of the 3D tissue without any change in its conditions from those at the time of culturing. For example, the initial length of the 3D tissue was not changed because the length of the 3D tissue was fixed by the fixtures in advance. After again confirming the 3D tissue was held by the MVCs, the tissue was strained in one direction at a constant speed of 0.01 mm/s for arbitrary distances (Figure 4c). Then, after finishing the strain, the beating force was measured. The operations to strain the 3D tissue and measure the beating forces were repeated for each strain stimulation. Beating forces were measured for each strain stimulation without detachment of the 3D tissue. The extension was considered to be ended when the 3D tissue became detached from the MVCs. This detachment was judged from a rapid decrease of beating force of the 3D tissue or a decrease of vacuum pressure. The tensile test was done in the culture medium (DMEM + 10% FBS) at about 25 °C. In this experiment, the beating of 3D tissues was not evoked by such stimulations as electrical or optical stimulation. The passive contractile forces were measured. The initial length of the 3D tissue (distance between the MVCs) was 5 mm. Maximum strain value was 0.9. The force was measured at various strains before reaching the maximum strain. The beating force was defined as the value at the peak amplitude. The beating forces were measured from 8 to 20 times for each strain stimulation. We adjusted the number of measurements according to beat frequency. When the beating was slow, in order to shorten the experiment time, the number of beats for measurement was decreased. The measured peak forces were averaged and standard errors were also calculated. Three samples were used. Beating stress was calculated by force derived by the cross-sectional area of the 3D tissue. The slimmest parts of each 3D tissue were used for the cross-sectional area calculation. We assumed 89 μm as the thickness of all 3D tissues. The beating force measurement of the cardiac 3D tissues was conducted from 10 to 15 days after the seeding of cells in the culture inserts. The calculated beating stresses were averaged and standard errors were also calculated.

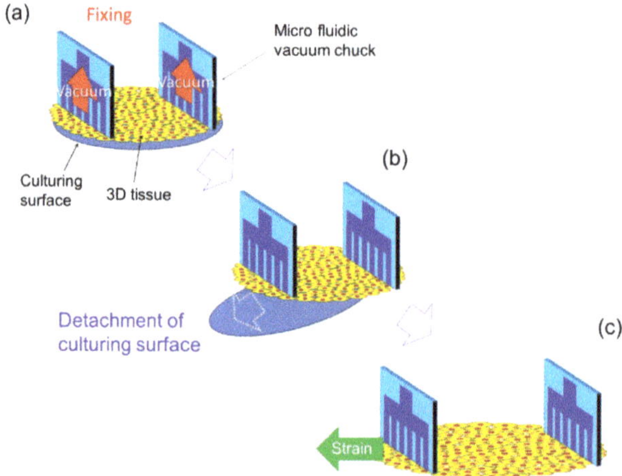

Figure 4. Schematic illustration of 3D tissue fixation by the micro vacuum chuck (MVC). (**a**) The 3D tissue, on a culturing surface that consisted of a culturing insert membrane, was fixed by the MVCs using the vacuum pressure. (**b**) The culturing surface was detached from the 3D tissue. (**c**) The 3D tissue was strained in one direction and beating force was measured.

3. Results

3.1. Results of Confirmation Experiment on Effects of Vacuum Hole Size and Vacuum Pressure

Figure 5 shows photos of a 3D tissue consisting of NHCF that was pulled into a single-hole MVC. As soon as the NHCF 3D tissue was attached on the MVC, part of it was pulled into the vacuum hole. Figure 6 shows the amount of NHCF 3D tissues which were pulled into the vacuum hole by various vacuum pressures for three sizes (side lengths) of the square-shaped holes, 50 μm, 100 μm, and 200 μm. The amount was defined as the distance between the tip of the pulled-in NHCF 3D tissue and the hole opening. We confirmed that the amount of pulled-in NHCF 3D tissues tended to increase with increment of vacuum pressure. Additionally, the amount of pulled-in NHCF 3D tissues also increased with increment of the vacuum hole size. When the hole was 50 μm × 50 μm, we observed collapse of the NHCF 3D tissue surface.

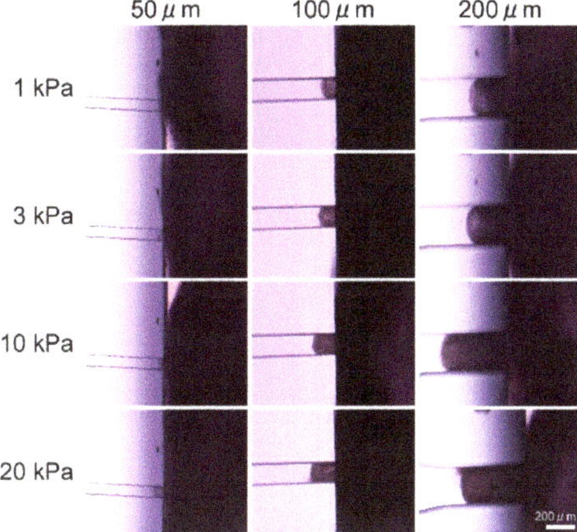

Figure 5. Photos of the normal human cardiac fibroblast (NHCF) 3D tissue that was pulled into the single-hole micro vacuum chuck (MVC). The cross-sectional shape of the vacuum hole was square and the side lengths were 50 μm, 100 μm and 200 μm. Applied vacuum pressures were 1 kPa, 3 kPa, 10 kPa, and 20 kPa. In each photo, a part of the NHCF 3D tissue was pulled into the hole. (Scale bar: 200 μm).

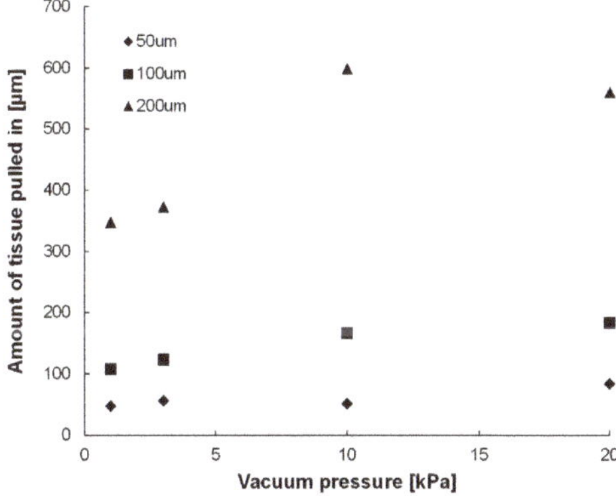

Figure 6. The amount of normal human cardiac fibroblast (NHCF) 3D tissues pulled into one vacuum hole by various vacuum pressures for three sizes of the square-shaped holes, 50 μm, 100 μm, and 200 μm. The amount was defined as the distance between the tip of the pulled-in NHCF 3D tissue and the hole opening. The amount of pulled-in 3D tissues tended to increase with increasing vacuum pressure, and it also increased with increasing size of the hole.

3.2. Results of Tensile Test of 3D Tissue Constructed of Human Induced Pluripotent Stem Cell-Derived Cardiomyocytes (hiPS-CM)

The photo of Figure 7 shows the MVCs could fix the 3D tissue constructed of hiPS-CM to the force measurement system successfully. Figure 8 shows photos taken during the tensile test. The cardiac 3D tissue was fixed by the MVCs. Figure 9 shows strain-beating force properties of the cardiac 3D tissue. The maximum beating force of the 3D tissue was 843 ± 111 μN (strain: 0.9). All beating forces increased with increasing strain. When making a statistical analysis of beating force, we used the smallest value of the maximum strain among three samples for the common maximum value of strain. Figure 10 shows strain-beating stress properties of the cardiac 3D tissue. The average maximum stress was 2.8 ± 0.5 kPa (strain: 0.9).

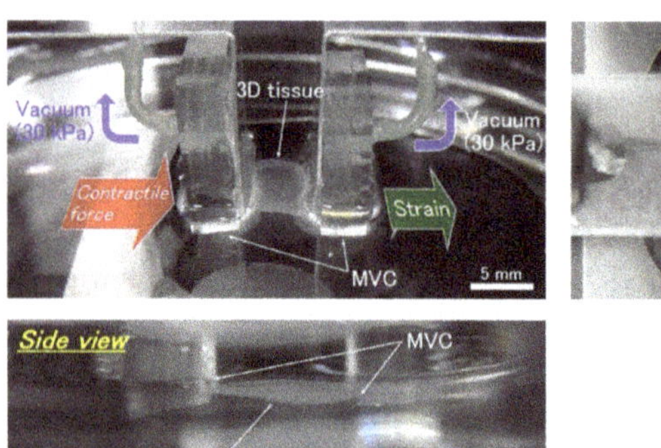

Figure 7. Photos of the cardiac 3D tissue fixed by the micro vacuum chuck (MVCs). Both ends of the cardiac 3D tissue were connected to the force transducer and the motorized stage with the MVCs which applied a vacuum pressure of 30 kPa.

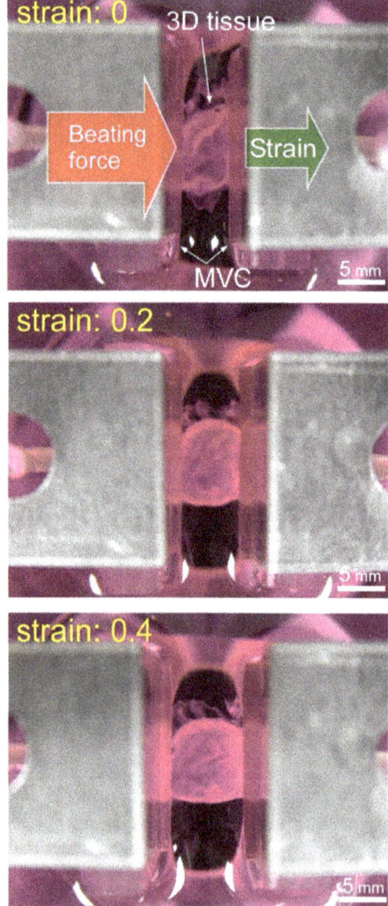

Figure 8. Photos of the cardiac 3D tissue during the tensile test.

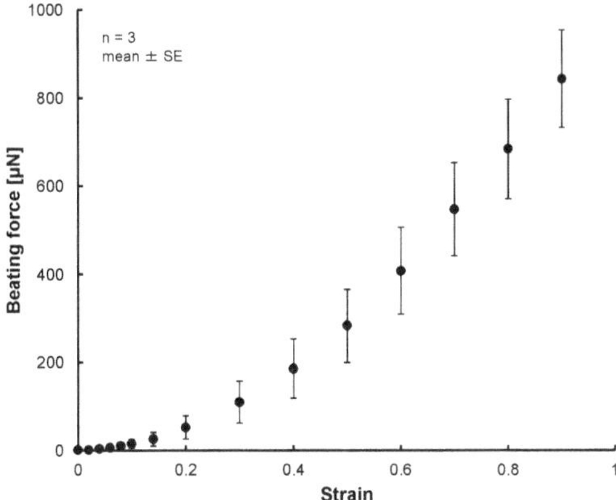

Figure 9. Strain-beating force properties of the cardiac 3D tissues. The beating forces were increased with the increment of strain.

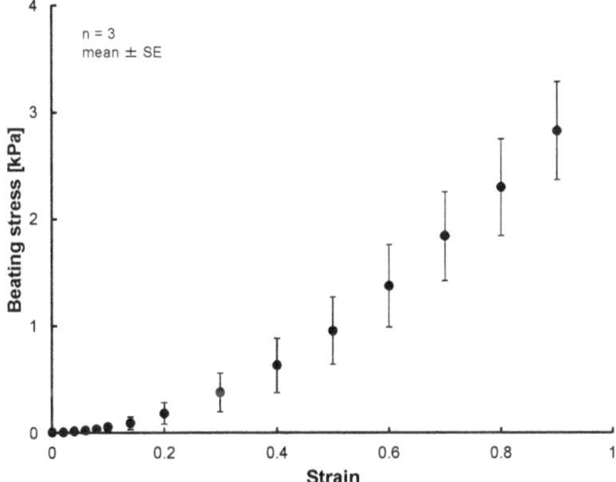

Figure 10. Strain-beating stress properties of the cardiac 3D tissues. The beating stress was increased with the increment of strain and the maximum beating stress was 2.8 ± 0.5 kPa (strain: 0.9).

4. Discussion

In evaluating 3D tissues, bio-mechanical properties are important. Therefore, we have applied the tensile test to the cardiac 3D tissue and measured beating force. In order to apply the tensile test to the 3D tissue, the tissue has to be connected to the tensile test system. However, there had been no fixture which was able to connect the 3D tissue to the tensile test system until our proposed system that held the tissue using vacuum pressure applied by an MVC. The MVC offers some advantages (see Section 2.1). In brief, these are the capabilities to: fix tissues in place without a complex operation; be applicable to variously sized and shaped 3D tissues; to mitigate damage caused to 3D tissue and uncertainty of the measurement; and to eliminate the chance of chemical damage.

We considered the detailed design of the MVC in an experiment which applied various vacuum hole sizes and various vacuum pressures to 3D tissues. In order to measure the beating properties of cardiac 3D tissues, a 3D tissue needs to be fixed using the MVCs even when tensile force is loaded to the 3D tissue. The MVCs must have sufficient vacuum pressure that can fix the 3D tissues. Because applying an excessive vacuum can cause damage to the tissue and unexpected length changes, a small amount of 3D tissue should be pulled into the vacuum hole by using the optimum hole size and shape for the MVC. By applying various sizes of vacuum holes, we observed that the amount of pulled-in

NHCF 3D tissues increased with increment of vacuum hole size (Figures 5 and 6). This indicated that the smaller size vacuum hole was better for fixing 3D tissues. The excessive vacuum condition for the smallest size MVC hole (50 μm) was smaller than that of other hole sizes. However, collapse of the NHCF 3D tissue surface was observed for this smallest size. The amount of excessive vacuum for the 200 μm hole MVC (max: 599) was 3.3 times larger than that of the 100 μm hole MVC (max: 183). Additionally, collapse of the NHCF 3D tissue surface was not observed for the 100 μm hole MVC. Thus, we concluded the MVC with the square hole size of 100 μm was the best for fixing 3D tissue.

Our developed MVC enables users to measure beating force of cardiac 3D tissue which is large sized and assembled from just cells by the LbL method. We observed that the cardiac 3D tissue was fixed by the MVC (Figures 7 and 8). By preparing multiple vacuum holes, the 3D tissue was not pulled into the holes excessively compared to the single-hole MVC (Uesugi et al. [71] described the single-hole MVC). In the tensile test, the applied vacuum pressure was 30 kPa. This value was larger than the maximum vacuum pressure (20 kPa) which was applied in the confirmation experiment on the effect of vacuum hole size (see Section 3.1). We consider the reason why vacuum pressure was larger than 20 kPa. When the cardiac 3D tissue had a loaded tensile strain, a gap occurred between the 3D tissue and the holes of the MVC due to the contractile force and beating force of the extended 3D tissue, and then, vacuum pressure was leaked. This meant that the fixing ability of the MVC was decreased. Thus, we had to increase the applied vacuum pressure to compensate for the leakage effect of vacuum pressure. We considered the high vacuum pressure of 30 kPa was not a problem because the amount of pulled-in NHCF 3D tissues became fairly steady above 10 kPa, from which we hypothesized that the amount of pulled-in cardiac 3D tissue at 30 kPa was close to that of 10 kPa and 20 kPa.

The beating forces of cardiac 3D tissue samples increased with increase of strain (Figure 9). The trend we saw has also been seen for other artificial cardiac 3D tissues which consisted of iPSC-derived cells [8,10,12,15,37,46] and native cardiac tissues and cardiomyocytes [73–76]. This behavior is described by the Frank–Starling relationship [73–76]. This could indicate the possibility for the beating properties of LbL constricted tissue to be close to those of other artificial tissues and native tissues. There has been no study which confirmed the relationship of 3D cardiac tissues whose size was large and which consisted of just cells assembled by the LbL method. By using the MVC, we could measure the contractile properties of the 3D tissues whose size was large and which consisted of just cells. Since our system could measure the cardiac 3D tissue beating properties which were similar to those of native heart tissue, we anticipate that our system will also be able to evaluate the native-like drug response of cardiac 3D tissues in drug screening.

The beating stress of cardiac 3D tissues was 2.8 ± 0.5 kPa (strain: 0.9) (Figure 10). Conventional studies also have reported beating stress of iPS-derived cardiac tissues. The beating stress of cardiac 3D tissues consisting mainly of iPSC-derived cells (μHM) was about 4 mN/mm^2 (beating of the tissues was controlled by applying a 1 Hz electrical stimulation without treatment of isoproterenol) [37]. The order of this stress value is close to our measured value. Because the beating force of μHM was increased with strain stimulation, the beating stress also increased further. The beating stresses of cardiac 3D tissues and cell sheet consisting of iPSC-derived cells and special gel have been reported as 0.08 [8], 0.62 [10], 1.34 [12], 3.3 [15] and 23.2 [11] mN/mm^2. Compared with these values, our measured beating stress was an intermediate value. The values of these conventional studies varied. We consider that the reason for varying beating stresses was the various experimental conditions such as shape of the 3D tissues, culturing time, number of cells, cell species content, amount of ECM, ECM species content, extent of maturation, and fixing method. The cardiac 3D tissues which were cultured in the conventional studies might be loaded with a static contractile stimulation by self-contraction (self-organization) during culturing. Therefore, we cannot discuss in greater detail the beating stresses which were reported by conventional studies of artificial cardiac 3D tissues.

The beating stresses of human myocardium and myocyte have been reported as 44.0 mN/mm^2 [77] and 51 mN/mm^2 [78]. These amounts of beating stresses of native tissue and cells are over 10 times larger than ours. We consider that these differences were due to the extent of maturation, orientation, cell density, and composition of the ECM of cardiomyocytes.

5. Conclusions

In order to evaluate the bio-mechanical properties of a 3D tissue, the tensile test is applied. When applying the tensile test, the 3D tissues have to be fixed to a tensile test system by a special fixture. However, there had been no fixture which was able to fix variously sized and shaped 3D tissues at a small stress concentration easily. Therefore, we developed a micro vacuum chuck (MVC) which could fix the 3D tissue by vacuum pressure and we also developed a special tensile test system for it. In order to optimize the size of the vacuum holes and vacuum pressure, we applied various sized vacuum holes and vacuum pressures to a NHCF 3D tissue. The results showed that a square hole shape with 100 μm sides was better for fixing the 3D tissue. Then, we carried out the tensile test with cardiac 3D tissue which was assembled from hiPS-CMs by the layer-by-layer (LbL) method. We confirmed that the MVC could fix the cardiac 3D tissue during the tensile test. By using the newly developed tensile test system, the beating stress of cardiac 3D tissues was measured (2.8 ± 0.5 kPa, strain: 0.9). Additionally, the tensile test results for the cardiac 3D tissue confirmed that it followed the Frank–Starling relationship, and that cardiac 3D tissue assembled by the LbL method was close to native cardiac tissue.

Based on these results, we consider this bio-mechanical evaluation method which uses the MVC and the developed tensile test system is useful for evaluation of 3D tissues, and it can contribute to the fields of regenerative medicine, drug development, pathology, bioactuator development, food industry, and BioArt.

Supplementary Materials: The following are available online at http://www.mdpi.com/2072-666X/10/7/487/s1, Figure S1: Schematic illustration showing assembly of the MVC.

Author Contributions: K.U., F.S., S.M., Y.S., T.A., M.A. and K.M. conceived and designed the experiments; K.U. performed the experiments; K.U. analyzed the data; K.U. contributed to development of the special fixture (MVC) and the special tensile test system; K.F, A.H. and Y.T. contributed to preparation of the 3D tissues. K.U. wrote the initial draft of the manuscript and all authors have contributed to data interpretation and approved the final version of the manuscript.

Funding: This work was supported by "Development of Manufacturing Technology for Functional Tissues and Organs Employing Three-Dimensional Biofabrication" funded by the Japan Agency for Medical Research and Development (AMED) (P14028), and by Japan Society for the Promotion of Science (JSPS) KAKENHI Grant Number 17H01254.

Conflicts of Interest: The authors declare no conflict of interest. The funding sponsors had no role in the design of the study; in the collection, analyses, or interpretation of data; in the writing of the manuscript, and in the decision to publish the results.

References

1. Langer, R.; Vacanti, J.P. Tissue engineering. *Science* **1993**, *260*, 920–926. [CrossRef]
2. Gribova, V.; Liu, C.Y.; Nishiguchi, A.; Matsusaki, M.; Boudou, T.; Picart, C.; Akashi, M. Construction and myogenic differentiation of 3D myoblast tissues fabricated by fibronectin-gelatin nanofilm coating. *Biochem. Biophys. Res. Commun.* **2016**, *47*, 515–521. [CrossRef]
3. Zimmermann, W.H.; Schneiderbanger, K.; Schubert, P.; Didié, M.; Münzel, F.; Heubach, J.F.; Kostin, S.; Neuhuber, W.L.; Eschenhagen, T. Tissue engineering of a differentiated cardiac muscle construct. *Circ. Res.* **2002**, *90*, 223–230. [CrossRef]
4. Eschenhagen, T.; Didié, M.; Münzel, F.; Schubert, P.; Schneiderbanger, K.; Zimmermann, W.H. 3D engineered heart tissue for replacement therapy. *Basic Res. Cardiol.* **2002**, *97*, 146–152. [CrossRef]
5. Zimmermann, W.H.; Melnychenko, I.; Eschenhagen, T. Engineered heart tissue for regeneration of diseased hearts. *Biomaterials* **2004**, *25*, 1639–1647. [CrossRef]
6. Pillekamp, F.; Reppel, M.; Rubenchyk, O.; Pfannkuche, K.; Matzkies, M.; Bloch, W.; Sreeram, N.; Brockmeier, K.; Hescheler, J. Force Measurements of Human Embryonic Stem Cell-Derived Cardiomyocytes in an *In Vitro* Transplantation Model. *Stem Cells* **2007**, *25*, 174–180. [CrossRef]

7. Xi, J.; Khalil, M.; Shishechian, N.; Hannes, T.; Pfannkuche, K.; Liang, H.; Fatima, A.; Haustein, M.; Suhr, F.; Bloch, W.; et al. Comparison of contractile behavior of native murine ventricular tissue and cardiomyocytes derived from embryonic or induced pluripotent stem cells. *FASEB J.* **2010**, *24*, 2739–2751. [CrossRef]

8. Tulloch, N.L.; Muskheli, V.; Razumova, M.V.; Korte, F.S.; Regnier, M.; Hauch, K.D.; Pabon, L.; Reinecke, H.; Murry, C.E. Growth of engineered human myocardium with mechanical loading and vascular coculture. *Circ. Res.* **2011**, *109*, 47–59. [CrossRef]

9. Schaaf, S.; Shibamiya, A.; Mewe, M.; Eder, A.; Stöhr, A.; Hirt, M.N.; Rau, T.; Zimmermann, W.H.; Conradi, L.; Eschenhagen, T.; et al. Human Engineered Heart Tissue as a Versatile Tool in Basic Research and Preclinical Toxicology. *PLoS ONE* **2011**, *6*, e26397. [CrossRef]

10. Masumoto, H.; Nakane, T.; Tinney, J.P.; Yuan, F.; Ye, F.; Kowalski, W.J.; Minakata, K.; Sakata, R.; Yamashita, J.K.; Keller, B.B. The myocardial regenerative potential of three-dimensional engineered cardiac tissues composed of multiple human iPS cell-derived cardiovascular cell lineages. *Sci. Rep.* **2016**, *6*, 29933. [CrossRef]

11. Jackman, C.P.; Carlson, A.L.; Bursac, N. Dynamic culture yields engineered myocardium with near-adult functional output. *Biomaterials* **2016**, *111*, 66–79. [CrossRef]

12. Ruan, J.L.; Tulloch, N.L.; Razumova, M.V.; Saiget, M.; Muskheli, V.; Pabon, L.; Reinecke, H.; Regnier, M.; Murry, C.E. Mechanical Stress Conditioning and Electrical Stimulation Promote Contractility and Force Maturation of Induced Pluripotent Stem Cell-Derived Human Cardiac Tissue. *Circulation* **2016**, *134*, 1557–1567. [CrossRef]

13. Nakane, T.; Masumoto, H.; Tinney, J.P.; Yuan, F.; Kowalski, W.J.; Ye, F.; LeBlanc, A.J.; Sakata, R.; Yamashita, J.K.; Keller, B.B. Impact of Cell Composition and Geometry on Human Induced Pluripotent Stem Cells-Derived Engineered Cardiac Tissue. *Sci. Rep.* **2017**, *7*, 45641. [CrossRef]

14. Tiburcy, M.; Hudson, J.E.; Balfanz, P.; Schlick, S.; Meyer, T.; Chang Liao, M.L.; Levent, E.; Raad, F.; Zeidler, S.; Wingender, E.; et al. Defined Engineered Human Myocardium with Advanced Maturation for Applications in Heart Failure Modeling and Repair. *Circulation* **2017**, *135*, 1832–1847. [CrossRef]

15. Sasaki, D.; Matsuura, K.; Seta, H.; Haraguchi, Y.; Okano, T.; Shimizu, T. Contractile force measurement of human induced pluripotent stem cell-derived cardiac cell sheet-tissue. *PLoS ONE* **2018**, *13*, e0198026. [CrossRef]

16. Soliman, S.; Pagliari, S.; Rinaldi, A.; Forte, G.; Fiaccavento, R.; Pagliari, F.; Franzese, O.; Minieri, M.; Di Nardo, P.; Licoccia, S.; et al. Multiscale three-dimensional scaffolds for soft tissue engineering via multimodal electrospinning. *Acta Biomater.* **2010**, *6*, 1227–1237. [CrossRef]

17. Liu, X.; Shen, H.; Song, S.; Chen, W.; Zhang, Z. Accelerated Biomineralization of Graphene Oxide—Incorporated Cellulose Acetate Nanofibrous Scaffolds for Mesenchymal Stem Cell Osteogenesis. *Colloids Surf. B Biointerfaces* **2017**, *159*, 251–258. [CrossRef]

18. Janjanin, S.; Li, W.J.; Morgan, M.T.; Shanti, R.M.; Tuan, R.S. Mold-shaped, nanofiber scaffold-based cartilage engineering using human mesenchymal stem cells and bioreactor. *J. Surg. Res.* **2008**, *149*, 47–56. [CrossRef]

19. Bhumiratana, S.; Eton, R.E.; Oungoulian, S.R.; Wan, L.Q.; Ateshian, G.A.; Vunjak-Novakovic, G. Large, stratified, and mechanically functional human cartilage grown *in vitro* by mesenchymal condensation. *Proc. Natl. Acad. Sci. USA* **2014**, *111*, 6940–6945. [CrossRef]

20. Barber, J.G.; Handorf, A.M.; Allee, T.J.; Li, W.J. Braided nanofibrous scaffold for tendon and ligament tissue engineering. *Tissue Eng. Part A* **2013**, *19*, 1265–1274. [CrossRef]

21. Chainani, A.; Hippensteel, K.J.; Kishan, A.; Garrigues, N.W.; Ruch, D.S.; Guilak, F.; Little, D. Multilayered Electrospun Scaffolds for Tendon Tissue Engineering. *Tissue Eng. Part A* **2013**, *19*, 2594–2604. [CrossRef]

22. Orr, S.B.; Chainani, A.; Hippensteel, K.J.; Kishan, A.; Gilchrist, C.; Garrigues, N.W.; Ruch, D.S.; Guilak, F.; Little, D. Aligned multilayered electrospun scaffolds for rotator cuff tendon tissue engineering. *Acta Biomater.* **2015**, *24*, 117–126. [CrossRef]

23. Tosun, Z.; Villegas-Montoya, C.; McFetridge, P.S. The influence of early-phase remodeling events on the biomechanical properties of engineered vascular tissues. *J. Vasc. Surg.* **2011**, *54*, 1451–1460. [CrossRef]

24. Arasteh, S.; Kazemnejad, S.; Khanjani, S.; Heidari-Vala, H.; Akhondi, M.M.; Mobini, S. Fabrication and characterization of nano-fibrous bilayer composite for skin regeneration application. *Methods* **2016**, *99*, 3–12. [CrossRef]

25. Jayarama Reddy, V.; Radhakrishnan, S.; Ravichandran, R.; Mukherjee, S.; Balamurugan, R.; Sundarrajan, S.; Ramakrishna, S. Nanofibrous structured biomimetic strategies for skin tissue regeneration. *Wound Repair Regen.* **2013**, *21*, 1–16. [CrossRef]

26. Amano, Y.; Nishiguchi, A.; Matsusaki, M.; Iseoka, H.; Miyagawa, S.; Sawa, Y.; Seo, M.; Yamaguchi, T.; Akashi, M. Development of vascularized iPSC derived 3D-cardiomyocyte tissues by filtration Layer-by-Layer technique and their application for pharmaceutical assays. *Acta Biomater.* **2016**, *33*, 110–121. [CrossRef]

27. Matsusaki, M.; Sakaue, K.; Kadowaki, K.; Akashi, M. Three-dimensional human tissue chips fabricated by rapid and automatic inkjet cell printing. *Adv. Healthc. Mater.* **2013**, *2*, 534–539. [CrossRef]

28. Yoshida, H.; Matsusaki, M.; Akashi, M. Multilayered blood capillary analogs in biodegradable hydrogels for *in vitro* drug permeability assays. *Adv. Funct. Mater.* **2013**, *23*, 1736–1742. [CrossRef]

29. Nishiguchi, A.; Yoshida, H.; Matsusaki, M.; Akashi, M. Rapid construction of three-dimensional multilayered tissues with endothelial tube network by the cell-accumulation technique. *Adv. Mater.* **2011**, *23*, 3506–3510. [CrossRef]

30. Matsusaki, M.; Case, C.P.; Akashi, M. Three-dimensional cell culture technique and pathophysiology. *Adv. Drug Deliv. Rev.* **2014**, *74*, 95–103. [CrossRef]

31. Takeda, M.; Miyagawa, S.; Fukushima, S.; Saito, A.; Ito, E.; Harada, A.; Matsuura, R.; Iseoka, H.; Sougawa, N.; Mochizuki-Oda, N.; et al. Development of *In Vitro* Drug-Induced Cardiotoxicity Assay by Using Three-Dimensional Cardiac Tissues Derived from Human Induced Pluripotent Stem Cells. *Tissue Eng. Part C Methods* **2017**, *24*, 56–67. [CrossRef]

32. Eschenhagen, T.; Fink, C.; Remmers, U.; Scholz, H.; Wattchow, J.; Weil, J.; Zimmermann, W.; Dohmen, H.H.; Schäfer, H.; Bishopric, N.; et al. Three-dimensional reconstitution of embryonic cardiomyocytes in a collagen matrix: A new heart muscle model system. *FASEB J.* **1997**, *11*, 683–694. [CrossRef]

33. Vandenburgh, H.; Shansky, J.; Benesch-Lee, F.; Barbata, V.; Reid, J.; Thorrez, L.; Valentini, R.; Crawford, G. Drug-screening platform based on the contractility of tissue-engineered muscle. *Muscle Nerve* **2008**, *37*, 438–447. [CrossRef]

34. Hansen, A.; Eder, A.; Bönstrup, M.; Flato, M.; Mewe, M.; Schaaf, S.; Aksehirlioglu, B.; Schwoerer, A.P.; Uebeler, J.; Eschenhagen, T. Development of a drug screening platform based on engineered heart tissue. *Circ. Res.* **2010**, *107*, 35–44. [CrossRef]

35. Zhang, D.; Shadrin, I.Y.; Lam, J.; Xian, H.Q.; Snodgrass, H.R.; Bursac, N. Tissue-engineered cardiac patch for advanced functional maturation of human ESC-derived cardiomyocytes. *Biomaterials* **2013**, *34*, 5813–5820. [CrossRef]

36. Turnbull, I.C.; Karakikes, I.; Serrao, G.W.; Backeris, P.; Lee, J.J.; Xie, C.; Senyei, G.; Gordon, R.E.; Li, R.A.; Akar, F.G.; et al. Advancing functional engineered cardiac tissues toward a preclinical model of human myocardium. *FASEB J.* **2014**, *28*, 644–654. [CrossRef]

37. Huebsch, N.; Loskill, P.; Deveshwar, N.; Spencer, C.I.; Judge, L.M.; Mandegar, M.A.; Fox, C.B.; Mohamed, T.M.; Ma, Z.; Mathur, A.; et al. Miniaturized iPS-Cell-Derived Cardiac Muscles for Physiologically Relevant Drug Response Analyses. *Sci. Rep.* **2016**, *6*, 24726. [CrossRef]

38. Pati, F.; Gantelius, J.; Svahn, H.A. 3D Bioprinting of Tissue/Organ Models. *Angew. Chem. Int. Ed.* **2016**, *55*, 4650–4665. [CrossRef]

39. Griffith, L.G.; Swartz, M.A. Capturing complex 3D tissue physiology *in vitro*. *Nat. Rev. Mol. Cell Biol.* **2006**, *7*, 211–224. [CrossRef]

40. Knowlton, S.; Onal, S.; Yu, C.H.; Zhao, J.J.; Tasoglu, S. Bioprinting for cancer research. *Trends Biotechnol.* **2015**, *33*, 504–513. [CrossRef]

41. Ki-Hwan Nam, K.H.; Smith, A.S.; Lone, S.; Kwon, S.; Kim, D.H. Biomimetic three-dimensional tissue models for advanced high-throughput drug screening. *J. Lab. Autom.* **2015**, *20*, 201–215.

42. Mironov, V.; Trusk, T.; Kasyanov, V.; Little, S.; Swaja, R.; Markwald, R. Biofabrication: A 21st century manufacturing paradigm. *Biofabrication* **2009**, *1*, 022001. [CrossRef]

43. Fink, C.; Ergün, S.; Kralisch, D.; Remmers, U.; Weil, J.; Eschenhagen, T. Chronic stretch of engineered heart tissue induces hypertrophy and functional improvement. *FASEB J.* **2000**, *14*, 669–679. [CrossRef]

44. Zimmermann, W.H.; Fink, C.; Kralisch, D.; Remmers, U.; Weil, J.; Eschenhagen, T. Three-dimensional engineered heart tissue from neonatal rat cardiac myocytes. *Biotechnol. Bioeng.* **2000**, *68*, 106–114. [CrossRef]

45. Hinds, S.; Bian, W.; Dennis, R.G.; Bursac, N. The role of extracellular matrix composition in structure and function of bioengineered skeletal muscle. *Biomaterials* **2011**, *32*, 575–583. [CrossRef]

46. Tchao, J.; Kim, J.J.; Lin, B.; Salama, G.; Lo, C.W.; Yang, L.; Tobita, K. Engineered Human Muscle Tissue from Skeletal Muscle Derived Stem Cells and Induced Pluripotent Stem Cell Derived Cardiac Cells. *Int. J. Tissue Eng.* **2013**, 98762. [CrossRef]

47. Horiguchi, H.; Imagawa, K.; Hoshino, T.; Akiyama, Y.; Morishima, K. Fabrication and Evaluation of Reconstructed Cardiac Tissue and Its Application to Bio-actuated Microdevices. *IEEE Trans. Nanobiosci.* **2009**, *8*, 349–355. [CrossRef]

48. Akiyama, Y.; Terada, R.; Hashimoto, M.; Hoshino, T.; Furukawa, Y.; Morishima, K. Rod-shaped Tissue Engineered Skeletal Muscle with Artificial Anchors to Utilize as a Bio-Actuator. *J. Biomech. Sci. Eng.* **2010**, *5*, 236–244. [CrossRef]

49. Hoshino, T.; Imagawa, K.; Akiyama, Y.; Morishima, K. Cardiomyocyte-driven gel network for bio mechano-informatic wet robotics. *Biomed. Microdevices* **2012**, *14*, 969–977. [CrossRef]

50. Kabumoto, K.; Hoshino, T.; Akiyama, Y.; Morishima, K. Voluntary Movement Controlled by the Surface EMG Signal for Tissue-Engineered Skeletal Muscle on a Gripping Tool. *Tissue Eng. Part A* **2013**, *19*, 1695–1703. [CrossRef]

51. Cvetkovic, C.; Raman, R.; Chan, V.; Williams, B.J.; Tolish, M.; Bajaj, P.; Sakar, M.S.; Asada, H.H.; Saif, M.T.; Bashir, R. Three-dimensionally printed biological machines powered by skeletal muscle. *Proc. Natl. Acad. Sci. USA* **2014**, *111*, 10125–10130. [CrossRef]

52. Catts, O.; Zurr, I. Growing for different ends. *Int. J. Biochem. Cell Biol.* **2014**, *56*, 20–29. [CrossRef]

53. Bhat, Z.F.; Bhat, H.F. Animal-free Meat Biofabrication. *Am. J. Food Technol.* **2011**, *6*, 441–459. [CrossRef]

54. Arshad, M.S.; Javed, M.; Sohaib, M.; Saeed, F.; Imran, A.; Amjad, Z. Tissue engineering approaches to develop cultured meat from cells: A mini review. *Cogent Food Agric.* **2017**, *3*, 1320814. [CrossRef]

55. McHugh, S. Real artificial: Tissue-cultured Meat, Genetically Modified Farm Animals, and Fictions. *Configurations* **2010**, *18*, 181–197. [CrossRef]

56. Catts, O.; Zurr, I. Semi-Living Art. In *Signs of Life: Bio Art and Beyond*, 1st ed.; Kac, E., Ed.; The MIT Press: Cambridge, MA, USA, 1932; pp. 231–247.

57. Catts, O.; Zurr, I. Growing Semi-Living Sculptures: The Tissue Culture & Art Project. *Leonardo* **2002**, *35*, 365–370.

58. Rees, J. Exhibition: Cultures in the capital. *Nature* **2008**, *451*, 891. [CrossRef]

59. Yetisen, A.K.; Davis, J.; Coskun, A.F.; Church, G.M.; Yun, S.H. Bioart. *Trends Biotechnol.* **2015**, *33*, 724–734. [CrossRef]

60. Vaage, N.S. What Ethics for Bioart? *Nanoethics* **2016**, *10*, 87–104. [CrossRef]

61. Stelarc. Extra Ear: Ear on the Arm Blender. *Diacritics* **2006**, *36*, 117–119.

62. Art Imitates Life: Artist Bioengineers Replica of van Gogh's Ear. CNN: Carina Storrs. 2015. Available online: https://edition.cnn.com/2015/11/13/health/van-gogh-ear-art-science/index.html (accessed on 11 July 2018).

63. Hayashi, R.; Yamato, M.; Takayanagi, H.; Oie, Y.; Kubota, A.; Hori, Y.; Okano, T.; Nishida, K. Validation system of tissue-engineered epithelial cell sheets for corneal regenerative medicine. *Tissue Eng. Part C Methods* **2009**, *16*, 553–560. [CrossRef]

64. Even-Ram, S.; Artym, V.; Yamada, K.M. Matrix Control of Stem Cell Fate. *Cell* **2006**, *126*, 645–647. [CrossRef]

65. Engler, A.J.; Sen, S.; Sweeney, H.L.; Discher, D.E. Matrix Elasticity Directs Stem Cell Lineage Specification. *Cell* **2006**, *126*, 677–689. [CrossRef]

66. Uesugi, K.; Akiyama, Y.; Yamato, M.; Okano, T.; Hoshino, T.; Morishima, K. Micro handling tool of cell sheet for measurement of adhesion force. In Proceedings of the μTAS 2009, Jeju, Korea, 27–31 October 2009; pp. 1844–1846.

67. Uesugi, K.; Akiyama, Y.; Yamato, M.; Okano, T.; Hoshino, T.; Morishima, K. Development of cell-sheet handling tool for measurement of cell sheet adhesion force. In Proceedings of the MHS 2009, Nagoya, Japan, 8–11 November 2009; pp. 614–619.

68. Uesugi, K.; Akiyama, Y.; Hoshino, T.; Akiyama, Y.; Yamato, M.; Okano, T.; Morishima, K. Measuring adhesion force of a cell sheet by the ninety-degree peel test using a multi hook type fixture. *J. Biomech. Sci. Eng.* **2013**, *8*, 129–138. [CrossRef]

69. Uesugi, K.; Akiyama, Y.; Hoshino, T.; Akiyama, Y.; Yamato, M.; Okano, T.; Morishima, K. Measuring mechanical properties of cell sheet by tensile test using self-attachable fixture. *J. Robot. Mechatron.* **2013**, *25*, 603–610. [CrossRef]

70. Uesugi, K.; Akiyama, Y.; Hoshino, T.; Akiyama, Y.; Yamato, M.; Okano, T.; Morishima, K. Measurement system for biomechanical properties of cell sheet. In Proceedings of the IROS 2013, Tokyo, Japan, 3–7 November 2013; pp. 1010–1015.

71. Uesugi, K.; Nishiguchi, A.; Matsusaki, M.; Akashi, M.; Morishima, K. Evaluation system for mechanobiology of three-dimensional tissue multilayered *in vitro*. In Proceedings of the MHS 2015, Nagoya, Japan, 23–25 November 2015; pp. 336–338.

72. Uesugi, K.; Fukumoto, K.; Shima, F.; Miyagawa, S.; Sawa, Y.; Akashi, M.; Morishima, K. Micro fluidic vacuum chuck system for handling of regenerative three dimensional tissue. In Proceedings of the μTAS 2016, Dublin, Ireland, 9–13 October 2016; pp. 671–672.

73. Allen, D.G.; Kentish, J.C. The cellular basis of the length-tension relation in cardiac muscle. *J. Mol. Cell. Cardiol.* **1985**, *17*, 821–840. [CrossRef]

74. Fuchs, F.; Smith, S.H. Calcium, cross-bridges, and the Frank-Starling relationship. *News Physiol. Sci.* **2001**, *16*, 5–10. [CrossRef]

75. Moss, R.L.; Fitzsimons, D.P. Frank-Starling relationship: Long on importance, short on mechanism. *Circ. Res.* **2002**, *90*, 11–13. [CrossRef]

76. Shiels, H.A.; White, E. The Frank-Starling mechanism in vertebrate cardiac myocytes. *J. Exp. Biol.* **2008**, *211*, 2005–2013. [CrossRef]

77. Hasenfuss, G.; Mulieri, L.A.; Blanchard, E.M.; Holubarsch, C.; Leavitt, B.J.; Ittleman, F.; Alpert, N.R. Energetics of isometric force development in control and volume-overload human myocardium. Comparison with animal species. *Circ. Res.* **1991**, *68*, 836–846. [CrossRef]

78. van der Velden, J.; Klein, L.J.; van der Bijl, M.; Huybregts, M.A.; Stooker, W.; Witkop, J.; Eijsman, L.; Visser, C.A.; Visser, F.C.; Stienen, G.J. Force production in mechanically isolated cardiac myocytes from human ventricular muscle tissue. *Cardiovasc. Res.* **1998**, *38*, 414–423. [CrossRef]

 *micromachines*

Article

Study Effects of Drug Treatment and Physiological Physical Stimulation on Surfactant Protein Expression of Lung Epithelial Cells Using a Biomimetic Microfluidic Cell Culture Device

Ting-Ru Lin [1], Sih-Ling Yeh [1], Chien-Chung Peng [1], Wei-Hao Liao [1] and Yi-Chung Tung [1,2,*]

[1] Research Center for Applied Sciences, Academia Sinica, Taipei 11529, Taiwan; b98504030@ntu.edu.tw (T.-R.L.); sihlingyeh@gmail.com (S.-L.Y.); vp@gate.sinica.edu.tw (C.-C.P.); jamliao@gate.sinica.edu.tw (W.-H.L.)

[2] College of Engineering, Chang Gung University, Taoyuan 33302, Taiwan

* Correspondence: tungy@gate.sinica.edu.tw; Tel.: +886-2-2787-3138; Fax: +886-2-2787-3122

Received: 10 May 2019; Accepted: 14 June 2019; Published: 16 June 2019

Abstract: This paper reports a biomimetic microfluidic device capable of reconstituting physiological physical microenvironments in lungs during fetal development for cell culture. The device integrates controllability of both hydrostatic pressure and cyclic substrate deformation within a single chip to better mimic the in vivo microenvironments. For demonstration, the effects of drug treatment and physical stimulations on surfactant protein C (SPC) expression of lung epithelial cells (A549) are studied using the device. The experimental results confirm the device's capability of mimicking in vivo microenvironments with multiple physical stimulations for cell culture applications. Furthermore, the results indicate the critical roles of physical stimulations in regulating cellular behaviors. With the demonstrated functionalities and performance, the device is expected to provide a powerful tool for further lung development studies that can be translated to clinical observation in a more straightforward manner. Consequently, the device is promising for construction of more in vitro physiological microenvironments integrating multiple physical stimulations to better study organ development and its functions.

Keywords: microfluidic device; cell culture; organ-on-chips; lung epithelial cell; surfactant protein

1. Introduction

Physical microenvironments play important roles in regulating various biological activities. In order to systematically investigate the effects of the physical microenvironments, various in vitro cell models have been developed and analyzed [1–5]. In a human body, the lung is an important respiratory organ, and various physical forces are involved during its development and operation. The cells within the lung experience different types of physical forces such as cyclic substrate deformation, surface tension, and hydrostatic pressure. In addition, during lung development in fetal stage, the physical stimulations are key factors regulating cell differentiation for proper lung function development. For instance, the stretch of tissues play a determinant role during lung development. It has been shown that intermittent and repetitive stretching of lungs induced by upper airway relaxation and the diaphragmatic contractions is an important factor for normal lung development in utero [6–8]. In addition, the larynx regulates the efflux of fetal intrapulmonary fluid produced by pulmonary epithelium from the trachea to the amniotic space to form positive pressure within the developing lungs [6,7]. However, due to the complicated anatomical organization of lung tissues, it is challenging to study in detail the effects of physical stimulations on the cells in vivo. The extensive study on lung development in vitro corresponding to physical stimulations is desired to decipher the underlying mechanisms for biomedical researches.

Recently, due to advantages provided by microfluidics, various microfluidic cell culture devices capable of reconstituting physiological microenvironments in organs, named organ-on-chips, have been developed for in vitro studies. Among the developed devices, several of them focus on reconstruction of physiological physical microenvironments in lungs for in vitro cell culture applications. For example, several devices are designed to apply cyclic strain and hydrostatic pressure on the cells to investigate mechanotransduction during breathing movement. The architecture of the devices emphasized on influences of application of substrate stretching on pneumocyte cells to investigate mechanisms behind the lung function [5,9–13], and analyses of cellular phenotypes, including orientation, attachment, and proliferation, are also performed after the cell culture [5,9,12]. Although the existing devices successfully provide physiological physical microenvironments for in vitro cell culture, and the results confirm the importance of physical stimulation for cell development, the existing devices mainly focus on providing single-type physical stimulations. In contrast, in vivo, multiple physical stimulations are often intertwined in physiological microenvironments. As a result, it is desired to integrate multiple physical stimulations within physiological ranges into single devices to construct in vitro cell culture models.

In this paper, we develop a microfluidic cell device capable of generating two physical stimulations, including hydrostatic pressure and cyclic substrate deformation, mimicking microenvironments during lung development. The device incorporates hydrostatic pressure and cyclic substrate deformation as physical stimulations for cell culture. The physical stimulations, including a hydrostatic pressure value, and the amplitude and frequency of the deformation, are designed according to the physiological conditions reported in previous studies. For the hydrostatic pressure, the device is designed based on the measured results from Nicolini et al., showing that the intra-amniotic pressure found in pregnancies with normal amniotic fluidic volume is 1–14 mmHg [14]. In terms of cyclic substrate deformation, Boyce et al. reported that the human fetal breathing rate is 30–90 times per minute [15]. In addition, Huh et al. indicated that the level of applied strain ranged from 5% to 15% matches normal levels observed in alveoli within whole lung in vivo [13]. As a result, the developed device is designed to provide the cyclic strain applied on cells at frequency of 30 stretches per minute and maximum surface strain of 9% under approximately 3 mmHg hydrostatic pressure to mimic physiological conditions.

To demonstrate the functionality provided by the designed biomimetic microfluidic cell culture device, we study expression of surfactant protein C (SPC) on type II pneumocytes under the steroid treatments and physical stimulations in this paper. SPC is an essential pulmonary surfactant protein maintaining normal operation of lungs. People lacking SPC tend to develop progressive interstitial lung disease [16]. In the experiments, the generated hydrostatic pressure and cyclic substrate deformation are numerically simulated and experimentally characterized. In the cell experiments, the biocompatibility of the device is first confirmed, and the device is further exploited to study the cell viability under combinations of drug treatments and physical stimulations. Furthermore, cell culture under various combinations of drug treatment and physical stimulations are performed to evaluate theirs effects on SPC expression of the cells. The results successfully confirm the capability of generating multiple physical stimulations and demonstrate the essentialness of physical stimulations on cellular responses. The developed device is believed to be able to pave the way to understand the cell behaviors under various physiological physicochemical stimulations.

2. Materials and Methods

2.1. Device Design and Fabrication

The microfluidic device is designed to generate physical stimulations mimicking physiological conditions in a lung to study surfactant protein expression of lung epithelial cells. The microfluidic device designed in this paper is made of an elastomeric material polydimethylsiloxane (PDMS), due to its cell compatibility, optical transparency, and mechanical deformability [17]. The microfluidic device is constructed by two PDMS channel layers separated by a flexible PDMS membrane as illustrated in Figure 1a. The top layer is used for cell culture in a cellular microenvironment with well-defined physical stimulations, including hydrostatic pressure and cyclic surface strain. In order to generate hydrostatic pressure within a physiological range, growth medium is introduced into the top channel layer from an inlet, and flows through a cell culture chamber (width: 500 μm; height: 85 μm; length: 1 cm) followed by a serpentine-shape channel with relative high flow resistance towards an outlet. The bottom layer is exploited for actuation to deform the sandwiched membrane on which the cells are cultured. Deformation of the membrane is precisely controlled by infusing and withdrawing liquid in the actuation channel (width: 2.5 mm) with a designated volume and flow rate using a computer-controlled syringe pump [12].

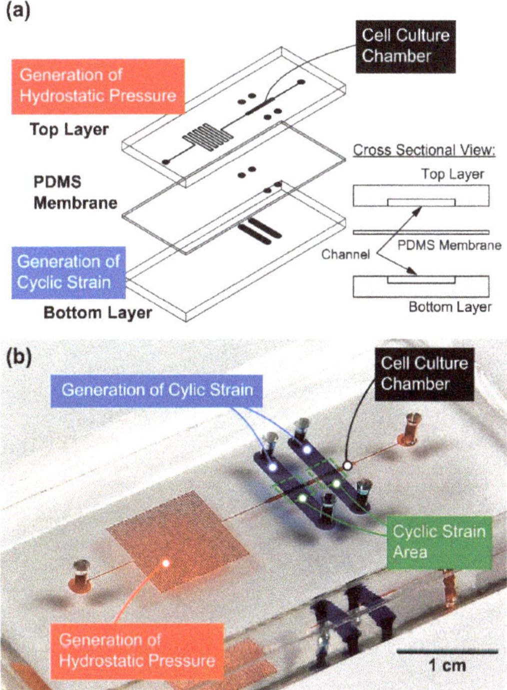

Figure 1. (**a**) Schematic of the designed microfluidic device to generate physical stimulations for lung epithelial cell surfactant protein expression studies. (**b**) Photo the fabricated device filled with colored food dyes (red: cell culture channel; blue: actuation channel.

The designed PDMS microfluidic device can be fabricated using soft lithography replica molding process [18,19]. For the fabrication of the two PDMS layers, the master molds fabricated on four-inch silicon wafers with 85 μm thick-negative photoresist (SU-8 2050, MicroChem Co., Newton, MA, USA) channel patterns are first prepared by photo lithography techniques. In order to prevent undesired adhesion between PDMS and the molds, their surfaces are silanized with 1H,1H,2H,2H-perfluorooctyltrichlorosilane (78560-45-9, Alfa Aesar, Ward Hill, MA, USA) in a desiccator at room temperature overnight. Degassed PDMS pre-polymer with base and curing agent (10:1 by weight) is then poured onto the molds and cured in a 60 °C oven overnight. The PDMS membrane with a thickness of 100 μm is prepared by spin coating the aforementioned PDMS pre-polymer on a four-inch silicon wafer at 100 rpm, and the membrane is then cured in the oven overnight.

To assemble the device, the inlet and outlet to the cell culture channel on the top layer are first made using a 2 mm-diameter biopsy punch (Miltex, York, PA, USA). The top layer is then irreversibly bonded to the membrane using oxygen plasma surface treatment at 90 W for 40 s (PX-250, Nordson MARCH Co., Concord, CA, USA). The inlets and outlets to the actuation channels on the bottom layer are then made by the biopsy punch through the assembled layer. The bottom layer is then bonded to the assembled top layer using the same surface treatment. Last, the bonded device is placed in the 60 °C oven overnight to promote adhesion and cell compatibility. A photo of the fabricated microfluidic device filled with colored food dyes is shown in Figure 1b.

2.2. Experimental Setup

To perform cell culture under well-controlled physical microenvironments using the developed microfluidic device, experimental setup as shown in Figure 2 is constructed. In the setup, a syringe pump (Fusion 400 classical Syringe Pump, Chemyx Inc., Stafford, TX, USA) is connected to the cell culture chamber inlet to continuously pump cell growth medium for perfusion of the cell culture. In order to generate cyclic substrate deformation, a computer-controlled syringe pump (Fusion 200 classical Syringe Pump, Chemyx Inc., Stafford, TX, USA) and a three-way valve are connected to the inlet and outlet of the actuation channel, respectively. To generate a precise cyclic surface strain for the cell culture, the syringe pump is controlled by a LabVIEW program (Ver. 2012, National Instruments, Co., Austin, TX, USA) to periodically infuse and withdraw deionized water with specific volume in and from the actuation channel. To assure the consistent membrane deformation during the experiments, the three-way valve connected to the outlet is exploited to eliminate the air bubbles trapped inside the actuation channel.

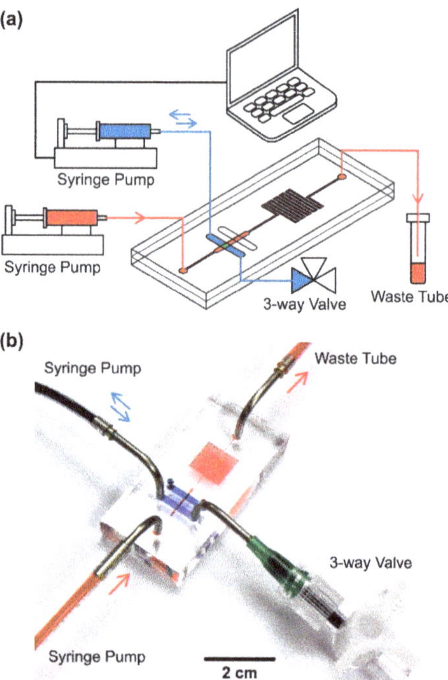

Figure 2. (**a**) Schematic and (**b**) photo of the experimental setup of the microfluidic cell culture device capable of mimicking physical stimulations within lungs.

2.3. Device Simulation and Characterization

In order to confirm the agreement of the physical microenvironments established within the microfluidic device with the physiological ones, numerical simulation and experimental characterization are performed in this paper. First, a three-dimensional (3D) finite element analysis (FEA) model is constructed using a commercially available FEA simulation software (COMSOL Ver. 4.3b, COMSOL Inc., Burlington, MA, USA) to estimate the hydrostatic pressure inside the cell culture chamber in the top layer during the cell perfusion culture. In the model, tetrahedral elements (number of iterations: 4; maximum element depth to process: 4) with auto mesh, and the material properties of water are exploited for flow simulation. The channel geometries are set to be identical to the mask design. For boundary conditions, no-slip conditions are applied on all the channel walls, and the inlet and outlet are set to be a flow with a uniform flow rate of 0.1 μL/min and atmosphere pressure, respectively. In the FEA software, Navier–Stokes equations are utilized as governing equations to solve the constructed fluidic model.

In order to mimic the physical stimulations of fetal breathing movements of lungs, the cyclic substrate deformation frequency at 30 stretches per minute and maximum surface strain of 9% are tested in the experiments. The deformation and the surface strain are experimentally characterized and numerically simulated to estimate the strain generated on the membrane top surface. For the characterization, fluorescein sodium salt solution (F6377, Sigma- Aldrich Co., St Louis, MO, USA) with concentration of 50 μg/mL is introduced into both cell culture and actuation channels located in the top and bottom layers. The solution is infused into and withdrawn from the actuation channel with a flow rate of 1.5 mL/min for 1 s to deform the membrane, and the deformation is observed through cross-sectional imaging using confocal microscopy (TCS SP5, Leica Microsystems, Wetzlar, Germany).

To estimate surface strain on the membrane top surface with the specific actuation conditions, the FEA simulation based on the governing equation stating divergence of stress equals the volume force is also performed. A 3D model simulating the membrane (0.5 mm × 2 mm × 0.1 mm in width × length × thickness) with the auto-meshed tetrahedral elements and boundary conditions of fixed edges is constructed in the software. In the simulation, the Young's modulus, Poisson's ratio, and density of the PDMS membrane are set as 7.5×10^5 Pa, 0.49, and 9.2×10^2 kg/m^3, respectively [5]. A series of uniformly distributed force is applied on the membrane bottom surface in the simulation to yield the same membrane maximum displacement observed in the experimental characterization.

2.4. Cell Culture and Seeding

In the cell experiments, carcinomic human alveolar basal epithelial cells (A549, ATCC, Manassas, VA, USA) are utilized to study pulmonary surfactant protein expression using the developed microfluidic cell culture device. The A549 cell line has been broadly used as a model of the type II pneumocytes for in vitro experiments [20]. The A549 cells are cultured using growth medium based on F-12K medium (Gibco 21127, Invitrogen Co., Carlsbad, CA, USA) with 10% *v/v* fetal bovine serum (FBS) (Gibco 10082, Invitrogen) and 1% *v/v* antibiotic–antimycotic (Gibco 15240, Invitrogen). The stocks are cultured in T25 cell culture flasks (Nunc 156367, Thermo Scientific Inc., Rochester, NY, USA) maintained at 37 °C in a humidified incubator under 5% CO$_2$ in air, and the cells are passaged by dissociation using trypLE (Gibco 12604, Invitrogen). Before introducing the cells into the microfluidic devices, the devices are sterilized by UV irradiation and then treated with O$_2$ plasma at 90 W for 40 s to make channel surfaces hydrophilic. The cell culture channels are then coated with extra-cellular matrix (ECM) protein, fibronectin (F2006, Sigma-Aldrich Co., St Louis, MO, USA), with concentration of 100 μg/mL in Dulbecco's Phosphate-Buffered Saline (DPBS) (Gibco 14190, Invitrogen) overnight inside the incubator to promote cell adhesion. For microfluidic cell culture experiments, A549 cell suspension with volume of 100 μL and density of 2×10^7 cells/mL in the growth medium is prepared. The cell suspension is then manually injected into the device using a 1 mL disposable syringe (Korea Vaccine Co. Ltd., Seoul, Korea) with a 14-gauge stainless steel needle (Jensen Global Inc., Santa Barbara, CA, USA). To assure the A549 cells well attach onto the PDMS membrane for the experiments, the cells are cultured in the device under static conditions in the cell incubator overnight before the drug treatment and physical stimulation experiments.

In order to compare effects of conventional drug treatments with physical stimulation on surfactant protein expression, a steroid drug, dexamethasone, is also tested in the experiments for comparison. Dexamethasone has been reported to be capable of stimulating fetal lung maturation and promoting surfactant secretion [21,22]. In the experiments, 1 µM dexamethasone (D4902, Sigma-Aldrich) in the growth medium is exploited to treat the cells in the microfluidic device to study the surfactant secretion resulted from the steroid [10]. In addition, to systematically investigate effects of the dexamethasone treatment, physical stimulation, and their combinations on the pulmonary surfactant protein expression, four sets of experiments are conducted in this study: (1) Device A: cell culture with neither supplement of dexamethasone nor physical stimulation as control; (2) Device B: cell culture with growth medium containing 1 µM dexamethasone; (3) Device C: cell culture with the physical stimulation; and (4) Device D: cell culture with 1 µM dexamethasone and the physical stimulation. The experiments are conducted inside a conventional cell incubator, and the growth medium is continuously perfused with flow rates of 0.1 µL/min during the experiments. All the experiments are repeated three times for statistical analysis. In addition, viabilities of the A549 cells cultured in the experiments under various conditions are characterized to investigate the cell compatibility of the device and effects of the treatments on cell viability. The viabilities are quantitatively evaluated using a fluorescence LIVE (green)/DEAD (red) Viability/Cytotoxicity Kit (L3224, Invitrogen) containing Calcein AM (2 µM) and ethidium homodimer-1 (2 µM). The fluorescence nuclei staining, bisbenzimide H33342 trihydrochloride (1 µg/mL) (B2261, Sigma-Aldrich) in DPBS, is also performed for cell quantification.

2.5. Analysis of Surfactant Protein Expression

Surfactant protein C (SPC) is one of the pulmonary surfactant proteins. SPC uniquely expressed in the alveolar type II cells, which are responsible for production and secretion of pulmonary surfactant [23]. In the cell experiments, immunocytochemical staining of SPC on the A549 cells is performed to study surfactant protein expression of the cells under drug treatment and physical stimulation. The immunocytochemical staining is conducted by first washing cells using DPBS and fixing the cells within the microfluidic devices by 4% paraformaldehyde (PFA) (158127, Sigma-Aldrich Co., St. Louis, MO, USA) for 20 min. The fixed cells are then washed with DPBS containing 0.1% *v/v* Tween 20 (#161-0781, Bio-Rad Laboratories, Inc., Hercules, CA, USA) and permeabilized using 0.1% Triton X-100 (T9284, Sigma-Aldrich) for 5 min. The cells are then blocked in 3% bovine serum albumin (BSA) (A7906, Sigma-Aldrich) in DPBS for 2 h at room temperature. The cells are then stained with anti-prosurfactant protein C antibody (ab40879, Abcam, Cambridge, UK) at a ratio of 1:250 in DPBS at 4 °C overnight. Last, donkey anti-rabbit IgG secondary antibody (A21206, Invitrogen), at a ratio of 1:1000, is used with an additional 1-h incubation at ambient temperature after washing the first antibody with PBST (0.1% (*v/v*) Tween 20 in D-PBS) for visualization. The nuclei stain with 2 µg/mL DAPI is also performed on the cells for the following analysis.

An inverted fluorescence microscope (AF7000, Leica Microsystems Ltd.) equipped with a charge-coupled device (CCD) camera (ORCA-R2, Hamamatsu Photonics, Shizuoka, Japan) is exploited to capture brightfield and fluorescence images of the cells. To quantify the protein expression level from the fluorescence images, a monochromatic channel of the green fluorescence images is process by a code written in MATLAB R2014a (MathWorks, Inc., Natick, MA, USA). In the code, average fluorescence intensity per pixel from all detected pixels within the cells, which are identified by pixels with intensities above a threshold, is calculated. The calculated average intensity of each experiment is then normalized to the average intensity of the control experiment (Device A) conducted at the same time to eliminate possible experimental bias from different experiments. For statistical analysis, three independent sets of experiments are conducted, and the analysis of variance (ANOVA), followed by Holm–Sidak tests, is exploited to compare the normalized quantitative fluorescence intensities corresponding to the four conditions.

3. Results and Discussion

3.1. Device Characterization

The physical microenvironments established within the microfluidic device are characterized by numerical simulation and experimental measurements. First, the hydrostatic pressure within the cell culture chamber during the perfusion cell culture with flow rate of 0.1 µL/min is simulated using the aforementioned FEA model. Figure 3a shows the simulated hydrostatic pressure distributions within the entire top layer channel. The average hydrostatic pressure within the cell culture chamber along the flow direction in the cell culture chamber is 2.91 mmHg with variation less than 0.01 mmHg, which is within the measured physiological values from previous research [14]. In addition, the shear stress within the cell culture chamber was also estimated using the simulation. The result show that the average shear stress on the chamber walls on which the cells are cultured is approximately 1.28×10^{-2} dyn/cm^2, which is negligible for the cells. The results suggest that well-controlled hydrostatic pressure with value mimicking physiological ones can be established for perfusion cell culture using the designed device.

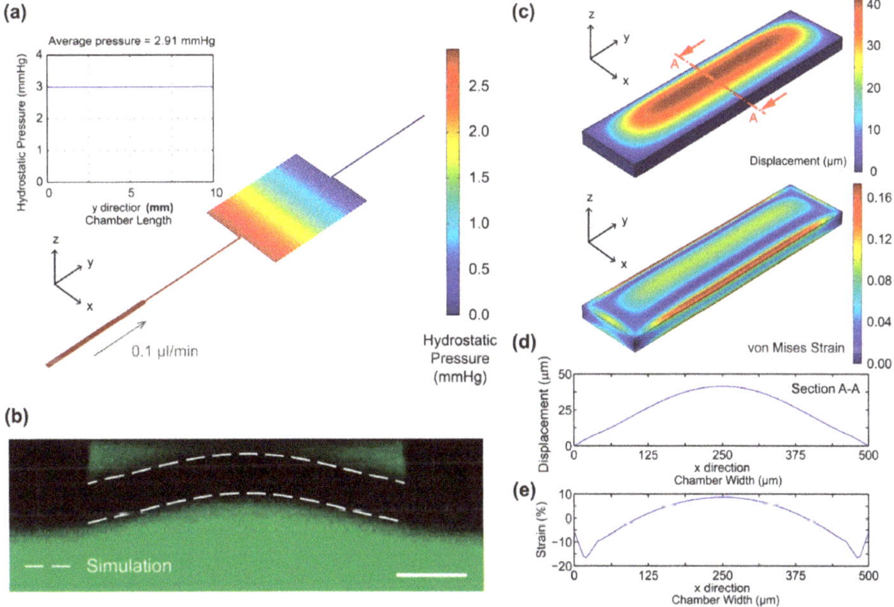

Figure 3. (**a**) The simulated hydrostatic pressure within the entire device. The inset demonstrates the uniform hydrostatic pressure along the flow direction in the cell culture chamber. (**b**) Confocal microscopic image showing the cross-section of the device with fluorescein solutions filled in the cell culture chamber and the actuation channel along the width of the cell culture chamber. The scale bar is 200 µm. The simulated deformations of the membrane are plotted in dotted lines showing the good agreement between the simulated and experimental results. (**c**) Contour plots of the displacement and von Mises strain distributions of the entire membrane across the cell culture chamber. (**d**) Plots of the displacement and strain along the top surface of the membrane across the width of the cell culture chamber.

In addition, deformation of the membrane within the microfluidic device is experimentally characterized using the confocal microscopy. Figure 3b shows the cross-sectional fluorescence confocal microscopic image of the deformed membrane when both top and bottom channels are filled with the fluorescein solution. To validate the numerical simulation results, the displacements of the top and bottom surfaces of the membranes are also plotted as white dotted lines and overlaid with the fluorescence image. The simulated deformation curves agree well with the experimental observation suggesting that the numerical simulation provides reasonably accurate prediction and can be used for the surface strain estimation. Figure 3c shows the simulated out of plane displacement and von Mises strain distribution of the entire membrane. The displacements of four membrane edges are zero due to the fixed boundary conditions, and the largest deformation greater than 40 μm occurs in the central part of the membrane. For the von Mises strain, the value decreases from a maximum value (16%) to zero and then increases from zero to 10% from the edge to the center along the channel width. Figure 3d shows the out of plane displacement (z direction) of the membrane at a cross-section located at the middle of the cell culture chamber along the flow direction (Section A-A). The average and maximum displacements are 24.7 μm and 41.6 μm, respectively. The results confirm that the membrane does not attach to the top surface of cell culture chamber resulting in cell damage because the channel height (85 μm) is larger than the sum of cell thickness and maximum z-displacement of the membrane.

To investigate the mechanical strain on the surface on which the cells are cultured, the normal strain in the direction along the width of the cell culture chamber (x-direction) is plotted as Figure 3e. From the edge to the center, the normal strain decreases slight to a minimum (−16%), and then gradually increases to a maximum (8.7%). Most of the simulated normal strains range from 2% to 8.7% across the width of the cell culture chamber (x = 125–375 μm), where the stretched cells are investigated. The designed strains ranges are within reported values from previous literatures. For instance, Liu et al. and Quinn et al. adopted cyclic biaxial 5–25% stretch to simulate the fetal breathing movement [9,13,24]. The simulated results confirmed that the device is capable of providing cultured cells physical stimulation of surface strain within the physiological ranges.

3.2. Cell Viability Characterization

To confirm the cell compatibility of the device and the cell experiments, the cells cultured inside the microfluidic device are observed using the microscope before and after the experiments. Figure 4a,b show phase contrast microscope images of the A549 cells right after seeding and after seeding overnight, respectively. The images demonstrate that the A549 cells can well adhere onto the fibronectin-coated PDMS membrane. To further estimate the cell viabilities after the experiments, the cells cultured in the device are stained with live(green)/dead(red) fluorescence dyes after 24-h culture under the combinations of the 1 μM dexamethasone and physical stimulations. As shown in Figure 4c, the fluorescence image indicates majority of the cells (>95%) show green fluorescence after the 24-h culture under the four combinations of 1 μM dexamethasone and physical stimulations within the device (Device A, B, C, and D), and there are no significant differences between the cell viabilities under different experimental conditions. The results suggest that the device possesses great biocompatibility, and the drug treatment and the physical stimulation do not significantly affect the cell viability.

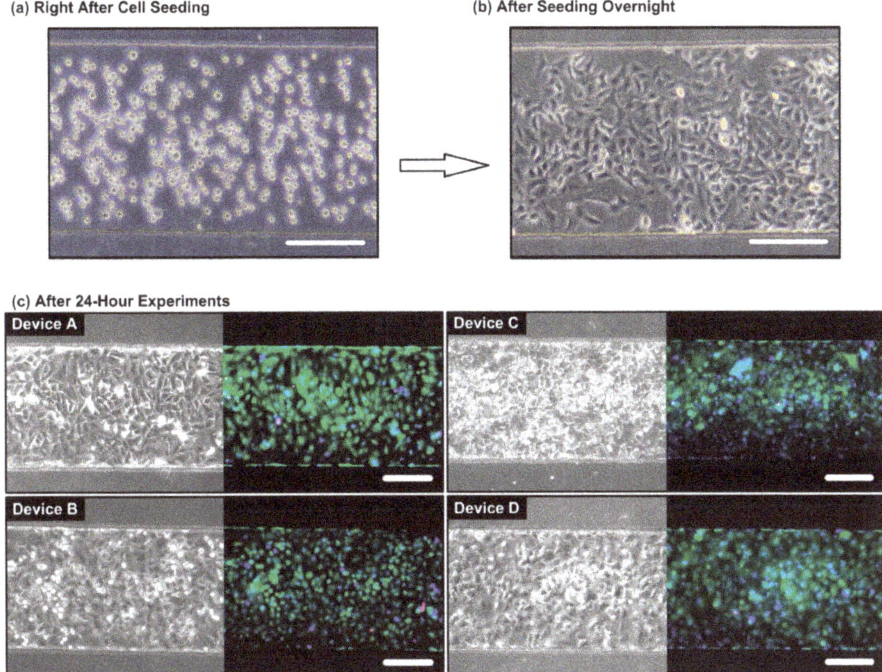

Figure 4. The microscopic images of the A549 cells (**a**) right after seeding into the microfluidic device, and (**b**) after culturing in the microfluidic device overnight. After the overnight culture, the A549 cells attach well onto the cell culture chamber substrate. (**c**) Bright field phase and live (green)/dead (red)/nuclei (blue) fluorescence images of the A549 cells inside the culture chamber after 24-h cell experiments under the 4 different conditions. Device A: control experiment in the growth medium; Device B: cell culture in the growth medium containing 1 µM dexamethasone; Device C: cell culture in the growth medium with the physical stimulations; and Device D: cell culture under the combination of dexamethasone treatment and physical stimulations. The scale bars are 200 µm.

3.3. Expression of SPC

To investigate the SPC expression of the A549 cells, fluorescence staining with microscopic imaging is performed in the experiments. Figure 5a shows the brightfield phase contrast images and fluorescence images stained with pulmonary SPC under the four different conditions after the 12-h experiments. The A549 cells cultured in the microfluidic device for 12 h show similar morphologies under the four different conditions. The results suggest that neither the physical strain nor the steroid treatment affect the cell morphology. The insensitivity of the A549 cell morphology to the physical strain and steroid treatment has been discussed in previous literature. For example, Liu et al. showed that the cell membranes, nuclei, lamellar bodies, and other fine structures were maintained under periodic mechanical stretch [9]. In addition, Rucka et al. reported that the morphological heterogeneity of A549 cells is not sensitive to dexamethasone treatment [25].

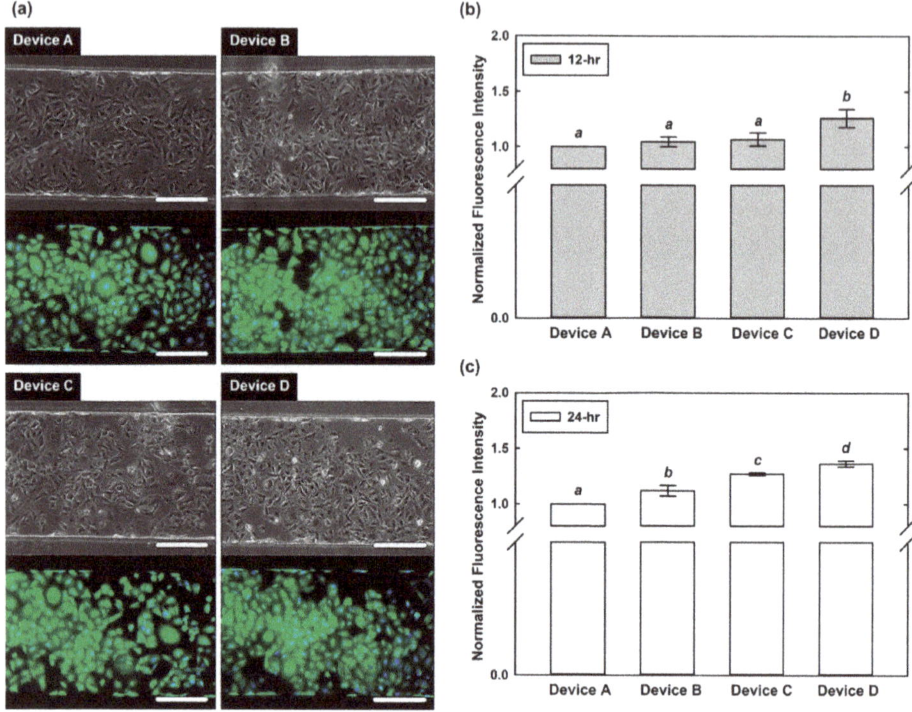

Figure 5. (**a**) Bright field phase and fluorescence images of the A549 cells cultured within the microfluidic devices with different combinations of steroid treatment and physical stimulation. Device A: neither steroid treatment nor physical stimulation (control); Device B: only steroid treatment; Device C: only physical stimulation; Device D: steroid treatment and physical stimulation. (**b**,**c**) Normalized average fluorescence intensity showing SPC expression of the A549 cells after 12- and 24-h experiments under the four different conditions, respectively. The fluorescence intensities showing SPC expression levels with statistically significantly different are designated with different letters (a, b, c, d = $p < 0.05$).

In order to quantitatively study the cells under the steroid treatment and under the designated physical stimulation mimicking lung physical microenvironment with specific strain range (from 2–9%), the cells on central chamber (x = 125–375 µm) are selected for SPC expression characterization by analyzing the fluorescence images as shown in Figure 5a. Figure 5b,c shows the normalized average intensity for A549 cells cultured in the microfluidic devices under the 4 different conditions after 12 h and 24 h, respectively. In Figure 5b, the normalized average intensities of A549 cells in Device B, C, and D are 8%, 6%, and 18% higher than that in Device A after 12-h experiment, respectively. The ANOVA analysis results indicate that no statistically significant differences between the results obtained from Device A, B, and C, although the normalized intensities are slightly higher for the cells cultured in Device B and C comparing to those cultured in Device A. In contrast, the intensities of the cells in Device D are statistically higher than those in Device A, B, and C. The results suggest that the treatment of dexamethasone or the physical stimulation on A549 cells for 12 h promotes the expression of SPC; however, the effects are limited comparing to the control experiments. In addition, the steroid treatment and the physical stimulation have similar effects on SPC expression of the A549 cells. In comparison, the SPC expression is significantly higher for the cells under the combination of the steroid treatment and physical stimulation indicating the synergic effects of both treatments on SPC expression of the A549 cells for 12-h experiments.

To further demonstrate the SPC expression of the A549 cells for longer treatments, Figure 5c shows the results of the 24-h cell experiments. The results indicate that the normalized average intensities of the A549 cells in Device B, C, and D are 10%, 27%, and 37% higher than that in Device A after the 24-h experiments, respectively. The results of the ANOVA analysis reveal there are significant differences among the results obtained from all four sets of experiments (Device A, B, C, and D). In the 24-h experiments, the SPC expression levels of the cells in the experimental conditions are all higher than those obtained in the control experiments. In the single treatment (either steroid treatment of physical stimulation) experiments, the cells under physical stimulation (Device C) have higher normalized average fluorescence intensities, indicating higher SPC expression level, comparing to those treated with the steroid (Device B). The results suggest that for 24-h treatment, the physical stimulation plays a more important role in promoting SPC expression of the A549 cells. Furthermore, the normalized average fluorescence intensities of the cells under the combined treatments (Device D with both steroid treatment and physical stimulation) are even higher than other conditions indicating even higher SPC expression level of the cells comparing to the cells under single treatments. The results again demonstrate the synergic effects of both treatments on SPC expression of the A549 cells. The similar experimental results have been mentioned in the previous studies. For example, Nakamura et al. reported that the increased SPC mRNA expression under the intermittent mechanical strain (5% elongation for 24-h) with dexamethasone treatment is higher than that without dexamethasone [10].

The experimental characterization results have confirmed that the developed device is capable of generating multiple well-controlled physical stimulations, including: cyclic strain and hydrostatic pressure with neglectable shear stress, which mimic in vivo physical microenvironments in lungs for in vitro cell culture. Furthermore, the cell experiment observation confirms the importance of the biomimetic microfluidic cell culture devices capable of reconstituting key physical microenvironments for in vitro cell studies. Further studies of cellular responses under more different physical stimulation combinations using the developed device as demonstrated in this paper can greatly help biologists elucidate influences of breathing movement under different normal or disease states. In addition, the experiments combined with drugs and other soluble factors can also be conducted using the developed device to investigate the effects of interactions between chemical and physical microenvironments on cellular behaviors. These characteristics warrant the reported biomimetic device a favorite candidate to conduct in vitro biomedical research on extensive pulmonary cells.

4. Conclusions

This paper reports a simple, but powerful, biomimetic microfluidic device capable of generating hydrostatic pressure and cyclic strain to provide a platform mimicking the lung physiological physical microenvironment for in vitro cell studies. The device is capable of providing an in vitro model with more in vivo-like microenvironments compared to conventional cell culture methods, and the device can be exploited for various pulmonary studies. For demonstration, in the experiments, the hydrostatic pressure applied to pneumocyte cells is used to mimic the microenvironment in fetal trachea during apneic periods on the saccular stag, and the physical surface strain applied upon a flexible PDMS membrane to stretch cells within the luminal cell culture chamber is used for simulating fetal breathing movement. The results demonstrate that the developed device can successful construct physical microenvironments similar to the physiological ones, and the cell experimental results suggest that physical stimulation may play a more important role in regulating the SPC expression level than the steroid treatment. The developed device can pave the way to better understand the cellular behaviors under various lung physiological conditions, and is promising for further translational studies.

Author Contributions: Y.-C.T. conceived and designed the experiments; T.-R.L., S.-L.Y., C.-C.P., and W.-H.L. performed the experiments; T.-R.L. and Y.-C.T. analyzed the data; T.-R.L., S.-L.Y., and Y.-C.T. wrote the paper.

Funding: This paper was funded by the Taiwan Ministry of Science and Technology (MOST 107-2627-M-001-007 and 107-2628-E-001-003-MY3), and the Academia Sinica Career Development Award (AS-CDA-106-M07).

Conflicts of Interest: The authors declare no conflict of interest.

References

1. Tarbell, J.M. Shear Stress and the Endothelial Transport Barrier. *Cardiovasc. Res.* **2010**, *87*, 320–330. [CrossRef] [PubMed]
2. Liu, M.C.; Shih, H.C.; Wu, J.G.; Weng, T.W.; Wu, C.Y.; Lu, J.C.; Tung, Y.C. Electrofluidic Pressure Sensor Embedded Microfluidic Device: A Study of Endothelial Cells under Hydrostatic Pressure and Shear Stress Combinations. *Lab Chip* **2013**, *13*, 1743–1753. [CrossRef] [PubMed]
3. Federico, V.; Bianchi, F.; Ahluwalia, B.; Domenici, C. Hydrostatic Pressure and Shear Stress Affect Endothelin-1 and Nitric Oxide Release by Endothelial Cells in Bioreactors. *Biotechnol. J.* **2013**, *9*, 146–154.
4. Collingsworth, A.M.; Torgan, C.E.; Nagda, S.N.; Rajalingam, R.J.; Kraus, W.E.; Truskey, G.A. Orientation and Length of Mammalian Skeletal Myocytes in Response to a Unidirectional Stretch. *Cell Tissue Res.* **2000**, *302*, 243–251. [CrossRef]
5. Kamotani, Y.; Bersano-Begey, T.; Kato, N.; Tung, Y.C.; Huh, D.; Song, J.W.; Takayama, S. Individually Programmable Cell Stretching Microwell Arrays Actuated by a Braille Display. *Biomaterials* **2008**, *29*, 2646–2655. [CrossRef] [PubMed]
6. Kotecha, S. Lung Growth for Beginners. *Paediatr. Respir. Rev.* **2000**, *1*, 308–313. [CrossRef]
7. Laudy, J.A.; Wladimiroff, J.W. The Fetal Lung. 1: Developmental Aspects. *Ultrasound Obstet. Gynecol.* **2000**, *16*, 284–290. [CrossRef]
8. Cosmi, E.; Anceschi, M.M.; Piazze, J.; La Torre, R. Ultrasonographic Patterns of Fetal Breathing Movements in Normal Pregnancy. *Int. J. Gynecol. Obstet.* **2003**, *80*, 285–290. [CrossRef]
9. Liu, M.; Skinner, S.J.M.; Xu, J.; Han, R.N.N.; Tanswell, A.K.; Post, M. Stimulation of Fetal Rat Lung Cell Proliferation in vitro by Mechanical Stretch. *Am. J. Physiol.* **1992**, *263*, L376–L383. [CrossRef]
10. Nakamura, T.; Liu, M.; Mourgeon, E.; Slutsky, A.; Post, M. Mechancial Strain and Dexamethasone Selectively Increase Surfactant Protein C and Tropelastin Gene Expression. *Am. J. Physiol. Lung Cell Mol. Physiol.* **2000**, *278*, L974–L980. [CrossRef]
11. Sanchez-Esteban, J.; Wang, Y.; Cicchiello, L.A.; Rubin, L.P. Cyclic Mechanical Stretch Inhibits Cell Proliferation and Induces Apoptosis in Fetal Rat Lung Fibroblasts. *Am. J. Physiol. Lung Cell Mol. Physiol.* **2002**, *282*, L448–L456. [CrossRef]
12. Douville, N.J.; Zamankhan, P.; Tung, Y.C.; Li, R.; Vaughan, B.L.; Tai, C.F.; White, J.; Christensen, P.J.; Grotberg, J.B.; Takayama, S. Combination of Fluid and Solid Mechanical Stresses Contribute to Cell Death and Detachment in a Microfluidic Alveolar Model. *Lab Chip* **2011**, *11*, 609–619. [CrossRef]
13. Huh, D.; Matthews, B.D.; Mammoto, A.; Montoya-Zavala, M.; Hsin, H.Y.; Ingber, D.E. Reconstituting Organ-Level Lung Functions on a Chip. *Science* **2010**, *328*, 1662–1668. [CrossRef] [PubMed]
14. Nicolini, U.; Fisk, N.M.; Rodeck, C.H.; Talbert, D.G.; Wigglesworth, J.S. Low Amniotic Pressure in Oligohydramnios—Is this the Cause of Pulmonary Hypoplasia? *Am. J. Obstet. Gynecol.* **1989**, *161*, 1098–1101. [CrossRef]
15. Boyce, E.S.; Dawes, G.S.; Gough, J.D.; Poore, E.R. Doppler Ultrasound Method for Detecting Human Fetal Breathing in Utero. *Br. Med. J.* **1976**, *2*, 17–18. [CrossRef] [PubMed]
16. Meyer, K.C. Diagnosis and Management of Interstitial Lung Disease. *Transl. Respir. Med.* **2014**, *2*, 4. [CrossRef]
17. Whitesides, G.M.; Ostuni, E.; Takayama, S.; Jiang, X.; Ingber, D.E. Soft Lithography in Biology and Biochemistry. *Ann. Rev. Biomed. Eng.* **2001**, *3*, 335–373. [CrossRef]
18. Lu, J.C.; Liao, W.H.; Tung, Y.C. Magnet-assisted Device-level Alignment for the Fabrication of Membrane-sandwiched Polydimethylsiloxane Microfluidic Devices. *J. Micromech. Microeng.* **2012**, *22*, 006–075. [CrossRef]
19. Chang, C.W.; Cheng, Y.J.; Tu, M.; Chen, Y.H.; Peng, C.C.; Liao, W.H.; Tung, Y.C. A Polydimethylsiloxane–polycarbonate Hybrid Microfluidic Device Capable of Generating Perpendicular Chemical and Oxygen Gradients for Cell Culture Studies. *Lab Chip* **2014**, *14*, 3762–3772. [CrossRef]
20. Nardone, L.L.; Andrews, S.B. Cell Line A549 as a Model of the Type II Pneumocyte: Phospholipid Biosynthesis from Native and Organometallic Precursors. *Biochim. Biophys. Acta Lipids Lipid Metab.* **1979**, *573*, 276–295. [CrossRef]
21. Kamath-Rayne, B.D.; DeFranco, E.A.; Marcotte, M.P. Antenatal Steroids for Treatment of Fetal Lung Immaturity After 34 Weeks of Gestation. *Obstet. Gynecol.* **2012**, *119*, 909–916. [CrossRef] [PubMed]

22. Sullivan, L.C.; Orgeig, S. Dexamethasone and Epinephrine Stimulate Surfactant Secretion in Type II Cells of Embryonic Chickens. *Am. J. Physiol. Regul. Integr. Comp. Physiol.* **2001**, *280*, R770–R777. [CrossRef] [PubMed]
23. Kalina, M.; Mason, R.J.; Shannon, J.M. Surfactant Protein C Is Expressed in Alveolar Type II Cells but Not in Clara Cells of Rat Lung. *Am. J. Respir. Cell Mol. Biol.* **1992**, *6*, 594–600. [CrossRef] [PubMed]
24. Quinn, T.P.; Schlueter, M.; Soifer, S.J.; Gutierrez, J.A. Cyclic Mechanical Stretch Induces VEGF and FGF-2 Expression in Pulmonary Vascular Smooth Muscle Cells. *Am. J. Physiol. Lung Cell. Mol. Physiol.* **2002**, *282*, L897–L903. [CrossRef]
25. Rucka, Z.; Vanhara, P.; Koutna, I.; Tesarova, L.; Potesilova, M.; Stejskal, S.; Simara, P.; Dolezel, J.; Zvonicek, V.; Coufal, O.; et al. Differential Effects of Insulin and Dexamethasone on Pulmonary Surfactant-Associated Genes and Proteins in A549 and H441 Cells and Lung Tissue. *Int. J. Mol. Med.* **2013**, *32*, 211–218. [CrossRef] [PubMed]

Brief Report

Competing Fluid Forces Control Endothelial Sprouting in a 3-D Microfluidic Vessel Bifurcation Model

Ehsan Akbari [1], Griffin B. Spychalski [2], Kaushik K. Rangharajan [1], Shaurya Prakash [1] and Jonathan W. Song [1,3,*]

[1]   Department of Mechanical and Aerospace Engineering, The Ohio State University, Columbus, OH 43210, USA
[2]   Department of Biomedical Engineering, The Ohio State University, Columbus, OH 43210, USA
[3]   Comprehensive Cancer Center, The Ohio State University, Columbus, OH 43210, USA
*   Correspondence: song.1069@osu.edu; Tel.: +1-614-292-8519

Received: 5 June 2019; Accepted: 2 July 2019; Published: 4 July 2019

Abstract: Sprouting angiogenesis—the infiltration and extension of endothelial cells from pre-existing blood vessels—helps orchestrate vascular growth and remodeling. It is now agreed that fluid forces, such as laminar shear stress due to unidirectional flow in straight vessel segments, are important regulators of angiogenesis. However, regulation of angiogenesis by the different flow dynamics that arise due to vessel branching, such as impinging flow stagnation at the base of a bifurcating vessel, are not well understood. Here we used a recently developed 3-D microfluidic model to investigate the role of the flow conditions that occur due to vessel bifurcations on endothelial sprouting. We observed that bifurcating fluid flow located at the vessel bifurcation point suppresses the formation of angiogenic sprouts. Similarly, laminar shear stress at a magnitude of ~3 dyn/cm^2 applied in the branched vessels downstream of the bifurcation point, inhibited the formation of angiogenic sprouts. In contrast, co-application of ~1 µm/s average transvascular flow across the endothelial monolayer with laminar shear stress induced the formation of angiogenic sprouts. These results suggest that transvascular flow imparts a competing effect against bifurcating fluid flow and laminar shear stress in regulating endothelial sprouting. To our knowledge, these findings are the first report on the stabilizing role of bifurcating fluid flow on endothelial sprouting. These results also demonstrate the importance of local flow dynamics due to branched vessel geometry in determining the location of sprouting angiogenesis.

Keywords: angiogenesis; shear stress; biomechanics; vessel branching

1. Introduction

Blood vessels comprise a hierarchical network that transports oxygen and nutrients throughout the body [1]. Expansion of this network occurs by angiogenesis, where endothelial cells (ECs) that line the inner surface of all blood vessels, sprout and branch to support tissue nourishment and growth [2,3]. Angiogenesis is necessary to help repair injured tissue [4], while uncontrolled angiogenesis is a prominent characteristic of rapidly growing solid tumors [5]. Therefore, the growth and remodeling of blood vessels by angiogenesis is critical throughout physiology [6,7], and increasing our fundamental understanding of angiogenesis is necessary to improve therapeutic strategies used for regenerative and cardiovascular medicine, and cancer therapy.

Research on angiogenesis has traditionally focused on identifying biochemical signaling factors that regulate EC function [8]. Yet in physiology, blood vessels are continuously exposed to fluid mechanical forces due to blood flow, including shear stress tangential to the endothelium, and normal flow across the vessel wall. Emerging research has highlighted the importance of the forces generated by blood flow in potently influencing the angiogenic process. These studies in the in vitro setting have been buoyed largely by the advancements in microfabrication techniques that enable the development of perfusable models that integrate 3-D tissue scaffolds for investigating angiogenesis in response to controlled fluid forces [9,10]. For instance, the application of 3 dyn/cm^2 laminar shear stress (LSS) was shown to suppress endothelial sprouting induced by vascular endothelial growth factor (VEGF) [11], while LSS values greater than 10 dyn/cm^2 have been shown to induce angiogenic sprouting [12]. Furthermore, previous studies have shown the pro-angiogenic role of transvascular flow (TVF) that is driven by a transmural pressure difference between the vasculature and the adjacent interstitium [11–13].

Endothelial sprouting has predominantly been examined in in vitro microfabricated devices with straight blood vessel models. However, the in vivo vasculature consists of a hierarchy of branching structures that generates bifurcating fluid flow (BFF) at the base of vessel bifurcations. BFF can be characterized by local stagnation pressure imparted normal to the endothelium and near-zero average shear stress, distinguishing BFF from previously studied hemodynamic factors (i.e., LSS and TVF). Recently, we reported an in vitro microfluidic model of a branching vessel that enables measurement of vascular permeability under application of BFF, LSS, and TVF [14] while also presenting an interface between fluid flow, ECs, and 3-D extracellular matrix (ECM). The findings from this study demonstrated the vascular permeability outcomes in response to the local flow dynamics at specific locations along the bifurcating vessel structure. However, the role of BFF in co-regulating endothelial sprouting has not been reported previously. Since elevated vascular permeability typically coincides with increased angiogenesis [15], the purpose of this paper was to investigate endothelial sprouting due to the flow dynamics produced by branching vessel geometry.

Here we report that application of BFF with ~38 dyn/cm^2 stagnation pressure at the vessel bifurcation point (BP) and ~3 dyn/cm^2 LSS in each branched vessel (BV) inhibits the formation of angiogenic sprouts. Furthermore, co-application of TVF with BFF and LSS induces the formation of angiogenic sprouts, thereby suggesting the presence of competing effects between TVF with BFF and LSS in regulating endothelial sprouting in a vessel bifurcation model. This work presents the first quantitative report on the stabilizing role of BFF for inhibiting endothelial sprouting. Moreover, these results demonstrate that sprout location can be determined by the local flow dynamics due to vessel bifurcations. Consequently, the findings reported here advance the current understanding of the fluid mechanical regulators of angiogenesis.

2. Materials and Methods

2.1. Microfluidic Model of Vessel Bifurcation

A microfluidic platform was developed as a 3-dimensional (3-D) in vitro analogue of a bifurcating vessel as previously described [14] (Figure 1A,B). The microchannels were fabricated by soft lithography of polydimethylsiloxane (PDMS), as reported previously [14]. Upon seeding, the microchannels were fully lined with mouse aortic endothelial cells (MAECs) to form an in vitro vessel analogue (Figure 1D). An important feature of the microfluidic model is a 3-D ECM compartment comprised of a mixture of collagen gel (3 mg/mL rat tail-type I) and fibronectin (10 μg/mL) that was positioned between the two parallel BV regions and intersects the base of the branching vessel or the BP (Figure 1D). Located at the BP and the two BVs are 100 μm wide openings (referred to as apertures) between two PDMS posts that help contain the pre-polymerized collagen gel solution in the 3-D ECM compartment [16]. These apertures enabled direct contact of the MAEC monolayer with the supporting collagen ECM where sprouts can infiltrate (Figure 1E).

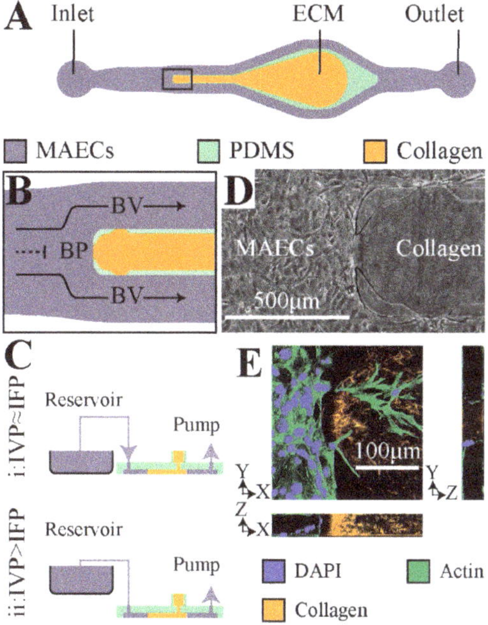

Figure 1. Schematic and characterization of the in vitro microfluidic device reported previously [14] used here for examination of sprouting angiogenesis. (**A**) Top-view for the microfluidic device. (**B**) Schematic showing a close-up for the local fluid flow at the bifurcation point (BP) aperture and the flow along apertures in the branched vessel (BV). (**C**) The schematic of the approach implemented to control the difference between intravascular pressure (IVP) and the interstitial fluid pressure (IFP) through controlled elevation of the reservoir. (**D**) A representative phase-contrast image of the BP fully lined with mouse aortic endothelial cells (MAECs) adjacent to the polymerized extracellular matrix (ECM) hydrogel. (**E**) A representative confocal image of the microfluidic device fully seeded with MAECs. Furthermore, the 3-D structure of the polymerized collagen matrix (orange) was resolved using total reflectance confocal imaging.

2.2. Cell Culture and Perfusion of the Microfluidic Model

MAECs were generously provided by the laboratory of Dr. Mike Ostrowski and cultured as previously described [17]. Briefly, MAECs were grown in DMEM-F12 cell culture media supplemented with 20% heat-inactivated fetal bovine serum, 1% penicillin-streptomycin, 10 U/mL heparin and 30 µg/mL endothelial cell growth supplement. The microchannels were pre-treated with fibronectin (10 µg/mL) for 2 h at 37 °C followed by treatment with cell culture growth media overnight at 37 °C. Subsequently, MAECs at 6–10 passage number were seeded at 20,000 cells/µl concentration, allowed to adhere overnight at 37 °C, and grown to confluence for 24 h. Controlled perfusion in the microfluidic model was applied using a programmable syringe pump as previously described [14]. A full-scale computational model of the microfluidic platform was developed using COMSOL Multiphysics (version 4.4) to determine: (a) the average LSS experienced by the MAECs that are adhered on the ECM at the BP and in each BV, (b) the level of TVF resulting in interstitial flow based on the measured hydraulic permeability of the ECM and the endothelial cell monolayer [14].

2.3. Immunofluorescence

Following treatment with each experimental condition, the microdevices were flushed 3 times with phosphate buffered saline (PBS) and fixed using 3% paraformaldehyde for 30 min at room temperature. Next, the devices were flushed 3 times with PBS and incubated with the blocking buffer (5% donkey serum with 0.1% Triton X-100 in PBS) for 1 hour at room temperature. The devices were then flushed 3 times with PBS and incubated with phalloidin (solution made by diluting the stock by 1:20 in PBS with 10% blocking buffer) for 30 min at room temperature to stain for actin. Next, the devices were flushed 3 times with PBS and incubated with DAPI (solution made by diluting the stock by 1:1000 in double distilled water) for 5 min at room temperature. Finally, the devices were flushed 3 times with PBS prior to epifluorescence imaging (TS-100, Nikon).

2.4. Quantification of Increase in Sprouting Area

Sprouting area was quantified using NIH ImageJ. A user-defined region of interest was defined for each aperture (i.e., BP or BV) and the total area covered by sprouting MAECs was quantified. The total sprouting area at 72 h after seeding was subtracted by the sprouting area at 24 h after seeding to report the total increase in sprouting area over 48 h.

2.5. Quantification of Sprout Elongation and Alignment

The morphological responses of MAEC sprouting were quantified using an elongation index and angle of orientation, which are commonly used parameters that quantify the extent that cells elongate and align in the direction of flow, respectively [18]. The elongation index for each sprouting MAEC was defined by the major axis (X) and minor axis (Y) of an ellipse fit to the cell area using MATLAB (Equation (1)) [19]. An elongation index of 0 represents a cell fit perfectly to a circle (non-elongated), and elongation index of 1 represents a cell fit perfectly to a line (fully elongated).

$$\text{Elongation Index} = \frac{X - Y}{X + Y} \tag{1}$$

The angle of orientation between each sprouting MAEC and interstitial flow was defined by the absolute angle between the major axis of the ellipse fit to the cell area and the interstitial flow vector at the cell area centroid. An angle of orientation of 0° is a cell aligned perfectly with the direction of interstitial flow and 90° is a cell aligned perpendicular to the direction of interstitial flow.

2.6. Statistical Analysis

Each experimental test condition was conducted at least in triplicate to report statistical analysis. The values reported for the sprouting area represent the average ± standard error of the mean. Two-sided Student's t-test was used to report statistical significance between each two pair of test conditions. The following symbols were used to report statistical significance: * indicates p-value < 0.05, *** indicates p-value < 0.001.

3. Results

3.1. Bifurcating Fluid Flow (BFF) and Laminar Shear Stress (LSS) Attenuate Endothelial Sprouting

The branching vessel geometry of our 3-D microfluidic model enables evaluation of angiogenic sprouting in response to BFF and LSS within the same fluidic circuit. In addition, our microfluidic model enables control of TVF levels independent of the perfusion rate, thereby enabling us to decouple the effects of TVF on endothelial sprouting from the effects of BFF and LSS. This capability is achieved by controlling the pressure difference via a static pressure head column between the intravascular pressure (IVP) and interstitial fluid pressure (IFP) domains (Figure 1C). Consequently, TVF can be either induced or restricted based on whether there is a difference in IVP and IFP [14]. When IVP is equal to IFP (IVP = IFP), TVF is ~0 (Figure 2A). Therefore, under these experimental conditions, sprouting angiogenesis responses were due to BFF and LSS and in the absence of TVF. When IVP is greater than IFP (IVP > IFP), a 1.5 cm H_2O hydrostatic pressure difference (~147 Pa or ~1470 dyn/cm^2) results in a TVF of ~1 µm/s oriented from the MAEC-lined intravascular region, across the endothelium, and into the interstitial ECM compartment (Figure 2A). TVF is a form of interstitial flow across the blood vessel wall [20], and in normal physiology, interstitial flow is thought to be on the order of 0.1–1 µm/s [21,22]. In our microfluidic model, introduction of ~147 Pa pressure head resulted in an average interstitial flow of 0.75 µm/s. Therefore, under these conditions, sprouting angiogenesis was evaluated in response to co-application of BFF and LSS alongside TVF.

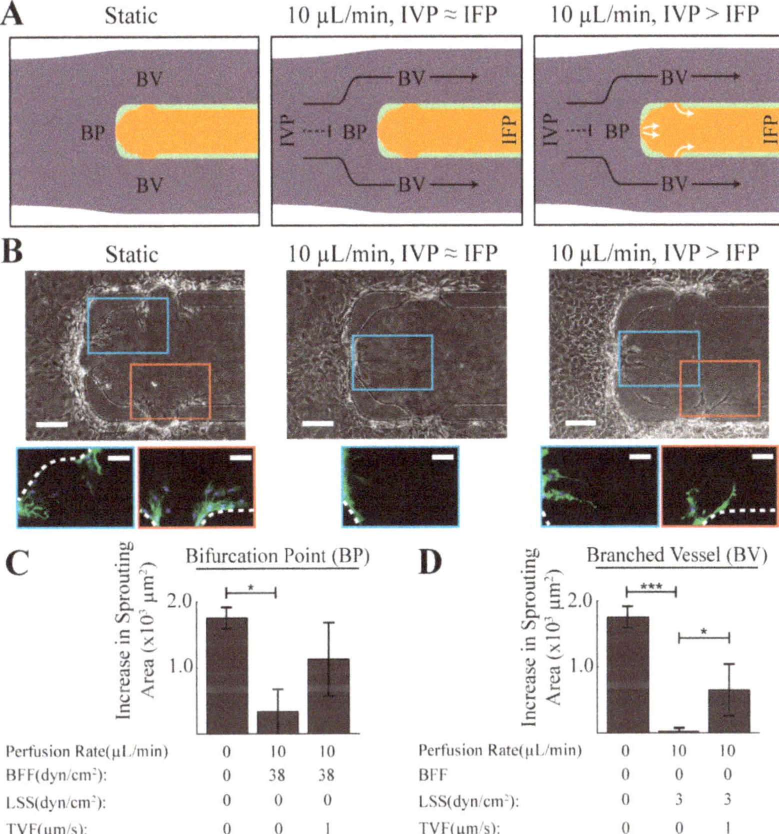

Figure 2. Application of bifurcating fluid flow (BFF) and laminar shear stress (LSS) attenuated endothelial sprouting. (**A**) The top-view schematics of the microdevice depicting the fluid mechanical factors under: (i) static, (ii) perfusion with equilibrated IVP and IFP resulting in BFF (dashed black line) at the BP, LSS (solid black line) in each BV and negligible transvascular flow (TVF), and (iii) perfusion with elevated IVP resulting in BFF at the BP, LSS in each BV and 1 µm/s TVF (white solid lines). (**B**) Representative phase contrast and epifluorescence images of sprouting MAECs in response to perfusion with ~10 µL/min after 48 h under equilibrated and elevated IVP compared to static control condition. Dashed white lines mark the location of the polydimethylsiloxane (PDMS) posts. (**C,D**) Quantitative representation of the level of increase in sprouting area in response to treatment with each experimental test condition: at the BP, and in BV. *: *p_value* < 0.05. ***: *p_value* < 0.001.

The microfluidic devices were perfused with complete MAEC growth medium at a volumetric flow rate of 10 µL/min to generate ~3 dyn/cm^2 LSS in BV as previously described [14]. This LSS level is within the physiological range of shear stress in the microcirculation in vivo [23]. In addition, previously reported computational estimations showed that 10 µL/min perfusion flow rate in this microfluidic model generates ~38 dyn/cm^2 stagnation pressure at the BP imparting a normal force due to stagnation flow at the BP along with near-zero shear stress [14]. To our knowledge, there are no previous reports of direct measurements of stagnation pressure at microcirculatory vessel bifurcations in vivo. However, estimates for the bulk flow rate of a representative large arteriole in vivo are ~2 µL/min based on the typical parameters for diameter (60 µm) and mean velocity (12 mm/s) [24]. Therefore, we estimate that the described level of stagnation pressure in this study (~38 dyn/cm^2) due to the 10 µL/min perfusion rate and the geometrical characteristics of the utilized microfluidic model to be within the physiological range of BFF experienced in the microcirculation in vivo.

Application of ~38 dyn/cm^2 BFF for 48 h, in the absence of TVF, significantly attenuated endothelial sprouting area by ~80% compared to the static control condition where MAECs undergo spontaneous sprouting (Figure 2B,C). Similarly, treatment with ~3 dyn/cm^2 LSS for 48 h, in the absence of TVF, completely inhibited sprouting area (i.e., reduced by ~98% compared to the static control condition) (Figure 2B,D). The observed suppression of MAEC sprouts by ~3 dyn/cm^2 LSS is in line with previously reported results in vitro with human umbilical vein endothelial cells (HUVECs) in response to approximately the same LSS levels [11]. However, to our knowledge, the findings here are the first report that endothelial sprouting is attenuated by BFF.

3.2. TVF Competes with BFF and LSS to Induce Formation of Angiogenic Sprouts

After establishing the effects of BFF and LSS on sprouting, the effects of systematically introducing TVF (~1 µm/s) via a static pressure head (~147 Pa) along with the simultaneous fixed flow rate of 10 µL/min that produces ~38 dyn/cm^2 BFF and 3 dyn/cm^2 LSS in our bifurcation model was evaluated. Co-application of TVF with BFF at the BP resulted in a ~3-fold increase in MAEC sprouting area compared to BFF alone, although this response was not statistically significant (Figure 2B,C). Similarly, co-application of TVF with LSS at the BV regions resulted in a ~28-fold increase in MAEC sprouting area compared to LSS alone (Figure 2B,D). The observed competing effects between TVF versus BFF and LSS on sprouting is in accordance with previous reports on the pro-angiogenic effect of TVF [11,12]. Furthermore, the level of endothelial sprouting induced by co-application of TVF alongside BFF and LSS was lower compared to the level of sprouting under static control condition. These findings suggest that the observed stabilizing effects of 38 dyn/cm^2 BFF and 3 dyn/cm^2 LSS counteracts the vessel sprouting responses triggered by a static pressure head of ~147 Pa inducing a TVF of ~1 µm/s.

3.3. Interstitial Flow Streamlines Coordinate Elongation of the Angiogenic Sprouts

After establishing the effects of BFF, LSS, and TVF in triggering or suppressing endothelial sprouting, the role for interstitial flow through the 3-D ECM compartment due to IVP > IFP (Figure 2A) supported the alignment and elongation of MAEC sprouts was evaluated. It is known that endothelial cells cultured in 2-D (two-dimensional) configurations align and elongate with the direction of LSS within 24 h [25]. In addition, it was previously shown that when exposed to interstitial flow, breast cancer cells aligned parallel to flow streamlines [26]. The MAEC sprouts that formed in response to co-application of TVF alongside BFF and LSS at the BP and each BV, respectively, were more elongated (estimated average elongation index of ~0.71) compared to static control condition (estimated average elongation index of ~0.41). These results suggest that interstitial flow through the ECM helps promote the extension of sprouting MAECs.

Next, it was examined whether the local interstitial streamlines help coordinate the direction of MAEC sprouting. Interstitial flow streamlines that were previously determined by computational modeling [14] were superimposed onto the optical microscopy images of MAEC sprouting into the 3-D ECM (Figure 3). Under an ideal or perfect alignment, the orientation of the MAEC sprouts would be 0° with respect to the interstitial flow streamlines. At the BP, MAEC sprouts that formed by co-application of TVF and BFF aligned with the direction of the local interstitial flow streamlines as visually demonstrated based on the superimposed images of the sprouting MAECs and the computationally estimated BP interstitial flow streamlines (Figure 3A). Our analysis by measuring the orientation angle provided an estimated average of ~7°, thereby suggesting a high level of influence of the interstitial flow on the sprout alignment. In contrast, MAEC sprouts at the BV and formed under co-application of LSS and TVF do not prominently align with the local interstitial flow streamlines (estimated average angle of orientation of ~49°), as visually depicted based on the overlapped image of sprouting MAECs and the local BV interstitial flow (Figure 3A). These outcomes were expected because at the BP aperture, the direction of TVF and interstitial flow in the supporting ECM are approximately parallel. In contrast, at the BV, the direction of interstitial flow changes prominently within ~50 μm away from the aperture interfaces (Figure 3A). Therefore, the MAEC sprout(s) that extend from the BV aperture interfaces are required to reorient in order to align with the direction of interstitial flow. Furthermore, it is noteworthy that the reported values for elongation index and alignment with interstitial flow serve as a descriptive assessment of the sprouting directionality. Thorough assessment of the role of interstitial flow in orchestrating sprouting directionality requires more advanced image analysis techniques that are beyond the scope of this study. Finally, it was confirmed that in the absence of MAECs, collagen matrix fiber orientation, i.e., the 3-D ECM, was not altered by perfusion and interstitial flow over 48 h (Figure 3B). Therefore, the observed alignment between angiogenic sprouts and the corresponding local interstitial flow streamlines is not mediated by flow-induced reorientation of the collagen fibers.

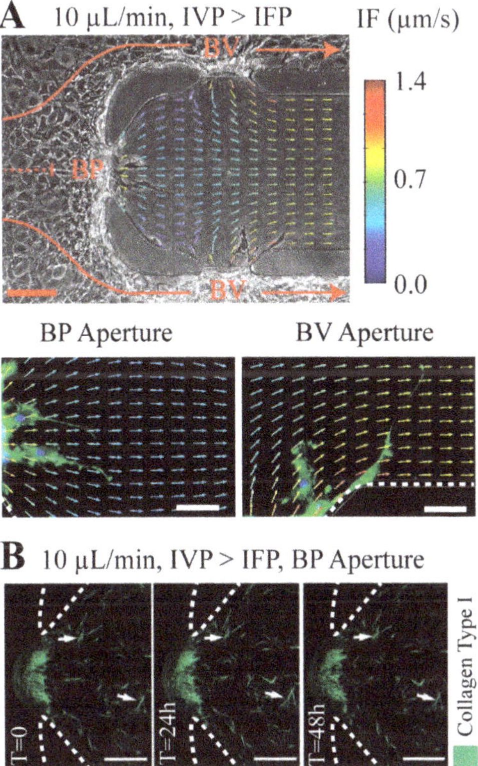

Figure 3. The angiogenic sprouts formed under elevated IVP were more elongated and showed alignment with the interstitial flow streamlines. (**A**) The computational interstitial flow streamline map superimposed on representative phase contrast and epifluorescence images of the angiogenic sprouts formed under elevated IVP condition. (**B**) Confocal reflectance microscopy images of collagen fibers under application of interstitial flow in the absence of MAECs. Arrowheads in the panels indicate landmark fibers to track among these images. Interstitial flow did not elicit a direct change in collagen fiber orientation. Red scale bar, 100 μm. White scale bars, 50 μm. Dashed white lines mark the location of the PDMS posts.

4. Discussion

Obtaining a detailed and mechanistic understanding of the main physical regulators of angiogenic vascular sprouting and remodeling is significant for advancing therapeutic strategies for modulating pathological angiogenesis. These studies can be facilitated by biomimetic platforms that reconstitute tissue-level function in vitro while also enabling controlled application of pressure and shear stress to cultured endothelial cells. In this work, the effect of stagnation pressure at vessel bifurcations and LSS on sprouting angiogenesis in presence and absence of TVF was evaluated. The studies were conducted in an in vitro microfluidic device that presents BFF in the presence of a collagen ECM. The results reported here show, for the first time, that endothelial sprouting was inhibited at the bifurcation point at 38 dyn/cm^2 stagnation pressure with nearly no shear stress. Application of 3 dyn/cm^2 LSS in the BV also imparted a stabilizing effect on the endothelium by suppressing the formation of angiogenic sprouts compared to static control conditions. While the LSS results agree with the previously reported role for tangential fluid forces in inhibiting VEGF induced endothelial sprouting [11], and decreasing endothelial permeability [14,27] the inhibition of sprouting due to a stagnation pressure is a new finding. It is worth noting that past in vivo observations on increased endothelial sprouting from vessels with low to no blood flow, such as damaged or occluded vessels [28] or blind-ending sprouts [29], also corroborate the reported stabilizing effect of 3 dyn/cm^2 LSS.

At the BP, the BFF represents negligible LSS but significant stagnation pressure arising from an impinging fluid flow. However, BFF did not show uncontrolled angiogenic sprouting. A previous in vivo study in the context of embryonic development showed that angiogenic sprouts form from points with local minimum shear stress except when the minimum shear occurs at the convergence of two blood vessels [30]. Interestingly, the fluid mechanical conditions at vessel convergences are comparable to flow characteristics at point of flow stagnation (locally stagnated flow with average negligible shear stress), which is based on the included angle of the incoming streams and the relative fluid shear of the converging fluid streams [31]. Furthermore, it was recently demonstrated that 38 dyn/cm^2 BFF at the BP for 6 h induced a significant decrease in endothelial permeability, thus causing stabilization of the endothelium [14]. In other words, as BFF stabilizes the endothelium, it is logical to expect a reduction in sprouting angiogenesis, which is the outcome we observed in this study.

Interestingly, simultaneous co-application of transvascular flow (TVF) alongside 38 dyn/cm^2 BFF and 3 dyn/cm^2 LSS induced the formation of angiogenic sprouts at BP and in each BV, respectively. Therefore, while BFF and LSS stabilize the endothelium, the TVF competes with the effects of BFF and LSS to induce formation of angiogenic sprouts. The sprouts formed under co-application of TVF alongside BFF and LSS were more elongated compared to static control condition. Furthermore, the angiogenic sprouts formed at the BP under elevated IVP (i.e., presence of BFF and TVF) were visually observed to be along the direction of interstitial flow streamlines at the BP, but not at the BV (i.e., presence of LSS and TVF) after 48 h. The visual alignment observations also find supporting evidence when taken together with the previous results on the pro-angiogenic effect of fluid flow across the endothelial monolayer [12,13].

Previous studies have attempted to discover the primary endothelial mechanosensory pathway that transduces the mechanical stimuli from LSS to the downstream endothelial biological response [32,33]. Since BFF is characterized by near-zero average shear stress at the stagnation point, it is plausible if the mechanotransduction pathway through which BFF leads to endothelial response is distinct from the previously reported pathways for LSS. Therefore, the detailed signaling pathways pertaining to BFF mechanotransduction requires further investigations. Moreover, while the results here were in the context of the blood vessel morphogenesis, it has previously been observed that intraluminal lymphatic valves preferentially form at the bifurcation points of collecting lymphatic vessels [34]. Our microfluidic model can be readily adapted to study the morphogenetic events coordinated by the local flow dynamics of branched vessels with applications for both the blood and lymphatic vasculature.

In summary, the results reported here introduce BFF as a potent regulator of endothelial sprouting while showing that both bifurcating flows and tangential shear flows are important determinants of angiogenic sprouting in an in vitro microfluidic model of vessel bifurcations.

Author Contributions: Conceptualization, E.A., K.K.R., S.P., and J.W.S.; Methodology, E.A. and K.K.K.; Software, E.A.; Validation, E.A, K.K.R. and G.B.S.; Formal Analysis, E.A. and G.B.S.; Investigation, E.A. and G.B.S.; Resources, S.P. and J.W.S.; Data Curation, E.A. and G.B.S.; Writing—Original Draft Preparation, E.A. and G.B.S.; Writing—Review & Editing, E.A., G.B.S., K.K.R., S.P. and J.W.S.; Visualization, E.A. and G.B.S.; Supervision, S.P. and J.W.S.; Project Administration, S.P. and J.W.S.; Funding Acquisition, S.P. and J.W.S.

Funding: This research was funded by The American Heart Association (15SDG25480000), NHLBI (R01HL141941), and Center for Emergent Materials, an NSF-MRSEC grant DMR-1420451. Images presented in this report were generated using the instruments and services at the Campus Microscopy and Imaging Facility, The Ohio State University. This facility is supported in part by grant P30 CA016058 from the NCI. Partial personnel support through the US Army Research Office through grant number W911NF-16-0278 and The Mark Foundation for Cancer Research (18-024-ASP) are also acknowledged. E.A. and K.K.R acknowledge OSU Presidential Fellowships. G.B.S. acknowledges funding from the Undergraduate Summer Research Program of The American Heart Association Great Rivers Affiliate, a Barry M. Goldwater Scholarship, and an Ohio State University (OSU) Comprehensive Cancer Center Pelotonia Fellowship.

Conflicts of Interest: The authors declare no conflicts of interest.

References

1. Pries, A.R.; Secomb, T.W. Making microvascular networks work: Angiogenesis, remodeling, and pruning. *Physiology (Bethesda)* **2014**, *29*, 446–455. [CrossRef]

2. Carmeliet, P.; Jain, R.K. Molecular mechanisms and clinical applications of angiogenesis. *Nature* **2011**, *473*, 298–307. [CrossRef]

3. Potente, M.; Gerhardt, H.; Carmeliet, P. Basic and therapeutic aspects of angiogenesis. *Cell* **2011**, *146*, 873–887. [CrossRef]

4. Greaves, N.S.; Ashcroft, K.J.; Baguneid, M.; Bayat, A. Current understanding of molecular and cellular mechanisms in fibroplasia and angiogenesis during acute wound healing. *J. Derm. Sci.* **2013**, *72*, 206–217. [CrossRef]

5. Kerbel, R.S. Tumor angiogenesis. *N. Engl. J. Med.* **2008**, *358*, 2039–2049. [CrossRef]

6. Cao, Y. Angiogenesis modulates adipogenesis and obesity. *J. Clin. Investig.* **2007**, *117*, 2362–2368. [CrossRef]

7. Virmani, R.; Kolodgie, F.D.; Burke, A.P.; Finn, A.V.; Gold, H.K.; Tulenko, T.N.; Wrenn, S.P.; Narula, J. Atherosclerotic plaque progression and vulnerability to rupture: Angiogenesis as a source of intraplaque hemorrhage. *Arter. Thromb. Vasc. Biol.* **2005**, *25*, 2054–2061. [CrossRef]

8. Shibuya, M. Vascular endothelial growth factor and its receptor system: Physiological functions in angiogenesis and pathological roles in various diseases. *J. Biochem.* **2013**, *153*, 13–19. [CrossRef]

9. Akbari, E.; Spychalski, G.B.; Song, J.W. Microfluidic approaches to the study of angiogenesis and the microcirculation. *Microcirculation* **2017**, *24*, e12363. [CrossRef]

10. Wong, K.H.; Chan, J.M.; Kamm, R.D.; Tien, J. Microfluidic models of vascular functions. *Annu. Rev. Biomed. Eng.* **2012**, *14*, 205–230. [CrossRef]

11. Song, J.W.; Munn, L.L. Fluid forces control endothelial sprouting. *Proc. Natl. Acad. Sci. USA* **2011**, *108*, 15342–15347. [CrossRef]

12. Galie, P.A.; Nguyen, D.H.; Choi, C.K.; Cohen, D.M.; Janmey, P.A.; Chen, C.S. Fluid shear stress threshold regulates angiogenic sprouting. *Proc. Natl. Acad. Sci. USA* **2014**, *111*, 7968–7973. [CrossRef]

13. Vickerman, V.; Kamm, R.D. Mechanism of a flow-gated angiogenesis switch: Early signaling events at cell-matrix and cell-cell junctions. *Integr. Biol.* **2012**, *4*, 863–874. [CrossRef]

14. Akbari, E.; Spychalski, G.B.; Rangharajan, K.K.; Prakash, S.; Song, J.W. Flow dynamics control endothelial permeability in a microfluidic vessel bifurcation model. *Lab Chip* **2018**, *18*, 1084–1093. [CrossRef]

15. Nagy, J.A.; Dvorak, A.M.; Dvorak, H.F. Vascular hyperpermeability, angiogenesis, and stroma generation. *Cold Spring Harb. Perspect. Med.* **2012**, *2*, a006544. [CrossRef]

16. Huang, C.P.; Lu, J.; Seon, H.; Lee, A.P.; Flanagan, L.A.; Kim, H.Y.; Putnam, A.J.; Jeon, N.L. Engineering microscale cellular niches for three-dimensional multicellular co-cultures. *Lab Chip* **2009**, *9*, 1740–1748. [CrossRef]

17. Srinivasan, R.; Zabuawala, T.; Huang, H.; Zhang, J.; Gulati, P.; Fernandez, S.; Karlo, J.C.; Landreth, G.E.; Leone, G.; Ostrowski, M.C. Erk1 and Erk2 regulate endothelial cell proliferation and migration during mouse embryonic angiogenesis. *PLoS ONE* **2009**, *4*, e8283. [CrossRef]

18. Song, J.W.; Gu, W.; Futai, N.; Warner, K.A.; Nor, J.E.; Takayama, S. Computer-controlled microcirculatory support system for endothelial cell culture and shearing. *Anal. Chem.* **2005**, *77*, 3993–3999. [CrossRef]

19. Boas, L.V.; Faustino, V.; Lima, R.; Miranda, J.M.; Minas, G.; Fernandes, C.S.V.; Catarino, S.O. Assessment of the Deformability and Velocity of Healthy and Artificially Impaired Red Blood Cells in Narrow Polydimethylsiloxane (PDMS) Microchannels. *Micromachines (Basel)* **2018**, *9*, 384. [CrossRef]

20. Rutkowski, J.M.; Swartz, M.A. A driving force for change: Interstitial flow as a morphoregulator. *Trends Cell Biol.* **2007**, *17*, 44–50. [CrossRef]

21. Chary, S.R.; Jain, R.K. Direct measurement of interstitial convection and diffusion of albumin in normal and neoplastic tissues by fluorescence photobleaching. *Proc. Natl. Acad. Sci. USA* **1989**, *86*, 5385–5389. [CrossRef]

22. Swartz, M.A.; Lund, A.W. Lymphatic and interstitial flow in the tumour microenvironment: Linking mechanobiology with immunity. *Nat. Rev. Cancer* **2012**, *12*, 210–219. [CrossRef]

23. Gray, K.M.; Stroka, K.M. Vascular endothelial cell mechanosensing: New insights gained from biomimetic microfluidic models. In *Seminars in Cell & Developmental Biology*; Academic Press: Cambridge, MA, USA, 2017; pp. 106–117.

24. Popel, A.S.; Johnson, P.C. Microcirculation and Hemorheology. *Annu. Rev. Fluid Mech.* **2005**, *37*, 43–69. [CrossRef]

25. Levesque, M.J.; Nerem, R.M. The elongation and orientation of cultured endothelial cells in response to shear stress. *J. Biomech. Eng.* **1985**, *107*, 341–347. [CrossRef]

26. Polacheck, W.J.; Charest, J.L.; Kamm, R.D. Interstitial flow influences direction of tumor cell migration through competing mechanisms. *Proc. Natl. Acad. Sci. USA* **2011**, *108*, 11115–11120. [CrossRef]

27. Buchanan, C.F.; Verbridge, S.S.; Vlachos, P.P.; Rylander, M.N. Flow shear stress regulates endothelial barrier function and expression of angiogenic factors in a 3D microfluidic tumor vascular model. *Cell Adhes. Migr.* **2014**, *8*, 517–524. [CrossRef]

28. Yamada, H.; Yamada, E.; Hackett, S.F.; Ozaki, H.; Okamoto, N.; Campochiaro, P.A. Hyperoxia causes decreased expression of vascular endothelial growth factor and endothelial cell apoptosis in adult retina. *J. Cell. Physiol.* **1999**, *179*, 149–156. [CrossRef]

29. Anderson, C.R.; Hastings, N.E.; Blackman, B.R.; Price, R.J. Capillary sprout endothelial cells exhibit a CD36 low phenotype: Regulation by shear stress and vascular endothelial growth factor-induced mechanism for attenuating anti-proliferative thrombospondin-1 signaling. *Am. J. Pathol.* **2008**, *173*, 1220–1228. [CrossRef]

30. Ghaffari, S.; Leask, R.L.; Jones, E.A. Flow dynamics control the location of sprouting and direct elongation during developmental angiogenesis. *Development* **2015**, *142*, 4151–4157. [CrossRef]

31. Prakash, S.; Akberov, R.; Agonafer, D.; Armijo, A.D.; Shannon, M.A. Influence of Boundary Conditions on Sub-Millimeter Combustion. *Energy Fuels* **2009**, *23*, 3549–3557. [CrossRef]

32. Hahn, C.; Schwartz, M.A. Mechanotransduction in vascular physiology and atherogenesis. *Nat. Rev. Mol. Cell Biol.* **2009**, *10*, 53–62. [CrossRef]

33. Davies, P.F. Flow-mediated endothelial mechanotransduction. *Physiol. Rev.* **1995**, *75*, 519–560. [CrossRef]

34. Udan, R.S.; Dickinson, M.E. The ebb and flow of lymphatic valve formation. *Dev. Cell* **2012**, *22*, 242–243. [CrossRef]

 *micromachines*

Article

Endothelial Cell Activation in an Embolic Ischemia-Reperfusion Injury Microfluidic Model

Danielle Nemcovsky Amar, Mark Epshtein and Netanel Korin *

Faculty of Biomedical Engineering, Technion-Israel Institute of Technology, Haifa 32000, Israel;
am@campus.technion.ac.il (D.N.A.); epshmark@campus.technion.ac.il (M.E.)
* Correspondence: korin@bm.technion.ac.il; Tel.: +972-4829-4116

Received: 29 October 2019; Accepted: 4 December 2019; Published: 6 December 2019

Abstract: Ischemia, lack of blood supply, is associated with a variety of life-threatening cardiovascular diseases, including acute ischemic stroke and myocardial infraction. While blood flow restoration is critical to prevent further damage, paradoxically, rapid reperfusion can increase tissue damage. A variety of animal models have been developed to investigate ischemia/reperfusion injury (IRI), however they do not fully recapitulate human physiology of IRI. Here, we present a microfluidic IRI model utilizing a vascular compartment comprising human endothelial cells, which can be obstructed via a human blood clot and then re-perfused via thrombolytic treatment. Using our model, a significant increase in the expression of the endothelial cell inflammatory surface receptors E-selectin and I-CAM1 was observed in response to embolic occlusion. Following the demonstration of clot lysis and reperfusion via treatment using a thrombolytic agent, a significant decrease in the number of adherent endothelial cells and an increase in I-CAM1 levels compared to embolic occluded models, where reperfusion was not established, was observed. Altogether, the presented model can be applied to allow better understanding of human embolic based IRI and potentially serve as a platform for the development of improved and new therapeutic approaches.

Keywords: ischemia/reperfusion injury; thrombolysis; organ-on-a-chip; endothelial cell activation

---

1. Introduction

Cardiovascular diseases are the leading cause of mortality in the Western world. According to the World Health Organization, (WHO), 17.9 million people die each year, representing 31% of all death worldwide [1]. Lack of proper blood supply, also known as ischemia, is associated with several medical conditions including myocardial infraction, thrombotic stroke, embolic vascular occlusion, angina pectoris, cardiac surgery and organ transplantation [2,3]. Although, restoration of blood supply is crucial to prevent further cellular damage, rapid reperfusion increases tissue damage, even in remote tissues that were not affected by the ischemia. This phenomenon where restoration of flow paradoxically increases tissue damage is termed ischemia reperfusion injury (IRI) [4,5].

In order to study IRI pathophysiology and to develop new therapeutic approaches a variety of IRI animal models have been employed [6–8]. These include both small animal models as well as large animal models, which better resemble human anatomy but are more costly. However, IRI animal models do not properly recapitulate critical components in the physiology of human ischemia reperfusion injury. As a result, despite the promising outcomes of some pharmacological treatments in IRI animal models, when these studies were implemented in human clinical trials the results were disappointing. For example, although statin treatments, which can reduce the inflammatory response in IRI, showed promising results in a rat IRI model [9], in human clinical studies, the results were less conclusive [10]. Thus, there is both a scientific need and an ethical motivation to develop alternative in vitro models of ischemia reperfusion injury, which would be relevant to human physiology.

Microfluidic cell culture devices, also known as organ-on-chips, offer an alternative system for recapitulation of certain features of human physiology and organ functionality in vitro. A variety of organ-on-chip platforms have been developed to study blood vessels, muscle, brain, liver and lung physiology [11]. Such microfluidic devices allow the culturing of human cells under a controlled microenvironment and well-defined dynamic conditions [12] and thus can be valuable for studying ischemia and IRI. A variety of microfluidic devices that precisely control oxygen levels and shear stress inside the device have been developed to study the role of these factors in health and disease conditions [13]. Abaci et al. studied the effect of these two factors under physiological conditions in a microfluidic vascular model and showed that the vasotoxic cancer drug, 5-Fluorouracil, induces vascular hyper-permeability and not apoptosis. Additionally, using their microfluidic model, they demonstrated the transition of endothelial cells from a quiescent to an activated state under hypoxic conditions. Khanal et al. studied IRI in cardiac tissue using a microfluidic device with porcine cardiomyocytes where the cells were subjected to nitrogen to induce hypoxia followed by normoxia; mitochondrial depolarization, cell adhesion and morphology were then assessed [14]. Despite the significant advances in the field, only limited work has focused on recapitulating IRI per se, where flow is rapidly restored following vascular occlusion. Additionally, thus far, to the best of our knowledge, no study has addressed embolic IRI, where the obstruction of flow is cause by a clot/embolus, and its lysis/removal may result in an IRI, as is the case in the treatment of ischemic stroke.

Here, we propose a microfluidic IRI model comprising a vascular compartment lined with human endothelial cells that can be obstructed with a human blood clot and then re-perfused via thrombolytic treatment. As vascular inflammation is an important symptom and contributor to IRI, we focused on endothelial cell activation, which is an important initial step in this process. Activated endothelial cells overexpress several receptors on their membrane, in order to attract immune cells, including E-selectin, which is responsible for leukocyte rolling, and I-CAM1, which is responsible for leukocytes firm adhesion to endothelial cells [15]. Using our model, we simulate treatment by a thrombolytic agent that, although restores flow to a clot-occluded vascular compartment, causes endothelial cell (EC) activation, as measured by the over-expression of the EC surface receptors E-selectin and I-CAM1. Thus, the presented model allows us to replicate, in a controlled manner, key features of IRI following restoration of flow upon removal of vascular embolic occlusion. Altogether, the suggested model may offer a better understanding of human IRI physiology in thromboembolic diseases and can also potentially serve as a novel platform to study new therapeutic approaches for treatment of IRI.

2. Materials and Methods

2.1. Microfluidic Device

The microfluidic device was designed using SolidWorks. The design contains two straight channels, 1 mm in height and width and 3 cm in length, connected by a narrowing, 200 μm in height and width and 2 mm in length, between them designed to allow a blood clot to occlude and obstruct the normal flow. The microfluidic device has one main inlet for cell seeding and perfusion. A second inlet, the clot inlet, was added for clot injection and its corresponding outlet is just before the narrowing, in order to avoid flow through the EC compartment. Another inlet was added at the EC channel post the narrowing, for measurement purposes when the pre-stenosis channel is occluded with a blood clot. The microdevice mold was 3D printed (Object®Eden500, Formlabs, Massachusetts, USA) using Formlabs Clear Resin v4. Polydimethylsiloxane (PDMS, sylgard®184, Silicone Elastomer) was poured over the mold and allowed to cure for 24 h, at room temperature. The PDMS was peeled from the mold and 1 mm holes were punched to create inlets and outlets. The PDMS and a glass slide were treated with plasma to create firm bonding and placed at 60 °C, overnight. A schematic view and image of the microdevice are shown in Figure 1a.

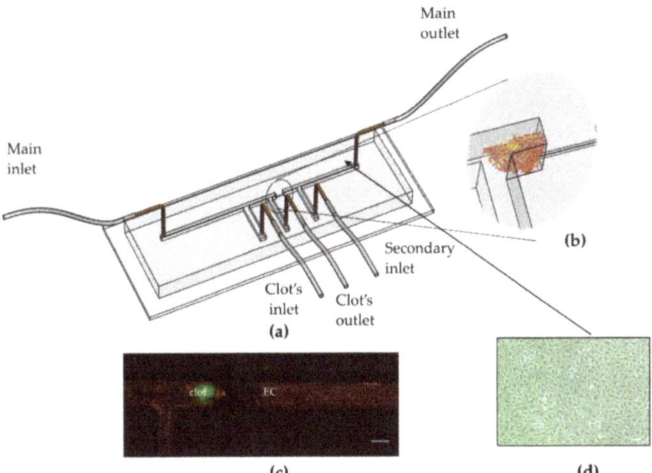

Figure 1. The microdevice recapitulating an embolic ischemia reperfusion injury. (**a**) Illustration of the microfluidic device: a narrowed section (200 μm × 200 μm) enables the occlusion by a clot prior to the vascular compartment, lined with endothelial cell. The different inlets and outlets allow independent perfusion of the cells and loading of the clot, as well as introduction of a thrombolytic solution to restore flow. (**b**) Schematic of the narrowed section in the microdevice where the blood clot obstructs the flow. (**c**) Live staining of endothelial cells (in red) and a blood clot occluding the channel (in green). Scale bar: 1 mm. (**d**) Phase microscopy image of the endothelial cells (ECs) inside the microfluidic device. Scale bar: 10 μm.

2.2. Cell Culture

Human umbilical vein endothelial cells (HUVECs, Lonza) were maintained in endothelial cell medium (ECM, ScienceCell, Carlsbad, CA, USA). Cells were cultured at 37 °C, in a humidified environment, with 5% CO_2. Cells were passaged every 4–5 days, i.e., when they reached 85% confluence, using Trypsin EDTA solution B (Biological Industries, Cromwell, CT, USA). The microdevice was coated with human fibronectin (100 μg/mL, Sigma), overnight, at 4 °C. HUVECs at passages 4–6 were seeded in the microdevice at a density of 1.5×10^6 cells/mL and incubated for two hours to allow the cells to adhere. The microdevice was connected to a programmable syringe pump for controlled perfusion overnight (flow rate 50 μL/h). Figure 1b shows a phase microscopy image of ECs grown under perfusion within the device.

2.3. Clot Fabrication

Calcium chloride (1 mol/L) was added to citrated human whole blood at a 5:1 ratio, to allow recalcification of the blood. Human fibrinogen plasminogen-depleted (5 mg/mL, Enzyme Research Laboratories) was added to stabilize the clot and increase its stiffness [16]. Human alpha-thrombin (1 u/mL, Enzyme Research Laboratories) was added to initiate clot formation. The solution was immediately injected into silicone tubes (Tygon, I.D. 0.79 mm) [17] and clots were incubated at 37 °C, overnight, to allow maturation. Clots were cut to 2 mm size in length and then incubated in saline for one hour prior to being inserted into model. Then a single clot was injected into the microfluidic device and allowed to occlude the entrance of the endothelialized vascular compartment (Figure 1c).

2.4. Thromblysis

A blood clot was prepared as described above and infused into the main microchannel via the clot inlet it produced an occlusion of the entrance to the narrowed section. A thrombolytic solution was prepared by mixing human Glu-Plasminogen (200 µg/mL, Enzyme Research Laboratories) with serum free medium and tissue plasminogen activator (tPA, 126 µg/mL, Actilyse), a thrombolytic drug. The solution was infused into the device through the inlet and to the pre-stenosis outlet. The pre-stenosis outlet was then blocked and a constant pressure (60 mmHg) was enforced using the pressure-controlled MFCS^{TM}-EZ system (Fluigent, Paris, France). The device was placed under an upright fluorescence microscope (Nikon SMZ25, Nikon Instruments Inc., Melveille, NY, USA) for time-lapse imaging. Flow rate was recorded during the experiment using a flow rate sensor (Fluigent, Paris, France).

2.5. Immunofluorescence

Following each experiment, the microdevice was washed three times with phosphate buffered saline (PBS) and then the cells were fixed with 4% paraformaldehyde (PFA), for 15 min, at room temperature. The cells were washed three times with PBS and incubated with blocking buffer (Bovine serum albumin 2%), for one hour. As E-selectin and I-CAM1 are receptors expressed on endothelial cell membrane and play an important role during inflammation. The E-selectin binding peptide (Esbp, CDITWDQLWDLMK–CONH2), that was synthesized by Shamay et al [18] and labeled with fluorescein isothiocyanate (FITC)-Lys. HUVECs were incubated with Esbp (50 µg/mL) for two hours at room temperature. To stain ICAM-1, cells were incubated at 4 °C, overnight with a mouse anti-ICAM-1 monoclonal antibody (5 µg/mL, ThermoFisher), which reacts with the human ICAM-1 receptor. The microdevice was then washed twice and incubated with a secondary antibody for labeling (Alexa Fluor 647 anti-mouse, ThermoFisher) for one hour. Lastly, cells were incubated for 5 min with 4′,6-Diamidino-2-Phenylindole, Dihydrochloride (DAPI), to stain the nucleus in all the channels. Cells were then washed three times with PBS.

2.6. Image Acquisition and Quantification

Images were taken with an inverted Eclipse Ti-E (Nikon Instruments Inc., Melville, NY, USA) microscope equipped with a 10× objective (N.A. 0.45, Nikon Instruments Inc., Melville, NY, USA) and an EMCCD camera (iXon Ultra, Andor-an Oxford Instruments company, Belfast, Ireland). ANDOR iQ3 Imaging software (Andor-an Oxford Instruments company, Belfast, Ireland) was used for stage movement control and image capture. Quantification of fluorescence intensity and number of cells was done with NIH ImageJ.

2.7. Statistical Analysis

Data analysis was performed with a *t*-test: two sample assuming unequal variance. The amount of variation within the data set is presented by the standard deviation (SD), and the hypothesis test was conducted using *p*-value to determine the significance of the results. p-value < 0.05 is presented using (*), p-value < 0.01 (**), p-value < 0.001 (***) and p-value < 0.0001 (****). All experiments were repeated at least three times.

3. Results

3.1. Establishing Embolic Vascular Occlusion in the Device

Microfluidic devices were coated with fibronectin and seeded with HUVECs. After seeding, cells were incubated for two hours to allow adherence. Cells were maintained under constant flow for 24 hours in order to achieve over 95% confluence and reduce any stress that can be induced by cell harvesting. The microfluidic device was designed to have a 200 µm narrowing between the two straight 1 mm channel compartments, thus allowing a pre-prepared human blood clot to obstruct and occlude the flow. Using separate inlet/out ports, we were able to incorporate the blood clot without harming the cell culture; >90% confluence was maintained (Figure 1).

3.2. Endothelial Activation upon Vascular Occlusion

Both ischemia upon vascular occlusion and IRI, which follows upon restoration of flow are characterized by an inflammatory response in the affected tissue [19]. One of the key steps in the inflammatory process, which has also been a therapeutic target in IRI, is EC activation [20]. During EC activation, numerous receptors are overexpressed on the cell membrane [21], including E-selectin and I-CAM1. E-selectin is a glycoprotein expressed on endothelial cells after activation by numerous cytokines such as TNF-α. It is the main receptor on EC responsible for the initial adhesion of leukocytes during the inflammatory response [22]. I-CAM1 is a member of the immunoglobulin family, whose levels increase significantly following endothelial activation. It is the key mediator of the firm adhesion of leukocytes to endothelial cells [23]. To monitor E-selectin and ICAM-1 expression levels we used a fluorescence staining procedure, as outlined in the Methods section.

As a first step we tested whether embolic occlusion in our device results in EC activation. Staining of E-selectin and I-CAM1 was done following 2 h under the following conditions (a) channels perfused with serum free medium (unblocked), (b) channels obstructed by an emboli and (c) channels perfused with the inflammatory cytokine TNF-α at (0.1 µg/mL) in serum free media. TNF-α is known to induce a long-term inflammatory response by stimulating inflammatory mediators and proteases, thereby promoting endothelial cell activation [24]. As shown in Figure 2, indeed exposure to TNF-α caused a profound increase in the expression of both E-selectin (green) and I-CAM1 (red). Importantly, a significant increase in both receptors was measured in the embolic occlusion model compared to the results in the perfused devices. These results confirm that upon embolic occlusion in the designed device, the basal levels of both receptors increase, indicating EC activation.

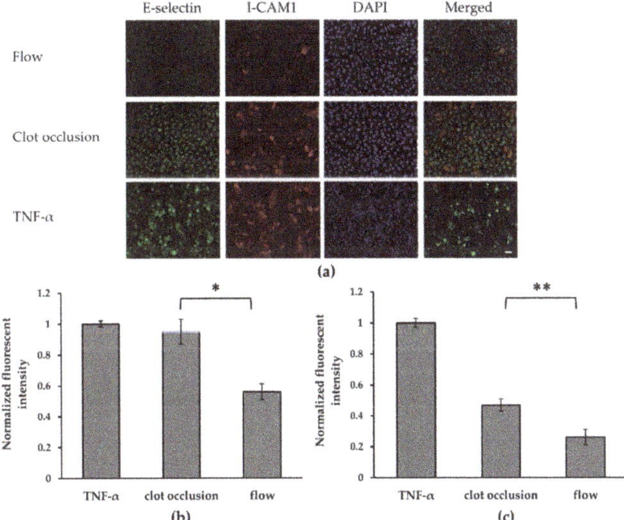

Figure 2. Endothelial cell (EC) activation following embolic occlusion in a microfluidic device. (**a**) Fluorescence confocal imaging of ECs stained for E-selectin, using Esbp (in green), I-CAM1, using a mouse-anti-ICAM1 antibody (in red) and nuclei (in blue), under various conditions (i) normal flow in the channel (FLOW), (ii) channel occluded with a blood clot (CLOT) or (iii) channel perfused with TNF-α, a proinflammatory stimulator (TNF-α). Scale bar: 10 µm. E-selectin and I-CAM1 intensity represents receptor expression, it elevates in correlation with cell activation. (**b,c**) Graph comparing between the normalized fluorescence intensity of (**b**) I-CAM1 and (**c**) E-selectin under the conditions mentioned in A. Both I-CAM1 and E-selectin levels in TNF- α channel and occluded channel increased compared to normally perfused channel with statistics significance. Significance determined by the unpaired Student's *t*-test, *: *p*-value < 0.05, **: *p*-value < 0.01 ($n = 3$).

3.3. Restoration of Flow via Thrombolysis

To regain perfusion, we prepared a thrombolytic solution composed of tPA and PLG supplemented with FITC-dextran for fluorescence labeling. tPA is a protease that converts PLG into plasmin (PLS). PLS is the final product in the thrombolysis cascade, that dissolves the fibrin mesh and ultimately the blood clot [25]. Currently the only Food and Drug Administration (FDA)-approved therapy for ischemic stroke is the alteplase, which is recombinant tPA [26]. The therapy involves the intravenous infusion of alteplase, to dissolve the clot. Although the treatment is critical it has a sever side effect, intracerebral hemorrhage (ICH) [27]. To simulate a thrombolytic treatment in our model, a blood clot was inserted, and occlusion of the vascular compartment was confirmed. Then to restore flow, a thrombolysis solution was introduced to the pre-occluded side; a physiological pressure gradient of 60 mmHg was maintained using a pressure-controlled system. Upon administration of the thrombolytic drug, within <10 min clot degradation occurred (Figure 3a), and flow was restored as measured using a flow sensor (Figure 3c). However, when the same saline solution with FITC-dextran excluding the tPA was injected, the clot remained stable (>2 h) under the same pressure gradient and no flow was re-established, see Figure 3b,c. Thus, these results demonstrated the ability to replicate thrombolysis-based reperfusion following an embolic occlusion.

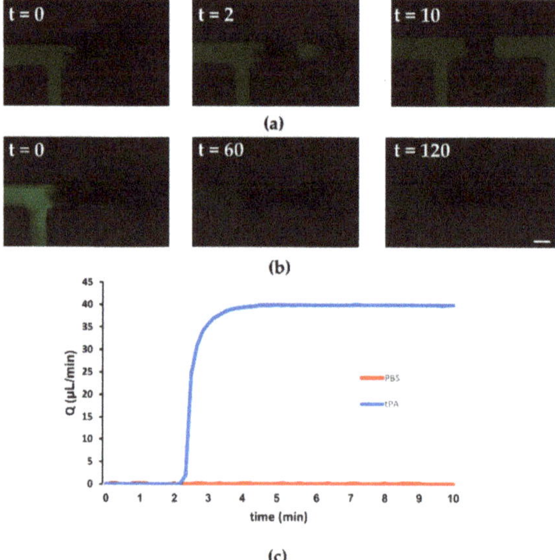

Figure 3. Thrombolytic reperfusion in the ischemia reperfusion injury (IRI) microfluidic device (**a,b**) Fluorescence time-lapse microscopy following thrombolysis and control treatment. (**a**) Treatment with a thrombolysis solution comprised of tissue plasminogen activator (tPA), plasminogen and FITC-dextran showing the reperfusion of a FITC-dextran solution upon treatment with tPA. (**b**) Treatment with phosphate buffered saline (PBS) with FITC-dextran (without tPA)—reopening did not occur even after two hours of experiment. Scale bar: 1 mm. (**c**) Flow rate measurements for the control (PBS) and thrombolysis (tPA) solutions, flow rate increased within several minutes of treatment in the channel treated with the thrombolytic solution, indicating reperfusion of the channel.

3.4. Endothelial Cell Activation upon Reperfusion

Following our demonstration that reperfusion of an embolic occluded channel can be induced via a thrombolytic agent, we examined whether such a re-perfusion can result in EC activation and injury. As a first indication of EC injury following reperfusion, there was a significant decrease in the number of adherent ECs normalized per area, compared to the occluded channel and normally perfused channel see Figure 4a,b. This can be attributed to reperfusion injury, as the sudden flux of fluid could cause excessive trauma to the cells, which might result in their detachment. When EC activation was examined by quantitative evaluation of the E-selectin and I-CAM1 levels, normalized per cell, the results showed that although E-selectin levels were elevated to the same extent in both the re-perfused channel and the occluded channel, while I-CAM1 exhibited statistically significant higher expression in the re-perfused channel as compared to the occluded channel. These results implicate that indeed re-perfusion elevates EC activation as well as possible EC injury.

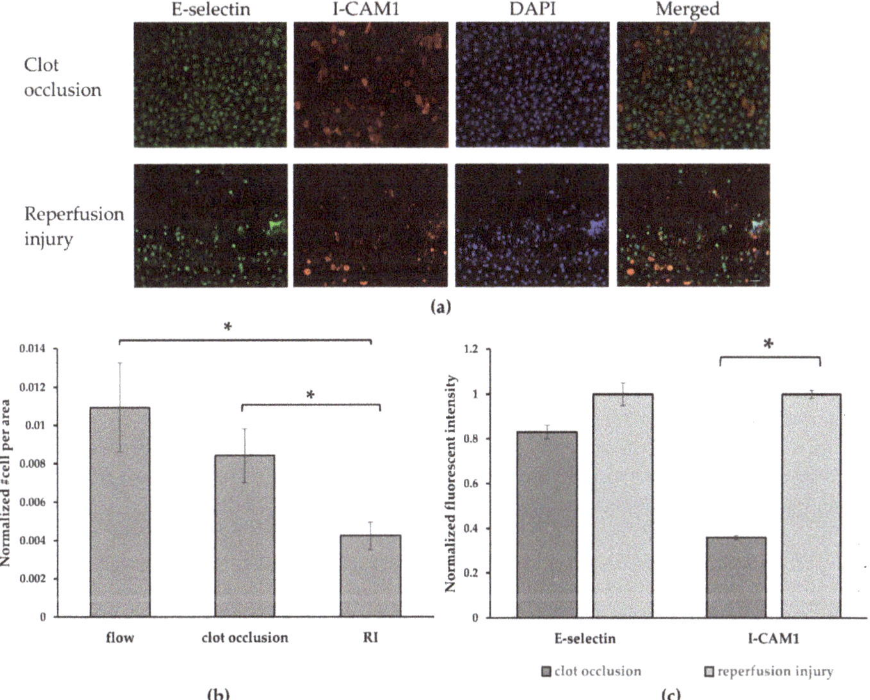

Figure 4. Reperfusion Injury and EC activation following reperfusion. (**a**) Confocal images of ECs stained for E-selectin (in green), I-CAM1 (in red) and nuclei (in blue). (i) Channel occluded with a blood clot. (ii) A re-perfused channel treated with tPA. Scale bar: 10 µm. (**b**) Normalized number of adherent cells per area. We can see significant decrease in number of adherent cells after reperfusion injury compared to normally perfused and occluded channels. (**c**) Normalized fluorescence intensity per cell of E-selectin and I-CAM1 in the occluded vs. the reperfused channels. Although not statistically significant is shown for E-selectin levels, for I-CAM the expression is significantly more pronounced. Statistical significance was determined by unpaired Student's *t*-test *: *p*-value < 0.05 ($n = 3$).

4. Discussion

In this study, we examined an alternative in vitro microfluidic model to investigate, in a controlled manner, IRI following an embolic event. The model focused on simulating the injury to the endothelium, using a vascular compartment comprising human endothelial cells, which was subjected to an embolic occlusion and then to IRI as a result of restoration of flow via thrombolytic treatment. IRI is a highly complex condition, that involves proinflammatory and various pathophysiology processes. In this study, we focused on EC receptor overexpression as an indicator of EC activation, a known condition that occurs in IRI [28,29]. We confirmed in the microfluidic model that embolic vascular occlusion (>2 h) increased EC activation as compared to the perfused channel (Figure 2), as shown by significant elevation of both E-selectin and I-CAM1 levels. These results might be attributable to hypoxia or to other changes in the cellular micro-environment as a result of the occlusion and lack of an adequate medium supply. Additionally, experiment where reperfusion was established using a thrombolytic drug confirmed that reperfusion induced further cellular damage then caused by ischemia/occlusion. Under reperfusion, we noticed significant cell detachment from the microfluidic model and significantly increased I-CAM1 levels as compared to their levels in the ischemic channel. Although, E-selectin levels were elevated compared to normal flow, they increased to the same extent as they did during the occlusion conditions. This could be attributed to the fact the E-selectin is more sensitive to culture conditions and is also activated more rapidly, while I-CAM1 overexpression is a process that occurs over a long-time range and is more stable.

Although we focused on EC activation in the current study, other known IRI effects can be examined in our embolic IR model, including reactive oxygen species levels (ROS) [30], EC permeability [31], cell adhesion [32] and oxygen levels [33,34]. Additionally, while the current model focused on EC damage, use of a more complex co-culture model [31,35] could enable assessment of organ-specific tissue damage, such as the liver, brain and heart. Furthermore, we could combine multiple microfluidic devices with different cell types to study IRI on multiple organs simultaneously. Moreover, future studies could integrate blood components, known to play an active role in IRI, such as: leukocytes and platelets, which could be perfused into the system and their role might be studied.

While our results aligned with previous animal and in vitro studies [36–38], studies using rodent as animal model for IRI also showed an increase in the levels of ICAM-1 on EC membrane [37] and in leukocyte adhesion to EC after IRI [36]. Hence, we were able to recapitulate certain feature of IRI as seen in in vivo studies. Other in vitro model of IRI also exhibited higher levels of ICAM-1 after reperfusion [38]. However, this study induced ischemia in artificial manner. In this study we induced occlusion/ischemia by introducing an occlusive blood clot into the microfluidic device, in contrast to many animal and in vitro models. For example, in animal studies clamping of the artery is a very common way to mimic ischemia and IRI is induced upon removal of the clamp [36]. On the other hand, new in vitro studies in microfluidic devices have modulated the microenvironment inside the model (pH, hypoxia chambers, etc.) to achieve ischemic conditions [34]. However, to the best of our knowledge, no microfluidic study before has directly addressed the embolic IRI scenario. Additionally, with our approach, we might take into consideration the influence of the components relevant to the thrombolysis pathway as well as the kinetics of the thrombolysis process. Such a model could be beneficial for developing new therapeutic approaches as well as for studying ways to better perfuse occluded vessels while minimizing IRI in different clinical settings.

Author Contributions: Conceptualization: D.N.A. and N.K.; Methodology: D.N.A., M.E. and N.K.; Formal analysis: D.N.A.; Investigation, D.N.A. and M.E.; Resources, N.K.; Data curation, D.N.A.; Writing—original draft preparation, D.N.A.; Writing—review and editing, N.K.; Supervision, N.K.; Funding acquisition, N.K.

Funding: This research was partly funded by Israel Ministry of Science and Technology, MOST grant # 3-14350.

Acknowledgments: The authors thanks Moran Levi, Maria Khoury and Hila Zukerman for their help with the study.

Conflicts of Interest: The authors declare no conflict of interest.

References

1. Federation, H. *Global Atlas on Cardiovascular Disease Prevention and Control*; World Health Organization: Geneva, Switzerland, 2011.

2. Maxwell, S.R.; Lip, G.Y. Reperfusion injury: A review of the pathophysiology, clinical manifestations and therapeutic options. *Int. J. Cardiol.* **1997**, *58*, 95–117. [CrossRef]

3. Hausenloy, D.J.; Yellon, D.M.; Hausenloy, D.J.; Yellon, D.M. Myocardial ischemia-reperfusion injury: A neglected therapeutic target Find the latest version: Review series Myocardial ischemia-reperfusion injury: A neglected therapeutic target. *J. Clin. Investig.* **2013**, *123*, 92–100. [CrossRef] [PubMed]

4. Eltzschig, H.K.; Eckle, T. review Ischemia and reperfusion—From mechanism to translation. *Nat. Med.* **2011**, *11*, 1391–1401. [CrossRef] [PubMed]

5. Frank, A.; Bonney, M.; Bonney, S.; Weitzel, L.; Koeppen, M.; Eckle, T. Myocardial ischemia reperfusion injury: From basic science to clinical bedside. *Semin. Cardiothorac. Vasc. Anesth.* **2012**, *16*, 123–132. [CrossRef]

6. Khan, A.; Waqar, K.; Shafique, A.; Irfan, R.; Gul, A. *Vitro and In Vivo Animal Models: The Engineering Towards Understanding Human Diseases and Therapeutic Interventions*; Elsevier Inc.: Amsterdam, The Netherlands, 2018.

7. Traystman, R.J. Animal Models of Focal and Global Cerebral Ischemia. *ILAR J.* **2003**, *44*, 85–95. [CrossRef]

8. Black, S.C. In vivo models of myocardial ischemia and reperfusion injury. *J. Pharmacol. Toxicol. Methods* **2000**, *43*, 153–167. [CrossRef]

9. Cowled, P.A.; Khanna, A.; Laws, P.E.; Field, J.B.; Fitridge, R.A. Simvastatin Plus Nitric Oxide Synthase Inhibition Modulates Remote Organ Damage Following Skeletal Muscle Ischemia-Reperfusion Injury. *J. Investig. Surg.* **2008**, *21*, 119–126. [CrossRef]

10. Stalenhoef, A.F.H. The benefit of statins in non-cardiac vascular surgery patients. *YMVA* **2009**, *49*, 260–265. [CrossRef]

11. Benam, K.H.; Dauth, S.; Hassell, B.; Herland, A.; Jain, A.; Jang, K.J.; Karalis, K.; Kim, H.J.; MacQueen, L.; Mahmoodian, R.; et al. Engineered In Vitro Disease Models. *Annu. Rev. Pathol.* **2015**, *10*, 195–262. [CrossRef]

12. Bhatia, S.N.; Ingber, D.E. Microfluidic organs-on-chips. *Nat. Biotechnol.* **2014**, *32*, 760–772. [CrossRef]

13. Abaci, H.E.; Shen, Y.; Tan, S.; Gerecht, S. Recapitulating physiological and pathological shear stree and oxygen to model vasculature in health and disease. *Sci. Rep.* **2014**, *4*, 1–9.

14. Khanal, G.; Chung, K.; Solis-Wever, X.; Johnson, B.; Pappas, D. Ischemia/reperfusion injury of primary porcine cardiomyocytes in a low-shear microfluidic culture and analysis device. *Analyst* **2011**, *136*, 3519–3526. [CrossRef] [PubMed]

15. Burne-Taney, M.J.; Rabb, H. The role of adhesion molecules and T cells in ischemic renal injury. *Curr. Opin. Nephrol. Hypertens.* **2003**, *12*, 85–90. [CrossRef] [PubMed]

16. Weisel, J.W.; Litvinov, R.I. *Fibrin Formation, Structure and Properties*; Springer: Berlin/Heidelberg, Germany, 2017.

17. Silva, C.F.; Weaver, J.P.; Gounis, M.J. Mechanical Characterization of Thromboemboli in Acute Ischemic Stroke and Laboratory Embolus Analogs. *Am. J. Neuroradiol.* **2011**, *32*, 1237–1244.

18. Shamay, Y.; Paulin, D.; Ashkenasy, G.; David, A. E-selectin binding peptide-polymer-drug conjugates and their selective cytotoxicity against vascular endothelial cells. *Biomaterials* **2009**, *30*, 6460–6468. [CrossRef] [PubMed]

19. Collard, C.D.; Gelman, S. Prevention of Ischemia—Reperfusion Injury. *Anesthesiology* **2001**, *94*, 1133–1138.

20. Loukogeorgakis, S.P.; Panagiotido, A.T.; Yellon, D.M.; Deanfield, J.E.; MacAllister, R.J. Postconditioning Protects Against Endothelial Ischemia-Reperfusion Injury in the Human Forearm. *Circulation* **2006**, *113*, 1015–1019. [CrossRef]

21. Tavares, J.C.; Nicola, M. Adhesion Molecules and Endothelium. In *Endothelium and Cardiovascular Diseases*; Academic Press: Cambridge, MA, USA, 2018; pp. 189–201.

22. Parekh, R.B.; Edge, C.J. Selectins-glycoprotein targets for therapeutic intervention in inflammation. *Trends Biotechnol.* **1994**, *12*, 339–345. [CrossRef]

23. Lawson, C.; Wolf, S. ICAM-1 signaling in endothelial cells. *Pharmacol. Rep.* **2009**, *61*, 22–32. [CrossRef]

24. Sethi, G.; Sung, B.; Aggarwal, B.B. TNF: A master switch for inflammation to cancer. *Front. Biosci.* **2008**, *13*, 5094–5107. [CrossRef]

25. Cesarman-Maus, G.; Hajjar, K.A. Molecular mechanisms of fibrinolysis. *Br. J. Haematol.* **2005**, *129*, 307–321. [CrossRef] [PubMed]

26. Adeoye, O.; Hornung, R.; Khatri, P.; Kleindorfer, D. Recombinant Tissue-Type Plasminogen Activator Use for Ischemic Stroke in the United States A Doubling of Treatment Rates Over the Course of 5 Years. *Stroke* **2011**, *42*, 1952–1955. [CrossRef] [PubMed]

27. Piebalgs, A.; Gu, B.; Roi, D.; Lobotesis, K.; Thom, S.; Xu, X.Y. Computational Simulations of Thrombolytic Therapy in Acute Ischaemic Stroke. *Sci. Rep.* **2018**, *8*, 1–13. [CrossRef] [PubMed]

28. Trepels, T.; Zeiher, A.M.; Fichtlscherer, S. The Endothelium and Inflammation. *Endothelium* **2006**, *13*, 423–429. [CrossRef]

29. Neri, M.; Riezzo, I.; Pascale, N.; Pomara, C.; Turillazzi, E. Ischemia/Reperfusion Injury following Acute Myocardial Infarction: A Critical Issue for Clinicians and Forensic Pathologists. *Mediat. Inflamm.* **2017**, *2017*, 14. [CrossRef]
30. Chin, L.K.; Yu, J.Q.; Fu, Y.; Yu, T.; Liu, A.Q.; Luo, K.Q. Production of reactive oxygen species in endothelial cells under different pulsatile shear stresses and glucose concentrations. *Lab Chip* **2011**, *11*, 1856–1863. [CrossRef]
31. Frost, T.S.; Jiang, L.; Lynch, R.M.; Zohar, Y. Permeability of Epithelial/Endothelial Barriers in Transwells and Microfluidic Bilayer Devices. *Micromachines* **2019**, *10*, 533. [CrossRef]
32. Ferrero, M.E.; Bertelli, A.E.; Fulgenzi, A.; Pellegatta, F.; Corsi, M.M.; Bonfrate, M.; Ferrara, F.; De Caterina, R.; Giovannini, L.; Bertelli, A. Activity in vitro of resveratrol on granulocyte and monocyte adhesion to endothelium. *Am. J. Clin. Nutr.* **1998**, *68*, 1208–1214. [CrossRef]
33. Zhu, F.; Baker, D.; Skommer, J.; Sewell, M.; Wlodkowic, D. Real-time 2D visualization of metabolic activities in zebrafish embryos using a microfluidic technology. *Cytom. Part A.* **2015**, *87*, 446–450. [CrossRef]
34. Abaci, H.E.; Devendra, R.; Soman, R.; Drazer, G.; Gerecht, S. Microbioreactors to manipulate oxygen tension and shear stress in the microenvironment of vascular stem and progenitor cells. *Biotechnol. Appl. Biochem.* **2012**, *59*, 97–105. [CrossRef]
35. Yeon, J.H.; Park, J. Microfluidic Cell Culture Systems for Cellular Analysis Microfluidic Cell Culture Systems for Cellular Analysis. *BioChip J.* **2007**, *1*, 17–27.
36. Thiele, J.R.; Goerendt, K.; Stark, G.B.; Eisenhardt, S.U. Real-time digital imaging of leukocyte-endothelial interaction in ischemia-reperfusion injury (IRI) of the rat cremaster muscle. *J. Vis. Exp.* **2012**, *66*, e3973. [CrossRef] [PubMed]
37. Farhood, A.; McGuire, G.M.; Manning, A.M.; Miyasaka, M.; Smith, C.W.; Jaeschke, H. Intercellular adhesion molecule 1 (ICAM-1) expression and its role in neutrophil-induced ischemia-reperfusion injury in rat liver. *J. Leukoc. Biol.* **1995**, *57*, 368–374. [CrossRef] [PubMed]
38. Lee, W.H.; Kang, S.; Vlachos, P.P.; Lee, Y.W. A novel in vitro ischemia/reperfusion injury model. *Arch. Pharm. Res.* **2009**, *32*, 421–429. [CrossRef]

 *micromachines*

Article

Tetrafluoroethylene-Propylene Elastomer for Fabrication of Microfluidic Organs-on-Chips Resistant to Drug Absorption

Emi Sano [1,†], Chihiro Mori [1,†], Naoki Matsuoka [2], Yuka Ozaki [1], Keisuke Yagi [2], Aya Wada [2], Koichi Tashima [2], Shinsuke Yamasaki [2], Kana Tanabe [2], Kayo Yano [1] and Yu-suke Torisawa [1,3,*]

[1] Department of Micro Engineering, Kyoto University, Kyoto 615-8540, Japan; e.sano1017@gmail.com (E.S.); mori.chihiro.5u@kyoto-u.ac.jp (C.M.); ozaki.yuka.6z@kyoto-u.ac.jp (Y.O.); yano.kayo.7z@kyoto-u.ac.jp (K.Y.)
[2] AGC Inc, Tokyo 100-8405, Japan; naoki-matsuoka@agc.com (N.M.); keisuke-yagi@agc.com (K.Y.); aya-wada@agc.com (A.W.); kouichi-tashima@agc.com (K.T.); shinsuke.yamasaki@agc.com (S.Y.); kana.tanabe@agc.com (K.T.)
[3] Hakubi Center for Advanced Research, Kyoto University, Kyoto 615-8540, Japan
* Correspondence: torisawa.yusuke.6z@kyoto-u.ac.jp; Tel.: +81-75-383-3701
† These authors contributed equally to this work.

Received: 26 August 2019; Accepted: 15 November 2019; Published: 19 November 2019

Abstract: Organs-on-chips are microfluidic devices typically fabricated from polydimethylsiloxane (PDMS). Since PDMS has many attractive properties including high optical clarity and compliance, PDMS is very useful for cell culture applications; however, PDMS possesses a significant drawback in that small hydrophobic molecules are strongly absorbed. This drawback hinders widespread use of PDMS-based devices for drug discovery and development. Here, we describe a microfluidic cell culture system made of a tetrafluoroethylene-propylene (FEPM) elastomer. We demonstrated that FEPM does not absorb small hydrophobic compounds including rhodamine B and three types of drugs, nifedipine, coumarin, and Bay K8644, whereas PDMS absorbs them strongly. The device consists of two FEPM layers of microchannels separated by a thin collagen vitrigel membrane. Since FEPM is flexible and biocompatible, this microfluidic device can be used to culture cells while applying mechanical strain. When human umbilical vein endothelial cells (HUVECs) were subjected to cyclic strain (~10%) for 4 h in this device, HUVECs reoriented and aligned perpendicularly in response to the cyclic stretch. Moreover, we demonstrated that this device can be used to replicate the epithelial–endothelial interface as well as to provide physiological mechanical strain and fluid flow. This method offers a robust platform to produce organs-on-chips for drug discovery and development.

Keywords: organs-on-chips; microfluidics; drug absorption; fluoroelastomer

1. Introduction

Predicting drug efficacy and toxicity before clinical trials is crucial for the drug discovery and development processes [1–3]. In vitro systems that can reliably predict responses to drugs in humans could be a powerful platform to test drugs and to discover new therapeutics. The organs-on-chips technology, specifically, has proven useful for studying physiological mechanisms and pharmacological modulation as well as for developing disease models [4–8]. By mimicking natural tissue architecture and microenvironmental chemical and physical cues within microfluidic devices, organs-on-chips enable the reconstitution of complex organ-level functionality that cannot be recapitulated with conventional culture systems. Since numerous of the physiological microenvironments inside the human body are microfluidic in nature (e.g., the pulmonary system, liver sinusoids, and vascular networks), the use of microfluidic devices facilitates engineering of cellular microenvironments [9–12].

Microfluidic devices allow for control of local chemical gradients and dynamic mechanical forces which govern the development and function of organs [13,14]. Microfluidic organs-on-chips demonstrated that mechanical forces are important to drive cellular differentiation and function and to faithfully recapitulate human organ-level physiology and pathophysiology; e.g., in lung [15–17], gut [18,19], kidney [20–22], and cancer models [23]. Thus, these microfluidic devices mimicking the mechanical microenvironment of living tissues have great potential to predict human responses to drugs and serve as an alternative to animal models.

Polydimethylsiloxane (PDMS) is commonly used to fabricate microfluidic devices because it is easy to use, biocompatible, highly gas permeable, optically clear, and flexible [24,25]. Using soft lithography, various types of culture devices have been developed and widely used as tools to study basic and applied scientific research. Although PDMS devices are very useful, one serious drawback is that small hydrophobic molecules are strongly absorbed into PDMS surrounding microchannels [25–31]. This limitation is critical for the development of in vitro systems to test drugs because many pharmaceutical compounds are small hydrophobic molecules. It has been demonstrated that the use of PDMS-based devices significantly reduces concentrations and the effectiveness of drugs including anticancer drugs due to absorption into PDMS microchannels [29,30]. Although several types of elastomers and coating methods have been developed to reduce absorption of small hydrophobic molecules [32–35], they have not been widely adopted and there is still no elastic device which resists absorption of small hydrophobic drugs. To overcome this challenge, we explored the possibility of using a fluoroelastomer which is known to have excellent chemical resistance as an alternative material [36–38]. We developed a compartmentalized microfluidic device [39] consisting of two layers of microchannels made from a tetrafluoroethylene-propylene (FEPM) elastomer and a thin collagen vitrigel membrane. Furthermore, we demonstrated that drug absorption by the FEPM elastomer is comparable with that by standard cell culture plates made from polystyrene. This FEPM-based microfluidic device can be used to culture cells while applying physiologically relevant mechanical signals.

2. Materials and Methods

2.1. Device Fabrication

The microfluidic devices consist of two layers of microchannels separated by a thin collagen vitrigel membrane (Figure 1). The microchannel layers were fabricated from tetrafluoroethylene-propylene (FEPM) compounds (AFLAS, AGC Inc., Tokyo, Japan) at a ratio of 10:4 base to vulcanizing agents at 60 °C on a two-roll mill ($\varphi8''\times18''$ test roll machine for chemical machine design and production, Yamatetsu Machinery Inc., Tokyo, Japan). Each layer was formed by compression molding with a custom-designed hard-chrome-plated two-layer mold on a 200-ton vacuum compression molding machine (TYC-V-2RT, Tung Yu Hydraulic Machinery Co., Ltd., Nantou, Taiwan), press-curing at 160 °C for 30 min, and post-curing at 200 °C for 2 h in an oven (DH612, Yamato Scientific Co., Ltd., Tokyo, Japan) [37,38]. The cross-sectional size of microchannels is 1 mm in width × 1 mm in height and the gaps between each microchannel are 1 mm. A 10-µm-thick collagen vitrigel membrane (VIT-C001, AGC Techno Glass Co., Ltd., Shizuoka, Japan) was placed between two channel layers. The FEPM layers and the collagen vitrigel membrane were bonded together for 5 s at room temperature by a self-adhesion system that was generated during the formation of each FEPM layer, allowing the channel layers to be assembled without the use of glue. The bonding strength of the FEPM device containing the membrane was measured using a digital spring scale (PS-01, Dr. meter, London, England) and a vertical drill guide (DS-70, SK11, Hyogo, Japan) by pulling the FEPM layer vertically until it detached (n = 5). The dimensions of the FEPM layer were 20 mm (width) × 40 mm (length) × 2 mm (thickness) (Figure S1). The FEPM microfluidic devices were sterilized by placing them under UV light for 2 h prior to cell culture.

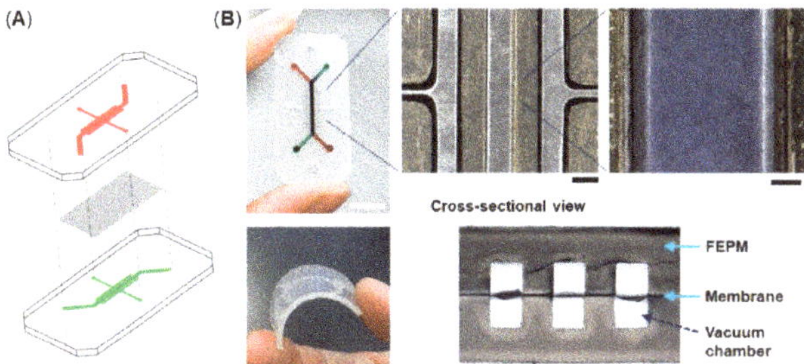

Figure 1. Compartmentalized microfluidic device fabricated from tetrafluoroethylene-propylene (FEPM) elastomer. (**A**) Schematic illustration of the microfluidic device. Two FEPM channel layers are separated by a collagen vitrigel membrane. (**B**) Photographic images of the device and the FEPM channel layer (bottom left). A cross-sectional view shows the top and bottom channels (1 mm in width and 1 mm in height) with the collagen vitrigel membrane (10 μm thick) in-between (bottom right). The gaps between each channel are 1 mm. Cells are cultured in the central channel and mechanical strain is exerted by applying vacuum to all the side vacuum chambers. Scale bar, 10, 1, and 0.2 mm for low, medium, and high magnification views, respectively.

To analyze the absorption of PDMS, slabs and microfluidic devices were fabricated from polydimetyisiloxane (Sylgard 184, Dow Corning, Midland, TX, USA) at a ratio of 10:1 base to curing agents using standard soft lithographic techniques [40]. The PDMS microfluidic device consists of a single microchannel similar to the FEPM device without side chambers. The microchannel layers were produced by casting PDMS prepolymer against the mold composted of SU-8 2150 (MicroChem, Westborough, MA, USA) patterns formed on a silicon wafer. PDMS was cured at 60 °C overnight. The channel layers were formed against the relief channel feature 300 μm in height and 1 mm in width. Two channel layers were bonded using oxygen plasma treatment (Covance MP, Femto Science, Hwaseong, Korea) at 500 mTorr pressure and 40 W power for 30 s followed by curing at 60 °C for 2 h.

2.2. Analysis of Absorption

To analyze absorption of a fluorescent dye, the inlet and outlet of the microfluidic devices were connected with silicon tubes (i.d. 1 mm, #1018-03, ARAM, Osaka, Japan) and a solution of 10 μM rhodamine B (Nacalai Tesque, Kyoto, Japan) in phosphate-buffered saline (PBS, Nacalai Tesque, Kyoto, Japan) was flowed through the microchannels in the devices using a syringe pump (NE-1000, New Era Pump Systems Inc, Farmingdale, NY, USA) at a flow rate of 1 μL·min^{-1} for up to 24 h. After washing with 10 mL of PBS, fluorescence images were obtained using an inverted microscope (IX-83, Olympus, Tokyo, Japan) with the 4× objective and a microscope digital camera (DP80, Olympus) and were analyzed with cellSens software (Olympus). We used the same exposure time (30 ms) to obtain fluorescent images. The fluorescent light source was a 100 W mercury lamp.

To evaluate drug absorption, slabs (2 mm thick × 6 mm in diameter) made from PDMS and FEPM were placed in polystyrene (PS) 96-well culture plates (AGC Techno Glass). Drug concentrations were determined by high-performance liquid chromatography (HPLC, Shimadzu Corporation, Kyoto, Japan) using a solvent delivery system equipped with an auto sampler. Three types of drugs, nifedipine (N0528, Tokyo Chemical Industry, Tokyo, Japan), Bay K8644 (B112 Sigma-Aldrich, St. Louis, MO, USA), and coumarin (031-16562, Fujifilm Wako Chemicals, Osaka, Japan), were used to test drug absorption. 1 μM of each drug in PBS was incubated for 0.5, 1, 2, 3, 6, and 24 h in PS 96-well plates with or without PDMS or FEPM slabs at 37 °C. An ultraviolet–visible (UV–Vis) absorbance detector (SPD-M20A, Shimadzu Corporation) was used to monitor the UV absorption maximum of the compounds under investigation. A ZORBAX 300SB-C18 column (4.6 × 150 mm, Agilent technologies, Santa Clara, CA, USA) was applied with a stationary phase of C18 (5 μm particles). As mobile phase, acetonitrile was used at 40–70% (depending on the compound) and 0.1% trifluoroacetic acid. Flow speed was

1.0 mL·min^{-1} and the injection volume was 100 µL. Recording and processing of data was performed using LabSolutions software (Shimadzu Corporation). In every experiment, a calibration curve was included to validate the system and the linearity of the compound UV absorption response.

2.3. Cell Culture

Green fluorescent protein expressing human umbilical vein endothelial cells (GFP-HUVECs, #cAP-0001GFP, Angio-Proteomie, Boston, MA, USA) and human umbilical vein endothelial cells (HUVECs, #C2517A, Lonza, Basel, Switzerland) were cultured in endothelial cell growth medium (EGM-2, Lonza) with 100 U·ml^{-1} penicillin and 100 U·ml^{-1} streptomycin (P/S, Nalacai Tesque, Kyoto, Japan) and cells between passage 5 and 7 were used for experiments. Caco-2 human intestinal epithelial cells (C2BBe1 clone of Caco-2 human colorectal adenocarcinoma cell line, #CRL-2102, ATCC, Manassas, VA, USA) were cultured in Dulbecco's Modified Eagle's Medium (DMEM 08458-45, Nacalai Tesque, Kyoto, Japan) containing 10% fetal bovine serum (FBS F7524, Sigma-Aldrich, St. Louis, MO, USA), MEM non-essential amino acid (Gibco, Dun Laoghaire, Ireland), 100 U·ml^{-1} penicillin, and 100 U·ml^{-1} streptomycin. Cells were cultured in an incubator at 37 °C and 5% CO_2.

2.4. Microfluidic Cell Culture

To culture HUVECs, the collagen vitrigel membranes in the FEPM devices were coated with fibronectin (100 µg·mL^{-1}) for 1 h prior to cell seeding. HUVECs (2.5×10^4 cells per 20 µL) were introduced into the top channel and incubated at 37 °C for 60 min to allow the cells to adhere to the membrane. The attached cells were then perfused with culture medium using a micro peristaltic pump (RP-TXP5F, Aquatech Co., Ltd., Osaka, Japan) at a flow rate of 10 µL·min^{-1} (0.008 dyne·cm^{-2}) [41]. To examine the effect of cyclic strain, HUVECs cultured in the devices were stretched with 5–10% strain at a frequency of 1 Hz (sinusoidal waveform) for 4 h by applying vacuum (30–40 kPa) to the side vacuum chambers of the devices using a vacuum controller (FX-500, Flexcell International Corporation, Burlington, VT, USA). Tubing (i.d. 1/16 inch, Tygon LMT-55, Saint-Gobain, La Defense, France) was connected from the vacuum source to the side chambers using needles (16G, Musashi Engineering, Inc., Osaka, Japan). The mechanical strain was evaluated by measuring the displacement between two points before and after stretching [15].

To form the epithelial-endothelial interface, the collagen vitrigel membranes in the FEPM devices were coated with Matrigel (400 µg·mL^{-1}, BD Biosciences, Franklin Lakes, NJ, USA) and rat type I collagen (50 µg·mL^{-1}, Gibco) for 2 h prior to cell seeding. Caco-2 cells (2.5×10^4 cells per 20 µL) were introduced into the top channel and incubated at 37 °C for 60 min to allow the cells to adhere to the membrane. The attached cells were then perfused with culture medium using a micro peristaltic pump at a flow rate of 10 µL·min^{-1}. After 4 days in culture to form a confluent monolayer of Caco-2 cells, the bottom side of the collagen vitrigel membrane was coated with fibronectin (100 µg·mL^{-1}) for 30 min by injecting the solution into the bottom channel. After washing with culture medium, HUVECs (2.5×10^4 cells per 20 µL) were introduced into the bottom channel and then the device was inverted and incubated at 37 °C for 30 min to make HUVECs adhere onto the bottom side of the membrane. The attached cells were then perfused at 10 µL·min^{-1} with DMEM and EGM-2 in the top and bottom channels, respectively. The devices were cultured for 3 more days to form a confluent monolayer of HUVECs.

2.5. Analysis of Cellular Viability

Cellular viability was assessed by fluorescence microscopy imaging. After 7 days in culture, the channels of the FEPM device were washed with PBS and then the upper channel was filled with a mixture solution containing Calcein-AM (2 µM in PBS) and ethidium homodimer-3 (4 µM in PBS) and incubated for 30 min according to standard protocol (Live/Dead Cell Staining Kit 2, PromoCell, Heidelberg, Germany). To test drug effect, cells were treated with 10 µg·mL^{-1} mitomycin C (Nacalai Tesque) for 24 h. HUVECs were cultured in the FEPM devices and in 96-well culture plates (#353072, Falcon, Corning, NY, USA) for 24 h prior to drug treatment. After 24 h treatment of mitomycin C, the devices and plates were washed with PBS and then were filled with a mixture solution containing calcein-AM and ethidium homodimer-3. Cellular viability was quantified by the percentage of calcein-AM-labeled cells (green) averaged over 3 different observation areas of the device from 3 independent experiments. All cellular images were taken using a microscope digital camera (DP80, Olympus) mounted on an inverted microscope (IX-83, Olympus) with the 4× and 10× objectives.

2.6. Analysis of Cellular Alignment

Morphological responses of cells to cyclic strain were quantified using angle of orientation [15,42]. The orientation angle is defined as the angle between the major axis of the best-fit ellipse around a cell and the axis of cyclic strain. The images were obtained by phase-contrast microscopy and analyzed with cellSens software (Olympus) to measure the angle of orientation. The data was analyzed statistically by the F-test at a 99% confidence level.

2.7. Immunofluorescence

Cells cultured in the FEPM devices were fixed with 4% paraformaldehyde (PFA, Nacalai Tesque, Kyoto, Japan) for 20 min, washed with PBS, and permeabilized with 0.1% Triton X-100 (Sigma-Aldrich, St. Louis, USA) for 15 min. The cells were then incubated with blocking buffer containing 1% bovine serum albumin (BSA, Sigma-Aldrich) for 30 min at room temperature and incubated with antibodies directed against vascular endothelial (VE)-cadherin (Abcam, ab33168, dilution 1:200, Cambridge, UK) or Claudin (Abcam, ab15098, dilution 1:200) overnight at 4 °C, followed by PBS washes. Subsequently, Alexa 568-conjugated secondary antibody (Abcam, ab175471, dilution 1:500) was introduced into the channels and incubated for 1 h at room temperature. The cells were co-stained with 4′,6-diamidino-2-phenylindole (DAPI, Invitrogen, D1306, Waltham, USA). Cross-sectional images were obtained at 2 μm intervals in the vertical direction using a confocal microscopy (FluoView FV1000 confocal, Olympus, Tokyo, Japan) with the 10× objective.

2.8. Statistics

Results are reported as mean ± standard deviation. The F-test was performed to analyze variance for two samples. The Levene test was performed for three samples to confirm that the variances are homogeneous. We confirmed that the data were normally distributed. The statistical significance of variance across groups was assessed by one-way analysis of variance (ANOVA) with a post-hoc analysis using the Bonferroni test. Significance level of $P < 0.01$ is denoted in graphs by an asterisk (*). Representative results from at least three independent biological replicates are shown.

3. Results

3.1. Characterization of Tetrafluoroethylene-Propylene (FEPM) Elastomer Devices

To explore whether our microfluidic devices fabricated from FEPM could prevent drug absorption, we compared absorption of a fluorescent dye into microchannels made either of FEPM or PDMS. A solution containing rhodamine B, which is a small hydrophobic fluorophore, was perfused into the microchannels and fluorescence intensity of the microchannels was monitored for 24 h (Figure 2). When the rhodamine B solution was placed in the FEPM and PDMS channels, fluorescence intensity of the FEPM channel was almost same as that of the PDMS channel (Figure S2), demonstrating that the FEPM devices can be used to evaluate this fluorophore. The fluorescence intensity of the PDMS microchannel gradually increased over time and the bright area also gradually spread out, indicating that the PDMS microchannel continuously soaked up the fluorophore. On the other hand, the fluorescence intensity of the FEPM microchannel almost did not change for 24 h and only weak fluorescence signals were observed due to adsorption of the fluorescent dye on the channel wall, demonstrating that the FEPM microchannel is resistant to absorption of this fluorophore. The cross-sectional images of the microchannels clearly showed that the fluorescent compound was absorbed inside the walls of the PDMS microchannel, whereas it was only observed on the channel surface and did not absorb into the FEPM microchannel. These results demonstrate that the FEPM devices are resistant to absorption of rhodamine B.

We then analyzed absorption of drugs quantitatively by measuring drug concentrations using HPLC. Three types of drugs, nifedipine, Bay K8644, and coumarin, were used to evaluate absorption by FEPM and analyzed by comparing with that by PDMS and polystyrene (PS) 96-well plates (Figure 3). These three drugs are small hydrophobic molecules known to absorb into PDMS [27,31]. The concentrations of these three drugs after incubation with PDMS decreased over time and reached almost half of the initial concentrations after 24 h of incubation. Importantly, the concentrations of all three drugs retrieved from wells containing PDMS slabs were significantly lower than those retrieved from standard PS wells after 24 h of incubation. On the other hand, the concentrations of all three drugs after incubation with FEPM did not change for 24 h. There was no significant difference in the residual concentrations of three drugs between the FEPM and PS well plates. The concentrations of Bay K8644 retrieved from wells containing FEPM slabs were even slightly higher than that in PS well plates over 6 h of incubation. These data clearly demonstrate that the FEPM elastomer is resistant to absorption of small hydrophobic drugs and is comparable with the standard cell-culture plates.

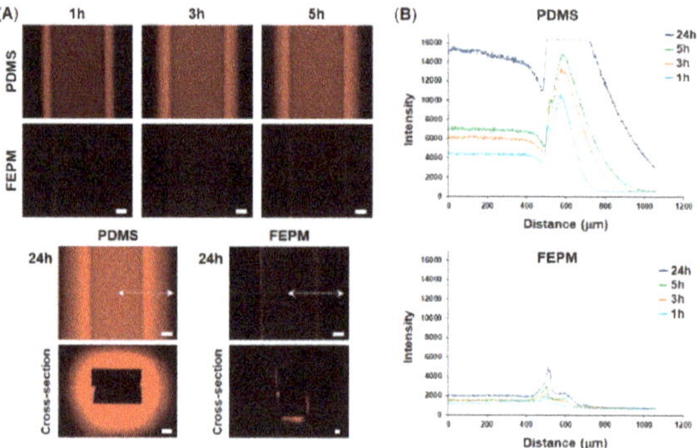

Figure 2. Absorption of rhodamine B into polydimethylsiloxane (PDMS) and FEPM. (**A**) Fluorescent images of rhodamine B absorbed into the PDMS and FEPM channels after 1, 3, 5, 24-h perfusion of rhodamine B solution. Bottom, cross-sectional images of the PDMS and FEPM channels after 24 h perfusion of rhodamine B solution. Scale bars, 200 µm. (**B**) Line profiles of the absorption of rhodamine B inside the walls of the PDMS and FEPM channels. The X-axis shows distance from the center of the channel (indicated by white dotted lines). (n = 3).

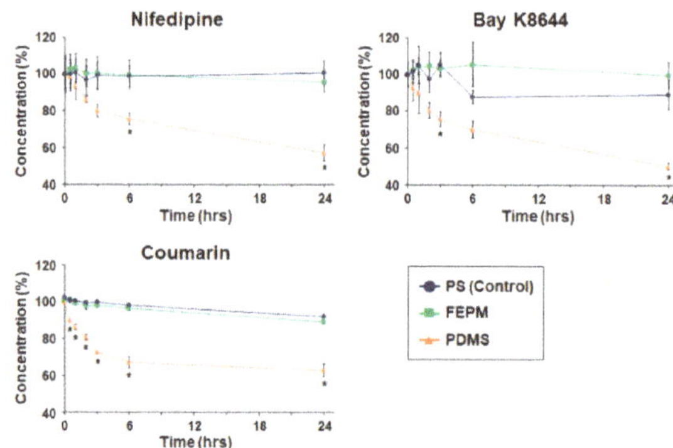

Figure 3. Time-course analysis of drug absorption into PDMS, FEPM, and polystyrene (PS) culture plates (control). Drug concentrations after 0.5, 1, 2, 3, 6, and 24-h incubation of nifedipine, Bay K8644, and coumarin evaluated by high-performance liquid chromatography (HPLC, n = 3, * $p < 0.01$).

3.2. Cell Culture within FEPM Devices

We utilized the FEPM elastomer to construct compartmentalized microfluidic devices as a cell culture system [39]. The device consists of two channel layers separated by a thin collagen vitrigel membrane. We first evaluated bond strength between the FEPM layers with the collagen vitrigel membrane in-between and obtained the minimum bond strength of 105 kPa (average value is 117 ± 13 kPa, n = 5). This bond strength is sufficient to maintain solutions inside channels without leakage during cyclic stretching although the value is lower than that of PDMS layers using plasma treatment (ca. 200 kPa) [43].

To examine cell culture in this device, we used human umbilical vascular endothelial cells (HUVECs) which are commonly used as a representative endothelial cell population. HUVECs were seeded into the top channel of the device and were cultured on the collagen vitrigel membrane under continuous medium perfusion for 7 days (Figure 4). HUVECs formed a monolayer cell sheet after 4 days in culture on-chip and maintained the GFP expression for 7 days. Importantly, almost all the cells were maintained viable for 7 days (viability = 98%, n = 3), indicating that the FEPM-based microfluidic devices can be used to culture cells and monitor cellular responses. We confirmed that HUVECs cultured in the devices are similar to those cultured in cell-culture plates (Figure S3). The cells were able to be cultured for at least 7 days within the microchannels without any leakage, suggesting that the bond strength of the FEPM layers is sufficient to culture cells under fluidic flow for 7 days. We then used this FEPM-based microfluidic device to produce cyclic stretching to mimic physiological movement inside the body. The level of applied mechanical strain ranged from 5% to 10% to match physiological levels of strain [44]. Membrane stretching causes distortion of cell shape as observed by increases in the length of adherent cells in the direction of applied tension (Figure 4C and Video S1). Microscopic analysis of cell shape confirmed that about 8% strain was exerted on the cells by applying suction to the vacuum chambers in the device. Thus, this device can be used to culture cells while applying physiological mechanical strain.

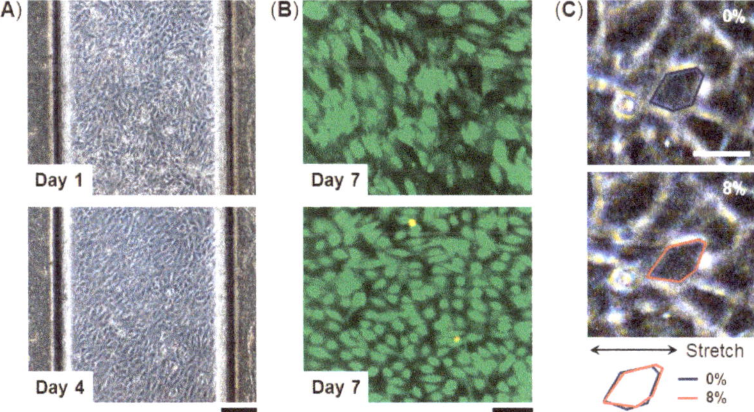

Figure 4. Microfluidic cell culture within FEPM devices. (**A**) Phase contrast micrographs of human umbilical vein endothelial cells (HUVECs) cultured within the device for 1 and 4 days. Scale bar, 200 μm. (**B**) Fluorescent images of green fluorescent protein (GFP)-expressing HUVECs cultured in the device for 7 days. Images were taken under different conditions. Top, green florescence shows GFP of the cells in the device. Bottom, green and red florescence represents live and dead cells, respectively. Scale bar, 100 μm. (n = 3) (**C**) Phase contrast images of HUVECs cultured within the FEPM device in the absence (top) and presence (bottom) of mechanical strain (8%) exerted by applying vacuum to the side chambers. Blue and red outlines indicate the shape of a single cell before (blue) and after (red) mechanical strain application. Scale bar, 50 μm.

We evaluated cellular responses induced by mechanical strain. HUVECs cultured in the FEPM devices were exposed to physiological cyclic strain (5–10%, 1Hz) for 4 h (Figure 5). Application of physiological cyclic strain induced the endothelial cells to reorient and align perpendicularly. Immunofluorescence analysis demonstrated that HUVECs elongated and aligned in the direction perpendicular to the applied strain (Figure 5B). Quantitative analysis showed a uniform distribution of the orientation angle in the absence of cyclic strain (Figure 5C). When the cells were exposed to cyclic strain for 4 h, the distribution of the orientation angle was biased toward 90°, as compared to control without strain (F-test, $p < 0.01$). Over 80% of the cells exhibited alignment with orientation angles between 60° and 120°. These results demonstrate that this device can be used to create physiological levels of mechanical strain and to monitor cellular responses to mechanical signals.

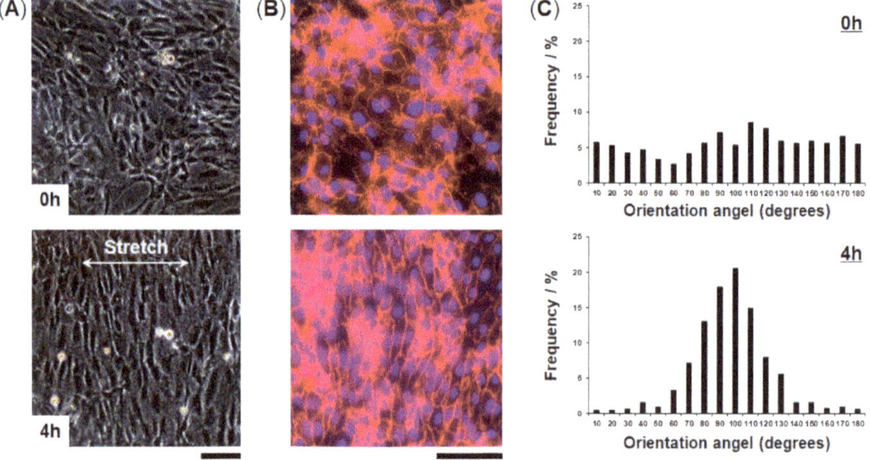

Figure 5. Alignment of vascular endothelial cells induced by application of cyclic strain. (**A**) Phase contrast micrographs of HUVECs cultured in the FEPM device before (top) and after (bottom) 4-h cyclic stretch with ~10% strain at a frequency of 1 Hz. (**B**) Fluorescent images of HUVECs cultured in the device in the absence (top) and presence (bottom) of cyclic strain. HUVECs were stained for vascular endothelial (VE)-cadherin (red) and 4′,6-diamidino-2-phenylindole (DAPI) (blue). Scale bars, 100 µm. (**C**) Histograms of cell orientation before (top) and after (bottom) 4 h of cyclic stretching. (n = 3).

3.3. Engineering of FEPM-Based Organs-on-Chips

To model complex physiological functions of epithelia, organs-on-chips typically recreate tissue-tissue interfaces separated by thin porous membranes coated with extracellular matrix. To explore whether our microfluidic device can be used to replicate tissue-tissue interfaces, we carried out microfluidic culture to form the interface between intestinal epithelial cells and vascular endothelial cells using HUVECs and Caco-2 cells which are commonly used as an intestinal epithelial model [18].

First we seeded Caco-2 cells into the top channel and cultured on the top side of the collagen vitrigel membrane within the FEPM device under continuous medium perfusion. After 4 days in culture to form a monolayer of epithelial cells, HUVECs were seeded into the bottom channel and cultured on the bottom side of the membrane for 3 more days. Immunofluorescence analysis confirmed that confluent cell layers formed on the both sides of the membrane; Caco-2 cells formed a confluent epithelium layer with well-developed tight junctions and HUVECs formed a confluent endothelium (Figure 6). Because the collagen vitrigel membrane swells when exposed to liquid, a curvature of the membrane was observed; however, confluent cell layers were able to form on the both side of the membrane. Thus, this device allows us to replicate a tissue–tissue interface between epithelium and vascular endothelium. These results demonstrate that the FEPM-based microfluidic devices could be used to build organs-on-chips for drug discovery and development.

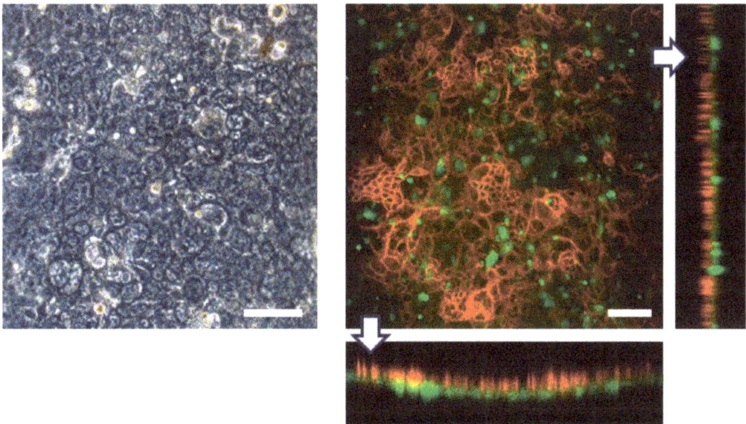

Figure 6. Engineering of tissue-tissue interfaces within FEPM devices. Human intestinal epithelial (Caco-2) cells are cultured on the upper surface of a collagen vitrigel membrane with vascular endothelial cells (HUVECs). Optical and confocal fluorescence micrographs of co-culture of Caco-2 cells stained for claudin (red) and GFP-HUVECs (green). Cross-sectional views (bottom and right) show the formation of monolayers of epithelium and endothelium on both sides of the membrane. Scale bars, 100 μm. (n = 3).

4. Discussion

Organs-on-chips produce tissue-level functionality not possible with standard in vitro culture methods by recapitulating tissue–tissue interfaces and physicochemical microenvironments. This technology has great potential to facilitate drug discovery and development and to replace animal models for drug testing. However, one significant drawback current organs-on-chips have is that devices are mostly made from PDMS which strongly absorbs small hydrophobic molecules. Since drug absorption into PDMS causes a reduction of drug concentrations and pharmacological activities, PDMS-based devices cannot widely be used as tools for drug discovery and development.

In this study, we investigated the possibility of using a tetrafluoroethylene-propylene (FEPM) elastomer to construct microfluidic organs-on-chips for drug testing. We demonstrated that the FEPM elastomer is resistant to absorption of three types of small hydrophobic drugs, whereas PDMS significantly absorbed these drugs, indicating that PDMS is not suitable for constructing devices to test drugs as previously reported [29,30]. By analyzing drug concentrations after 24-h incubation with FEPM, we found that the residual concentrations of these small hydrophobic drugs were almost the same level as those in standard culture plates, demonstrating that the FEPM elastomer is comparable to standard polystyrene culture plates in terms of absorption of small hydrophobic drugs.

We then developed compartmentalized microfluidic devices using the FEPM elastomer. The device comprised two FEPM channel layers separated by a collagen vitrigel membrane which is made from natural extracellular matrix (ECM). The cellular images demonstrated that the FEPM layers are optically clear and can be used for fluorescence imaging. The FEPM elastomer is hydrophobic and has Young's modulus of 0.8 MPa which is similar to PDMS [45]. Thus, the FEPM-based device can produce cyclic stretching of the ECM membrane with physiological levels of strain (~10%) by applying cyclic vacuum to the side chambers. We confirmed that there was no leakage between each layer and the collagen vitrigel membrane during cell culture with cyclic stretching. We also confirmed that this device can be used to culture cells and maintain cellular viability with continuous perfusion of culture medium for at least 7 days. By seeding cells into the both sides of the microchannels, the epithelial-endothelial interface can be formed in this device, suggesting that this device can be used to construct organs-on-chips. As an example, we demonstrated that an intestinal epithelial model can be constructed by replicating the interface between intestinal epithelial cells and vascular endothelial cells using Caco-2 cells and HUVECs. Furthermore, this device permitted imaging of cellular responses to environmental signals such as mechanical forces in real-time. When HUVECs were subjected to physiological cyclic strain (5–10%) for 4 h, the cells exhibited elongation and alignment on the ECM membrane in response to the cyclic strain. Moreover, we confirmed that this device can be used to test cytotoxic effect of mitomycin C (Figure S4). After 24 h treatment of mitomycin C (10 μg·mL^{-1}), there was no significant difference in the number of live and dead cells between the device culture and the conventional plate culture. Thus, this device provides a way to recapitulate cellular microenvironment including cyclic stretching and fluid flow for drug testing.

Since the FEPM elastomer is biocompatible, flexible, and optically clear, it has the potential to be an alternative to PDMS. However, there are some limitations of the FEPM elastomers. They are 100 times less gas permeable than PDMS and thus perfusion culture is required to maintain cellular viability. On the other hand, high permeation of water vapor of PDMS devices causes evaporation and osmolarity shifts that affect cell growth and development [25,46]. The FEPM devices may mitigate this problem because of low permeability of water vapor [47]. Because FEPM has hydrophobicity and low gas permeability, air bubbles may be generated inside channels when solutions are introduced. These bubbles can be removed by incubating the devices with culture medium in an incubator before cell seeding. Another limitation is that a large vacuum compression molding machine and precision-metal molds are required to fabricate channel layers. This is not user friendly; however, this may be suitable for mass production because the vacuum compression molding machine is commonly used for mass production of rubber products, whereas PDMS is incompatible with mass production due to the low speed of fabrication processes. Furthermore, this method can produce 100 μm features (Figure S5) although we used millimeter channels in this study.

In summary, we developed a novel microfluidic culture device using the FEPM elastomer with thin ECM membranes. Since this elastomer is resistant to drug absorption, the FEPM devices can be used to test the effects of drugs. This microfluidic device permits the formation of a tissue–tissue interface on a thin ECM membrane and the application of mechanical forces including cyclic strain and fluid shear stress. This device also allows real-time imaging of cellular responses. Therefore, this device could be a useful platform to construct organs-on-chips for drug discovery and development.

Supplementary Materials: The following are available online at http://www.mdpi.com/2072-666X/10/11/793/s1: Video S1: Cells cultured in the FEPM microfluidic device while applying cyclic strain. Figure S1: Design of the FEPM channel layer. Figure S2: Evaluation of fluorescence intensity of the FEPM device containing a rhodamine B solution. Figure S3: Images of HUVECs cultured on a cell culture plate. Figure S4: Evaluation of cellular viability after 24 h treatment of mitomycin C. Figure S5: Images of FEPM layers containing small features.

Author Contributions: Y.-s.T. conceived and designed the experiments; E.S., C.M., Y.O., K.T. (Koichi Tashima), S.Y. and K.T. (Kana Tanabe) performed the experiments and analyzed the data; N.M., K.Y. (Keisuke Yagi), A.W. and K.Y. (Kayo Yano) contributed to fabrication of the devices; C.M., N.M. and Y.-s.T. wrote the manuscript.

Funding: This research was funded by JSPS KAKENHI Grant No. JP17H02082, AMED under Grant No. JP18gm5810008, and the Kyoto University Hakubi Project.

Acknowledgments: We thank Kaori Kosugi, Hanako Mukai, Naoki Shizuka, Michio Jingu, Kazuo Iino, Yu-ki Kitamura, Shigeo Suetomi, Tomonori Sato, Tomomi Hagiwara, Hisayoshi Ishii, Tomohito Ohtsuki, Nahoko Yamada, Hiroko Motohashi, Masamichi Ide, Masahiro Ohkura, Takayuki Akasaka, Daisuke Taguchi, Norihito Miyagawa, Masakazu Ataku, Motoya Murase, Yoichiro Takahashi, Masahide Yodogawa, Toshiyuki Meguro and AGC Inc. for their technical assistance and advice.

Conflicts of Interest: E.S., C.M., Y.O., K.Y. (Kayo Yano) and Y.-s.T. has nothing to declare. N.M., K.Y. (Keisuke Yagi), A.W., K.T. (Koichi Tashima), S.Y., and K.T. (Kana Tanabe) are employees of AGC Inc.

References

1. Scannell, J.W.; Blanckley, A.; Boldon, H.; Warrington, B. Diagnosing the decline in pharmaceutical R&D efficiency. *Nat. Rev. Drug Discov.* **2012**, *11*, 191–200. [PubMed]
2. Muller, P.Y.; Milton, M.N. The determination and interpretation of the therapeutic index in drug development. *Nat. Rev. Drug Discov.* **2012**, *11*, 751–761. [CrossRef] [PubMed]
3. Bowes, J.; Brown, A.J.; Hamon, J.; Jarolimek, W.; Sridhar, A.; Waldron, G.; Whitebread, S. Reducing safety-related drug attrition: The use of in vitro pharmacological profiling. *Nat. Rev. Drug Discov.* **2012**, *11*, 909–922. [CrossRef] [PubMed]
4. Huh, D.; Torisawa, Y.; Hamilton, G.A.; Kim, H.J.; Ingber, D.E. Microengineered physiological biomimicry: Organs-on-chips. *Lab Chip* **2012**, *12*, 2156–2164. [CrossRef] [PubMed]
5. Bhatia, S.N.; Ingber, D.N. Microfluidic organs-on-chips. *Nat. Biotechnol.* **2014**, *32*, 760–772. [CrossRef]
6. Esch, E.W.; Bahinski, A.; Huh, D. Organs-on-chips at the frontiers of drug discovery. *Nat. Rev. Drug Discov.* **2015**, *14*, 248–260. [CrossRef]
7. Zhang, B.; Korolj, A.; Lai, B.F.L.; Radisic, M. Advances in organ-on-a-chip engineering. *Nat. Rev. Mater.* **2018**, *3*, 257–278. [CrossRef]
8. Yesil-Celiktas, O.; Hassan, S.; Miri, A.K.; Maharjan, S.; Al-kharboosh, R.; Quiñones-Hinojosa, A.; Zhang, Y.S. Mimicking human pathophysiology in organ-on-chip devices. *Adv. Biosyst.* **2018**, *2*, 1800109. [CrossRef]
9. Paguirigan, A.L.; Beebe, D.J. From the cellular perspective: Exploring differences in the cellular baseline in macroscale and microfluidic cultures. *Integr. Biol.* **2009**, *1*, 182–195. [CrossRef]
10. Torisawa, Y.; Mosadegh, B.; Bersano-Begey, T.; Steele, J.M.; Luker, K.E.; Luker, G.D.; Takayama, S. Microfluidic platform for chemotaxis in gradients formed by CXCL12 source-sink cells. *Integr. Biol.* **2010**, *2*, 680–686. [CrossRef]
11. Kim, S.; Kim, H.J.; Jeon, N.L. Biological applications of microfluidic gradient devices. *Integr. Biol.* **2010**, *2*, 584–603. [CrossRef] [PubMed]
12. Shemesh, J.; Jalilian, I.; Shi, A.; Heng Yeoh, G.; Knothe Tate, M.L.; Ebrahimi Warkiani, M. Flow-induced stress on adherent cells inmicrofluidic devices. *Lab Chip* **2015**, *15*, 4114–4127. [CrossRef] [PubMed]
13. Ingber, D.E. Cellular mechanotransduction: Putting all the pieces together again. *FASEB J.* **2006**, *20*, 811–827. [CrossRef] [PubMed]
14. Vining, K.H.; Mooney, D.J. Mechanical forces direct stem cell behaviour in development and regeneration. *Nat. Rev. Mol. Cell Biol.* **2017**, *18*, 728–742. [CrossRef] [PubMed]
15. Huh, D.; Matthews, B.D.; Mammoto, A.; Montoya-Zavala, M.; Hsin, H.Y.; Ingber, D.E. Reconstituting organ-level lung functions on a chip. *Science* **2010**, *328*, 1662–1668. [CrossRef] [PubMed]
16. Huh, D.; Leslie, D.C.; Matthews, B.D.; Fraser, J.P.; Jurek, S.; Hamilton, G.A.; Thorneloe, K.S.; McAlexander, M.A.; Ingber, D.E. A human disease model of drug toxicity-induced pulmonary edema in a lung-on-a-chip microdevice. *Sci. Transl. Med.* **2012**, *4*, 159ra147. [CrossRef]
17. Stucki, A.O.; Stucki, J.D.; Hall, S.R.; Felder, M.; Mermoud, Y.; Schmid, R.A.; Geiser, T.; Guenat, O.T. A lung-on-a-chip array with an integrated bio-inspired respiration mechanism. *Lab Chip* **2015**, *15*, 1302–1310. [CrossRef]
18. Kim, H.J.; Huh, D.; Hamilton, G.; Ingber, D.E. Human gut-on-a-chip inhabited by microbial flora that experiences intestinal peristalsis-like motions and flow. *Lab Chip* **2012**, *12*, 2165–2174. [CrossRef]
19. Kasendra, M.; Tovaglieri, A.; Sontheimer-Phelps, A.; Jalili-Firoozinezhad, S.; Bein, A.; Chalkiadaki, A.; Scholl, W.; Zhang, C.; Rickner, H.; Richmond, C.A.; et al. Development of a primary human Small Intestine-on-a-Chip using biopsy-derived organoids. *Sci. Rep.* **2018**, *8*, 2871. [CrossRef]
20. Jang, K.J.; Mehr, A.P.; Hamilton, G.A.; McPartlin, L.A.; Chung, S.; Suh, K.Y.; Ingber, D.E. Human kidney proximal tubule-on-a-chip for drug transport and nephrotoxicity assessment. *Integr. Biol.* **2013**, *5*, 1119–1129. [CrossRef]
21. Musah, S.; Mammoto, A.; Ferrante, T.C.; Jeanty, S.S.F.; Hirano-Kobayashi, M.; Mammoto, T.; Roberts, K.; Chung, S.; Novak, R.; Ingram, M.; et al. Mature induced-pluripotent-stem-cell-derived human podocytes reconstitute kidney glomerular-capillary-wall function on a chip. *Nat. Biomed. Eng.* **2017**, *1*, 0069. [CrossRef] [PubMed]

22. Homan, K.A.; Gupta, N.; Kroll, K.T.; Kolesky, D.B.; Skylar-Scott, M.; Miyoshi, T.; Mau, D.; Valerius, M.T.; Ferrante, T.; Bonventre, J.V.; et al. Flow-enhanced vascularization and maturation of kidney organoids in vitro. *Nat. Methods* **2019**, *16*, 255–262. [CrossRef] [PubMed]
23. Hassell, B.A.; Goyal, G.; Lee, E.; Sontheimer-Phelps, A.; Levy, O.; Chen, C.S.; Ingber, D.E. Human organ chip models recapitulate orthotopic lung cancer growth, therapeutic responses, and tumor dormancy in vitro. *Cell Rep.* **2017**, *21*, 508–516. [CrossRef] [PubMed]
24. Whitesides, G.M.; Ostuni, E.; Takayama, S.; Jiang, X.; Ingber, D.E. Soft lithography in biology and biochemistry. *Annu. Rev. Biomed. Eng.* **2001**, *3*, 335–373. [CrossRef] [PubMed]
25. Berthier, E.; Young, E.W.; Beebe, D. Engineers are from PDMS-land, Biologists are from Polystyrenia. *Lab Chip* **2012**, *12*, 1224–1237. [CrossRef] [PubMed]
26. Toepke, M.W.; Beebe, D.J. PDMS absorption of small molecules and consequences in microfluidic applications. *Lab Chip* **2006**, *6*, 1484–1486. [CrossRef]
27. Van Meer, B.J.; de Vries, H.; Firth, K.S.A.; van Weerd, J.; Tertoolen, L.G.J.; Karperien, H.B.J.; Jonkheijm, P.; Denning, C.; IJzerman, A.P.; Mummery, C.L. Small molecule absorption by PDMS in the context of drug response bioassays. *Biochem. Biophys. Res. Commun.* **2017**, *482*, 323–328. [CrossRef]
28. Auner, A.W.; Tasneem, K.M.; Markov, D.A.; McCawley, L.J.; Hutson, M.S. Chemical-PDMS binding kinetics and implications for bioavailability in microfluidic devices. *Lab Chip* **2019**, *19*, 864–874. [CrossRef]
29. Regehr, K.J.; Domenech, M.; Koepsel, J.T.; Carver, K.C.; Ellison-Zelski, S.J.; Murphy, W.L.; Schuler, L.A.; Alarid, E.T.; Beebe, D.J. Biological implications of polydimethylsiloxane-based microfluidic cell culture. *Lab Chip* **2009**, *9*, 2132–2139. [CrossRef]
30. Moore, T.A.; Brodersen, P.; Young, E.W.K. Multiple myeloma cell drug responses differ in thermoplastic vs PDMS microfluidic devices. *Anal. Chem.* **2017**, *89*, 11391–11398. [CrossRef]
31. Domansky, K.; Sliz, J.D.; Wen, N.; Hinojosa, C.; ThompsonII, G.; Fraser, J.P.; Hamkins-Indik, T.; Hamilton, G.A.; Levner, D.; Ingber, D.E. SEBS elastomers for fabrication of microfluidic devices with reduced drug absorption by injection molding and extrusion. *Microfluid. Nanofluid.* **2017**, *21*, 107. [CrossRef]
32. Domansky, K.; Leslie, D.C.; McKinney, J.; Fraser, J.P.; Sliz, J.D.; Hamkins-Indik, T.; Hamilton, G.A.; Bahinski, A.; Ingber, D.E. Clear castable polyurethane elastomer for fabrication of microfluidic devices. *Lab Chip* **2013**, *13*, 3956–3964. [CrossRef] [PubMed]
33. Borysiak, M.D.; Bielawski, K.S.; Sniadecki, N.J.; Jenkel, C.F.; Vogt, B.D.; Posner, J.D. Simple replica micromolding of biocompatible styrenic elastomers. *Lab Chip* **2013**, *13*, 2773–2784. [CrossRef] [PubMed]
34. Ren, K.; Zhao, Y.; Su, J.; Ryan, D.; Wu, H. Convenient method for modifying poly(dimethylsiloxane) to be airtight and resistive against absorption of small molecules. *Anal. Chem.* **2010**, *82*, 5965–5971. [CrossRef]
35. Sasaki, H.; Onoe, H.; Osaki, T.; Kawano, R.; Takeuchi, S. Parylene-coating in PDMS microfluidic channels prevents the absorption of fluorescent dyes. *Sens. Actuators B Chem.* **2010**, *150*, 478–482. [CrossRef]
36. Hiltz, J.A. Characterization of fluoroelastomers by various analytical techniques including pyrolysis gas chromatography/mass spectrometry. *J. Anal. Appl. Pyrol.* **2014**, *109*, 283–295. [CrossRef]
37. Kojima, G.; Kojima, H.; Tabata, Y. A new fluoroelastomer derived from tetrafluoroethylene and propylene. *Rubber Chem. Technol.* **1977**, *50*, 403–412. [CrossRef]
38. Kojima, G.; Wachi, H. Vulcanization of a fluoroelastomer derived from tetrafluoroethylene and propylene. *Rubber Chem. Technol.* **1978**, *51*, 940–947. [CrossRef]
39. Moraes, C.; Mehta, G.; Lesher-Perez, S.C.; Takayama, S. Organs-on-a-chip: A focus on compartmentalized microdevices. *Ann. Biomed. Eng.* **2012**, *40*, 1211–1227. [CrossRef]
40. McDonald, J.C.; Duffy, D.C.; Anderson, J.R.; Chiu, D.T.; Wu, H.; Schueller, O.J.; Whitesides, G.M. Fabrication of microfluidic systems in poly(dimethylsiloxane). *Electrophoresis* **2000**, *21*, 27–40. [CrossRef]
41. Bacabac, R.G.; Smit, T.H.; Cowin, S.C.; Van Loon, J.J.; Nieuwstadt, F.T.; Heethaar, R.; Klein-Nulend, J. Dynamic shear stress in parallel-plate flow chambers. *J. Biomech.* **2005**, *38*, 159–167. [CrossRef] [PubMed]
42. Sinha, R.; Le Gac, S.; Verdonschot, N.; van den Berg, A.; Koopman, B.; Rouwkema, J. Endothelial cell alignment as a result of anisotropic strain and flow induced shear stress combinations. *Sci. Rep.* **2016**, *6*, 29510. [CrossRef] [PubMed]
43. Eddings, M.A.; Johnson, M.A.; Gale, B.K. Determining the optimal PDMS–PDMS bonding technique for microfluidic devices. *J. Micromech. Microeng.* **2008**, *18*, 067001. [CrossRef]
44. Jufri, N.F.; Mohamedali, A.; Avolio, A.; Baker, M.S. Mechanical stretch: Physiological and pathological implications for human vascular endothelial cells. *Vasc. Cell* **2015**, *7*, 8. [CrossRef] [PubMed]

45. Liu, M.; Sun, J.; Sun, Y.; Bock, C.; Chen, Q. Thickness-dependent mechanical properties of polydimethylsiloxane membranes. *J. Micromech. Microeng.* **2009**, *19*, 035028. [CrossRef]
46. Heo, Y.S.; Cabrera, L.M.; Song, J.W.; Futai, N.; Tung, Y.C.; Smith, G.D.; Takayama, S. Characterization and resolution of evaporation-mediated osmolality shifts that constrain microfluidic cell culture in poly(dimethylsiloxane) devices. *Anal. Chem.* **2007**, *79*, 1126–1134. [CrossRef]
47. Laurenson, L.; Dennis, N.T.M. Permeability of common elastomers for gases over a range of temperatures. *J. Vac. Sci. Technol. A* **1985**, *3*, 1707. [CrossRef]

Article

A Cancer Spheroid Array Chip for Selecting Effective Drug

Jae Won Choi [1], Sang-Yun Lee [2,3] and Dong Woo Lee [1,*]

[1] Department of Biomedical Engineering, Konyang University, Daejeon 35365, Korea; zeak3659@naver.com
[2] Department of Health Sciences and Technology, SAIHST, Sungkyunkwan University, Seoul 06351, Korea; leesangyun316@gmail.com
[3] Medical & Bio Device (MBD), Suwon 16229, Korea
[*] Correspondence: dw2010.lee@gmail.com

Received: 7 September 2019; Accepted: 9 October 2019; Published: 12 October 2019

Abstract: A cancer spheroid array chip was developed by modifying a micropillar and microwell structure to improve the evaluation of drugs targeting specific mutations such as phosphor-epidermal growth factor receptor (p-EGFR). The chip encapsulated cells in alginate and allowed cancer cells to grow for over seven days to form cancer spheroids. However, reagents or media used to screen drugs in a high-density spheroid array had to be replaced very carefully, and this was a tedious task. Particularly, the immunostaining of cancer spheroids required numerous steps to replace many of the reagents used for drug evaluation. To solve this problem, we adapted a micropillar and microwell structure to a spheroid array. Thus, culturing cancer spheroids in alginate spots attached to the micropillar allowed us to replace the reagents in the microwell chip with a single fill of fresh medium, without damaging the cancer spheroids. In this study, a cancer spheroid array was made from a p-EGFR-overexpressing cell line (A549 lung cancer cell line). In a 12 by 36 column array chip (25 mm by 75 mm), the spheroid over 100 μm in diameter started to form at day seven and p-EGFR was also considerably overexpressed. The array was used for p-EGFR inhibition and cell viability measurement against seventy drugs, including ten EGFR-targeting drugs. By comparing drug response in the spheroid array (spheroid model) with that in the single-cell model, we demonstrated that the two models showed different responses and that the spheroid model might be more resistant to some drugs, thus narrowing the choice of drug candidates.

Keywords: organoid; 3D cell culture; spheroid array; high-throughput screening; drug efficacy

1. Introduction

When using conventional approaches for evaluating anticancer drugs, the 2-dimensional monolayer (2D) cell culture model is the gold standard. However, when cancer cells are cultured in plastic dishes, the cell morphology differs from the 3D growth occurring in animal cells in the living body. This environment also affects gene expression. It has been reported that animal cells grown in biocompatible 3D cell culture models exhibit different gene expression patterns than when they are grown in 2D cell culture models [1]. As a result, in vitro animal cell cultures have poor correspondence with in vivo animal cell cultures. Thus, many 3D cell culture models have been developed to overcome this poor correspondence [2]. Moreover, when animal cells are used for analyzing drug efficacy or toxicity, the drug reactivity in a 3D cell culture model differs greatly from what has been observed in conventional 2D cell culture models [3–6]. When cells derived from cancer patients are cultured in 3D, cell-cell interactions and the extracellular matrix (ECM) change the morphology of the cells in the culture, as well as the type and expression level of the major genes being expressed [3–6]. For these reasons, tools aiding in the development of 3D cell cultures are being studied, and some have even been commercialized. Generally, a 3D cell culture can be categorized into two models. A scaffold-free model allows cells to grow together without exogenous extracellular matrix and a scaffolding model allows cell cultivation in the ECM space. Recently, the scaffold method was used to form cancer organoids (spheroids over 100 μm in diameter), which are considered as near-physiological in vitro cell models [7–9]. Organoids could be used for biomedical research, genomic analysis of various diseases, and therapeutic studies [10–17]. Specifically, cancer organoid cultures could be a powerful tool for evaluating drug efficacy and toxicity during drug discovery studies [18], for conducting cytotoxicity investigations of new therapeutic compounds [19], as well as for personalizing cancer treatments [9,20]. Thus, many technologies such as hang drop technology [21], agarose microwells [22], and microfluidic chips [23] have been developed, which successfully demonstrate the performance of cancer organoid cultures. However, for commercial application of high throughput screening, the issue of automation needs to be resolved. Especially, while screening drugs in a high-density cancer spheroid array, the media need to be changed by careful pipetting, which is a tedious task, and a bottleneck in automation. To overcome this problem, we adopted a micropillar and microwell structure of the spheroid array, as shown in Figure 1. Previously, we have described a micropillar and microwell chip for culturing 3D cells and testing drug efficacy [24–26]; however, the drugs were exposed to single cells or small spheroids.

In this study, as shown by the spot images following the number of days in Figure 1, cancer spheroids over 100 μm in diameter began to form on a micropillar chip at day seven. After day seven, cancer spheroids maintained their size and overexpression of phosphor-epidermal growth factor receptor (p-EGFR). Culturing cancer spheroids in alginate spots attached to the micropillar allowed the media to be changed by replacing with new microwells filled with fresh media. Sixteen cancer cell lines successfully grew and formed spheroids over 100 μm in diameter in alginate spots on the micropillar chip, as shown in Figure 1b. We applied the cancer spheroid array to identify highly effective drugs targeting p-EGFR from among seventy drugs (including ten EGFR-targeting drugs as model drugs) by using a p-EGFR-overexpressing cell line (A549 cell line). The micropillar and microwell chip could support the single-cell and spheroid models and produced different drug responses according to the model.

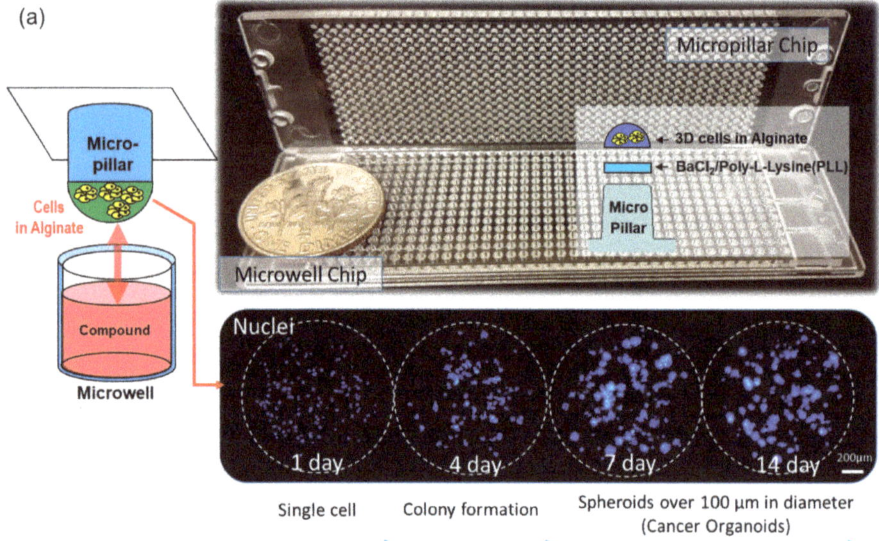

Figure 1. *Cont.*

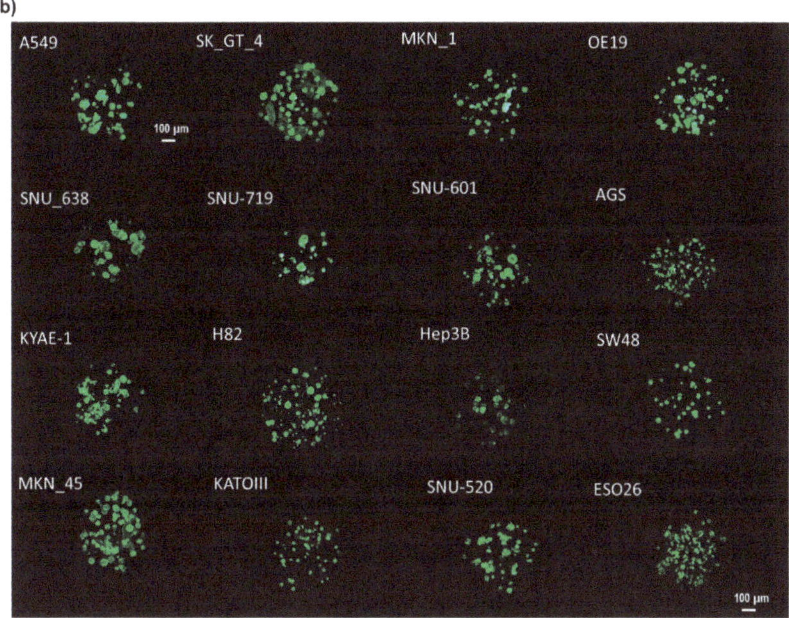

Figure 1. Cancer spheroid array chip based on micropillar and microwell chips. (**a**) Blue represents the nucleus. Spheroids over 100 µm in diameter were formed after seven days. (**b**) Sixteen cell lines (including A549) were grown for seven days in alginate spots on the micropillar chip.

2. Materials and Methods

2.1. Fabrication of Micropillar/Microwell Chips and Incubation Chamber for the Spheroid Array

The micropillar/microwell chip platform is typically used for short-term cell cultures to form spheroids. However, a technical issue arose when we applied this technique for long-term cell culture to form spheroids; the tightly combined chips lacked a CO_2 supply. Moreover, the low volume (1 µL) of media in the microwell evaporated very easily. To resolve these issues, the micropillar/microwell chips were modified; the micropillar was combined with the microwell chip very tightly to prevent evaporation. Nonetheless, this caused the combined chips to run out of CO_2 at a higher rate. Adding a spacer to the microwell chip (Figure 2c) created a gap between the chips so that CO_2 could easily penetrate the wells. The modified micropillar and microwell chips were manufactured by plastic injection molding. The micropillar chip was made of poly styrene-*co*-maleic anhydride (PS-MA) and contained 532 micropillars (with a 0.75 mm pillar diameter and a 1.5-mm pillar-to-pillar distance). PS-MA provides a reactive functionality to covalently attach poly-L-lysine (PLL), ultimately attaching alginate spots by their ionic interactions. Plastic molding was performed with an injection machine (Sodic Plustech Inc., Schaumburg, IL, USA). To prevent evaporation, an incubation chamber for micropillar/microwell chips (Figure 2d) was fabricated by cutting the cyclic olefin copolymer (COC) with a computer numerical control (CNC) machine. The COC was selected because of its high transparency, excellent biocompatibility, and adequate stiffness for physical machining. As shown in Figure 2d, four combined chips were placed in the reservoir, which was filled with distilled water (DI) to prevent evaporation. After 13 days, 5.3% of the media had evaporated from the incubation chamber.

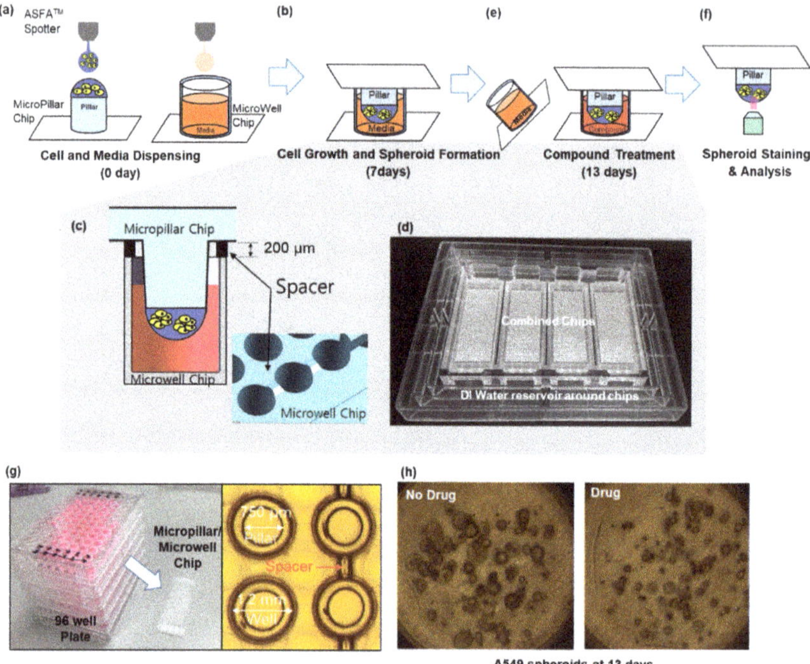

Figure 2. Experimental procedure for the micropillar/microwell chip system. (**a**) Cells and media were dispensed on a micropillar and in a microwell, respectively. (**b**) Cells were immobilized in alginate at the top of the micropillars and dipped into the microwells containing growth media for seven days to form spheroids with diameter over 100 µm. (**c**) The microchip was separated from other microwell chips by a spacer to unify penetration of CO_2 into the chips. (**d**) To prevent evaporation during incubation, the incubation chamber surrounds the four chips in water. (**e**) Compounds are dispensed into the microwells, and spheroids are exposed to the compounds by moving the micropillar chip to a new microwell chip. (**f**) Spheroids are stained with Calcein-AM, and the dried alginate spots on the micropillar chip are scanned for data analysis. (**g**) Comparison of the combined micropillar/microwell chip with a conventional 96-well plate. (**h**) Cancer spheroid images with and without drug treatment on day 13.

2.2. Cell Line Culture

A549 (lung), SK_GT_4 (esophagus), MKN_1 (stomach), OE19 (esophagus), SNU-638 (stomach), SNU-719 (stomach), SNU-601 (stomach), AGS (stomach), KYAE-1 (esophagus), H82 (lung), Hep3B (liver), SW48 (colorectal), MKN_45 (stomach), KATOIII (stomach), SNU-520 (stomach), and ESO26 (adenocarcinoma of the gastroesophageal junction) were cultured in RPMI 1640 medium (Gibco, Co Dublin, Ireland) supplemented with 10% fetal bovine serum (FBS) and 1% antibiotics (Gibco, Co Dublin, Ireland). All cell lines were purchased from the Korean Cell Line Bank (Seoul, South Korea). Cell lines were maintained at 37 °C in a 5% CO_2, in a humidified atmosphere and passaged every four days. Normally, we used cell lines under 20 passages after thawing the frozen cell stock. Under 20 passages, sixteen cell lines could easily form 3D cells in 0.5% (w/w) alginate on the chip platform.

2.3. Experimental Procedure

Approximately 100 cells in 50 nL of 0.5% (w/w) alginate were automatically dispensed onto a micropillar chip by using an ASFA™ Spotter ST (Medical & Bio Decision, Suwon, South Korea). ASFA™ Spotter ST uses a solenoid valve (The Lee Company, Westbrook, CT, USA) for dispensing the 50-nL droplets of the cell–alginate mixture and 1 µL of medium or drugs. The top of the micropillar was coated with 60 nL of 0.02 M $BaCl_2$. When the cell-alginate mixture was dispensed on the top of the micropillar, the barium ions replaced the sodium ions in the alginate; thus, the polymer strands in the alginate was cross-linked with the barium ions, resulting in an alginate gel. After dispensing the cells and media in the micropillar and microwell, respectively (Figure 2a), the micropillar chip containing the human cells in the alginate was combined (or "stamped") with the microwell chip filled with 1 µL of fresh media (Figure 2b). The micropillar and microwell chip in their combined form is shown in Figure 2b. After three days of incubation at 37 °C to stabilize the cells, the cells started forming spheres. We changed the media and allowed cells to grow up to seven days until the size of the spheres was larger than 100 µm. In order to uniformly infiltrate the CO_2 into the chip, the micropillar chips were spaced 200 µm apart from the microwell chip, as shown in Figure 2c. While incubating the cells on the chip, an incubation chamber (Figure 2d) was used to prevent evaporation of the media, even though the spacer made evaporation easier. On day seven, the drug was added to the spheroids over 100 µm in diameter, which were now cancer organoids Figure 2e. To evaluate the cancer spheroids, immunostaining was performed following the protocol described in the previous work [19]. Briefly, 3D cells cultured on the micropillar chip were fixed using a 4% paraformaldehyde solution (PFA, Biosesang, Seongnam, Korea) mixed with 2.5 mM $CaCl_2$ for 120 min. The amount of time required to fix 3D-cultured cells is considerably longer than the 60 min required to fix a 2D cultured cell. Experimentally, 120 min is the minimum amount of time required to prevent cell degradation in alginate. After fixation, the micropillars were transferred to a permeabilizing and blocking solution (1% bovine serum albumin (BSA) in phosphate-buffered saline (PBS) containing 0.3% Triton-X) for 1 h. Subsequently, each micropillar chip was incubated overnight at 4 °C with the antibody staining solution. The antibody staining solution was prepared by adding anti-p-EGFR (200:1, Abcam, Cambridge, UK, phosphor Y1092, Alexa Fluor 488, green fluorescent dye with excitation at 488-nm), Hoechst 33342 (1000:1, Thermo Fisher Scientific, Waltham, MA, USA, Hoechst 33342, blue fluorescent dye that could be excited using a 358-nm laser), and F-actin phalloidin (400:1, Thermo Fisher Scientific, Waltham, MA, USA, Alexa Fluor 594, red fluorescent dye that could be excited using 561-nm or 594-nm lasers) to the permeabilizing and blocking solution. The stained chip was then washed for 15 min in the staining buffer solution (MBD-STA500, Medical & Bio Device, Suwon, South Korea) and then dried completely in a dark environment. To image the stained cells, the micropillars were scanned using an optical scanner (ASFA™ Scanner HE, Medical & Bio Device, Suwon, Korea).

To evaluate the efficacy of 70 different drugs in inhibition of p-EGFR using a single chip, we dispensed 70 drugs at a concentration of 20 µM into the microwell. One alginate spot containing cells on a micropillar from among the 12 by 36 array was exposed to one drug (20 µM). One microliter of each drug was dispensed into a microwell using the ASFA™ Spotter ST (Medical & Bio Decision, Suwon, South Korea). One microliter of each drug could be dispensed using Echo Liquid Handlers (LABCYTE, San Jose, CA, USA). Each drug was administered to the six microwells (six replicates). In our previous study [27,28], even though cells were treated with drugs for one day (at almost a single-cell stage, before forming spheroids), most drugs showed high resistance in 3D-cultured cells. Thus, a high dose of 20 µM of drug was selected because the spheroids over 100 µm in diameter may show high resistance to drugs. The purpose of this paper was to demonstrate that, because of the high drug resistance, spheroids over 100 µm in diameter can help narrow down the options for drug candidates targeting p-EGFR. We expected the spheroids to have a higher resistance to drugs than the 3D single-cell model; therefore, we chose 20 µM as the drug concentration.

2.4. Comparison of Drug Response between the Single-Cell and Spheroid Models

To determine the differences in results obtained with each model, we compared the responses to drug treatment. For the single-cell response, cells were treated with drugs on day one, and cell viability was measured on day seven. For the spheroid response, spheroids were treated with drugs on day seven, and spheroid viability was measured on day 13, as shown in Figure 2. The drug treatment time and p-EGFR staining times (6, 24, 48, and 144 h) in the spheroid model were similar to those in the single-cell model. To compare p-EGFR expression and cell viability for the different chips in different days, we normalized cell viability and p-EGFR expression with those in the control (no drugs) in the same chip. Thus, p-EGFR was normalized by dividing its relative expression in the control sample with that in a drug treatment sample on the same chip. Chips 6, 24, 48, and 144 h were stained after administering the drug and normalized p-EGFR at each time point because p-EGFR inhibition occurred at different times depending on the drugs. Among the normalized p-EGFR samples taken at 6, 24, 48, and 144 h, a minimum p-EGFR normalization value was selected for evaluating drug inhibition of p-EGFR.

2.5. p-EGFR Measurement

The nucleus, p-EGFR, and F-actin in the 3D-cultured cells in alginate were stained with different color fluorescence (blue, green, and red). After fixing the 3D-cultured cells, the red fluorescent dye was used for identifying filamentous actin (F-actin)—one of the components of the cytoskeleton—because the cytoskeleton degrades when the 3D-cultured cells are affected by the drugs. The green fluorescent dye was used to identify p-EGFR in the cell membrane, whereas the blue fluorescent dye was used to identify the cell nucleus. An automatic optical fluorescence scanner (ASFA™ Scanner ST, Medical & Bio Device, Suwon, South Korea) was used to measure the red, green, and blue fluorescence intensities using an 8-bit code among the RGB codes (0–255); the 3D-cultured cells were identified according to intensity thresholds (20 green code). The relative p-EGFR expression levels (relative p-EGFR) were calculated by dividing the green area (p-EGFR) with the blue area (the nucleus of the cell) in one alginate spot, as shown below:

$$\text{Relative } p-\text{EGFR } [\%] = \frac{\textit{Total Green Area}}{\textit{Total Blue Area}} \times 100 \tag{1}$$

The cytoskeleton was measured by calculating the size of the red area (F-actin expression). F-actin and p-EGFR values were normalized to their corresponding controls (no drug treatment).

2.6. Viability Measurement

Calcein AM staining solution was prepared by adding 1.0 µL of calcein AM (4 mM stock from Invitrogen) in 8 mL of 140 mM NaCl supplemented with 20 mM CaCl$_2$. The non-fluorescent acetomethoxy derivate of calcein (calcein AM, AM = acetoxymethyl) is used because it can be transported through the cellular membrane into live cells. After transport into the cells, intracellular esterases remove the acetomethoxy group, the molecule gets trapped inside, and gives out strong green fluorescence. As dead cells lack active esterases, only live cells are labeled.

To check cell viability at day seven after drug treatment in the single and spheroid models, cells in the chips were stained with Calcein AM. The live cells were stained and produced green fluorescence. An automatic optical fluorescence scanner (ASFA™ Scanner ST, Medical & Bio Device, Suwon, South Korea) was used to measure green fluorescence intensities using an 8-bit code among the RGB codes (0–255). The area of the 3D-cultured cells was identified according to the intensity threshold (20 green codes) to reduce the background noise. The green area was used for calculating the cell viability. To determine relative viability, the green area for the cells exposed to the drug was divided by the control cell area without drug exposure. The relative viabilities are based on healthy cells without drug exposure. Six alginate spots were used for calculating the average and standard deviation of the relative viability, shown in Table 1.

Table 1. p-EGFR expression and relative cell viabilities when the single-cell and spheroids models were exposed to the drugs for seven days.

Drug	Target	Single-Cell Model				Cancer Spheroid Model				Drug	Target	Single-Cell Model				Cancer Spheroid Model			
		p-EGFR Expression [%]		Cell Viability [%]		p-EGFR Expression [%]		Cell Viability [%]				p-EGFR Expression [%]		Cell Viability [%]		p-EGFR Expression [%]		Cell Viability [%]	
		Average	SD	Average	SD	Average	SD	Average	SD			Average	SD	Average	SD	Average	SD	Average	SD
1_DMSO	-	100.0	0.0	100.2	6.5	100.0	0.0	100.0	8.8	37_AZD4547	FGFR1/2/3	76.4	9.5	0.0	0.0	74.9	22.6	111.6	20.6
2_AEE788	EGFR	30.0	8.5	0.0	0.1	56.9	16.1	3.6	1.6	38_BGJ398	FGFR1/2/3	100.0	0.0	66.2	12.7	100.0	0.0	109.8	11.7
3_Afatinib	EGFR	50.0	8.4	0.0	0.0	61.3	20.9	1.4	2.3	39_Dovitinib	Flt3, c-Kit, FGFR1/3, VEGFR1/2/3, PDGFRα/β	37.6	6.9	0.0	0.0	27.1	3.6	0.1	0.1
4_BMS-599626	EGFR	47.7	8.1	82.8	7.6	84.3	6.3	95.1	15.1	40_Bosutinib	dual Src/Abl	9.2	3.2	0.0	0.0	62.5	8.9	6.0	3.1
5_Erlotinib HCl	HER1/EGFR	67.1	12.7	60.0	10.1	78.6	3.9	59.4	7.1	41_Dasatinib	Bcr-Abl	25.4	6.8	25.0	11.1	22.0	2.6	80.3	8.0
6_Dacomitinib	EGFR	67.5	16.0	0.0	0.0	100.0	0.0	82.0	9.2	42_Nilotinib	Bcr-Abl	72.5	13.8	84.8	9.5	91.1	5.8	98.5	10.6
7_Gefitinib	EGFR	58.6	6.0	19.5	19.3	100.0	0.0	73.2	13.5	43_AZD6244	MEK1	78.0	20.6	6.1	1.4	100.0	0.0	20.6	6.3
8_Lapatinib	EGFR	63.9	12.3	73.9	5.8	100.0	0.0	84.5	8.5	44_Trametinib	MEK1/2	64.3	11.7	16.5	5.4	68.1	32.0	52.0	5.4
9_Neratinib	EGFR	59.8	7.2	54.9	5.0	100.0	0.0	73.1	6.0	45_Bortezomib	Proteasome	79.7	12.1	0.3	0.3	100.0	0.0	24.8	11.7
10_CI-1033	EGFR, HER2	48.9	12.9	66.6	6.8	100.0	0.0	69.2	10.7	46_Carfilzomib	Proteasome	100.0	0.0	0.2	0.2	100.0	0.0	39.2	9.5
11_CO-1686	EGFR	62.3	14.3	10.5	10.7	100.0	0.0	74.4	12.5	47_ABT-199	Bcl-2	72.7	8.7	3.1	0.9	76.8	13.4	14.7	3.8
12_BKM120	PI3K	100.0	0.0	1.9	0.8	100.0	0.0	36.1	4.4	48_ABT-888	PARP	100.0	0.0	82.2	5.4	100.0	0.0	112.3	8.7
13_BYL719	PI3K	100.0	0.0	13.0	4.2	100.0	0.0	49.0	9.1	49_AUY922	HSP (e.g. HSP90)	27.8	5.5	8.5	2.7	63.8	12.4	80.2	11.7
14_XL147	PI3K	65.1	11.1	7.7	2.1	38.0	8.6	28.0	3.8	50_Dabrafenib	BRAFV600	29.3	11.0	47.0	10.1	100.0	0.0	95.1	6.8
15_Everolimus	mTOR	100.0	0.0	57.5	4.0	67.6	16.3	86.6	12.1	51_Ibrutinib	Btk, modestly potent to Bmx, CSK, FGR, BRK, HCK	12.4	5.6	74.5	9.4	35.0	2.8	94.5	16.8
16_AZD2014	mTOR	73.1	9.6	6.9	2.1	100.0	0.0	33.4	7.6	52_LDE225	Smoothened	100.0	0.0	90.3	12.1	100.0	0.0	94.3	15.9
17_PF-05212384	P3k/mTOR	100.0	0.0	1.6	0.6	100.0	0.0	34.9	5.9	53_LDK378	ALK	100.0	0.0	0.0	0.0	64.2	19.9	0.0	0.0
18_XL765	P3k/mTOR	100.0	0.0	42.5	3.8	100.0	0.0	68.7	3.0	54_LGK-974	PORCN	100.0	0.0	72.8	4.5	100.0	0.0	81.9	7.3
19_BEZ235	P3k/mTOR	100.0	0.0	10.0	3.4	63.8	4.8	58.3	7.7	55_Olaparib	PARP1/2	100.0	0.0	59.2	5.3	67.8	18.0	74.5	17.3
20_AZD5363	Akt1/2/3	68.7	7.5	22.4	8.0	69.2	5.5	77.6	9.0	56_Panobinostat	HDAC	62.8	8.6	0.1	0.1	76.4	2.5	4.2	1.3
21_Axitinib	VEGFR1/2/3, PDGFRβ and c-Kit	100.0	0.0	34.4	12.0	100.0	0.0	119.0	9.6	57_PF-04449913	HSP90	100.0	0.0	80.6	15.1	100.0	9.2	87.8	15.2
22_Cediranib	VEGFR, Flt	100.0	0.0	0.0	0.0	72.9	14.9	32.7	44.7	58_Ruxolitinib	JAK1/2	29.9	12.2	6.5	2.3	100.0	0.0	67.8	8.1
23_Imatinib	v-Abl, c-Kit and PDGFR	45.3	9.3	86.4	3.7	49.9	10.9	87.9	7.8	59_Sotrastaurin	PKC	100.0	0.0	30.0	8.3	100.0	0.0	51.0	14.9

Table 1. *Cont.*

| Drug | Target | Single-Cell Model | | | | Cancer Spheroid Model | | | | Drug | Target | Single-Cell Model | | | | Cancer Spheroid Model | | | |
| | | p-EGFR Expression [%] | | Cell Viability [%] | | p-EGFR Expression [%] | | Cell Viability [%] | | | | p-EGFR Expression [%] | | Cell Viability [%] | | p-EGFR Expression [%] | | Cell Viability [%] | |
		Average	SD	Average	SD	Average	SD	Average	SD			Average	SD	Average	SD	Average	SD	Average	SD
24_Pazopanib HCl	VEGFR1/2/3, PDGFR, FGFR, c-Kit	89.0	14.2	46.2	11.5	60.4	23.1	95.8	8.5	60_Vemurafenib	B-RafV600E	34.8	5.5	44.5	2.8	97.1	5.8	70.7	24.4
25_Sunitinib	VEGFR2 and PDGFRβ	100.0	0.0	0.0	0.0	100.0	0.0	39.3	43.6	61_Vismodegib	Hedgehog/smothen	100.0	0.0	97.2	9.7	100.0	0.0	111.9	5.6
26_Tandutinib	FLT3, PDGFR, and KIT	100.0	0.0	8.0	5.0	100.0	0.0	23.6	4.2	62_PHA-665752	c-Met inhibitor	100.0	0.0	102.0	11.4	100.0	0.0	130.0	13.2
27_Tivozanib	VEGFR, c-Kit, PDGFR	19.8	4.6	3.2	2.2	64.0	22.2	56.7	5.9	63_TMZ	alkylating agent	88.5	6.3	80.7	7.8	49.6	21.1	97.3	13.2
28_Regorafenib	VEGFR1/2/3, PDGFRβ, Kit, RET and Raf-1	20.7	4.8	0.6	0.7	100.0	0.0	6.3	1.1	64_Amoral	morpholine antifungal drug	87.9	16.6	86.0	9.5	51.9	22.1	99.9	9.5
29_Vandetanib	VEGFR2	53.3	4.7	0.7	1.1	47.0	23.6	96.0	14.6	65_Mevas	HMG-CoA reductase inhibitor	100.0	0.0	88.9	7.2	84.0	11.4	107.4	8.1
30_Cabozantinib	VEGFR2,c-Met, Ret, Kit, Flt-1/3/4, Tie2, and AXL	28.7	7.0	8.0	5.1	29.3	13.3	83.0	16.8	66_Amio	antiarrhythmic medication	75.2	27.9	96.6	3.9	73.1	10.7	106.5	2.5
31_Foretinib	HGFR and VEGFR, mostly for Met and KDR	26.8	6.5	5.4	3.3	37.4	3.3	69.1	19.1	67_Flu	Anticholesterol agent. HMG-CoA inhibitor	100.0	0.0	87.4	15.2	100.0	0.0	105.7	5.3
32_Crizotinib	Met, ALK	70.4	4.4	0.0	0.0	78.1	23.7	0.2	0.2	68_Myco_acid	Inosine-5'-monophosphate dehydrogenase inhibitor	100.0	0.0	33.5	8.7	100.0	0.0	100.3	4.8
33_INCB28060	Met	62.5	13.5	88.9	7.4	52.2	10.6	115.4	12.9	69_Raloxi	Estrogen receptor inhibitor	100.0	0.0	95.1	4.9	100.0	0.0	103.0	13.4
34_LEE011	CDK4/6	100.0	0.0	64.2	9.0	100.0	0.0	86.0	5.7	70_Astemi	Histamine receptor ligand	100.0	0.0	78.6	8.5	75.4	14.9	93.5	11.4
35_PD 0332991	CDK4/6	100.0	0.0	1.2	0.9	100.0	0.0	85.7	10.1	71_Ferre	Retinoic acid receptor ligand	100.0	0.0	71.0	7.4	100.0	0.0	111.9	16.6
36_LY2835219	CDK4/6	66.6	8.5	0.0	0.0	75.1	25.9	82.1	11.2	-	-	-	-	-	-	-	-	-	-

2.7. Western Blot Assay

Total cell lysates from the A549 lung cancer cell lines were prepared using the Complete™ Lysis-M buffer solution (Roche Life Science, Penzberg, Germany). Protein extracts were resolved using 4–20% Mini-PROTEAN TGX™ Precast Protein Gels (Bio-Rad, Berkeley, CA, USA) and transferred onto iBlot®PVDF gel transfer stack membranes (Thermo Fisher Scientific, Waltham, MA, USA). After blocking non-specific binding sites for 1 h in 5% BSA in Tris-buffered saline containing 0.1% Tween-20 (TBS-T), the membranes were incubated overnight at 4 °C with specific primary antibodies. The antibodies included anti-p-EGFR (phospho Y1092) antibody (1:1000, Abcam, Cambridge, UK) and anti-beta actin (1:2000, Abcam, Cambridge, UK). These were used following the manufacturers' instructions.

3. Results and Discussion

3D cell cultures such as spheroids could be a highly useful tool for simulating the cancer microenvironment and predicting drug efficacy. To precisely predict the in vivo efficacy of targeting drugs, a spheroid cytoskeleton, as well as the expression levels of a target protein (p-EGFR), were measured by immunofluorescence staining in a high throughput manner. Cell viabilities were also measured by Calcein AM staining in a different chip. Using micropillar and microwell chips, the changes in expression levels of p-EGFR were measured in 3D-cultured cells to screen the targeting efficiency of 70 drugs. By comparing p-EGFR expression levels and cell viability in the single-cell model with those in the spheroid model, the possible drug candidate options could be narrowed down. Before drug administration, immunostaining for p-EGFR using the chip was verified by measuring p-EGFR in AGS [29] and A549 [30], which are well-known cell lines of p-EGFR overexpression. Figure 3 shows the relative expression levels of p-EGFR in these two cell lines. The A549 cell line showed a higher expression of p-EGFR than the AGS cell line.

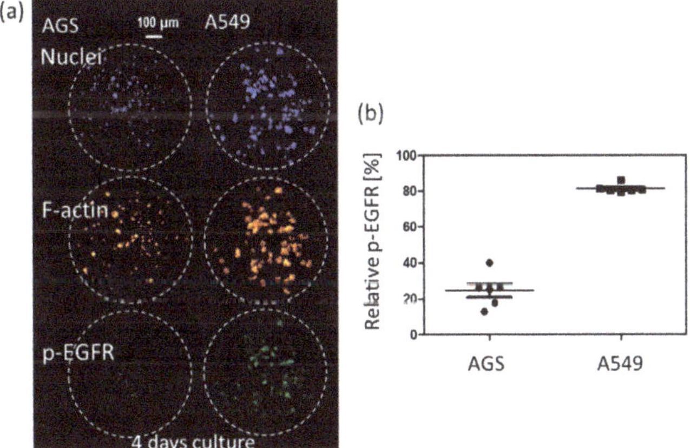

Figure 3. Relative phospho-epidermal growth factor receptor (p-EGFR) in AGS and A549 cell lines in a four day-culture. (**a**) Immunostaining images of two cell line. (**b**) Relative p-EGFR in AGS and A549 cell lines.

3.1. p-EGFR Expression in Spheroids

Figure 4 shows changes in p-EGFR expression levels over time. Figure 4a shows control images (no drug) according to days. As shown in the visualized images, cells grew for up to 14 days to form spheroids in 0.5% w/v alginate. Western blotting showed that the relative expression levels of p-EGFR in the spheroid model were higher than in the single-cell model, as shown in Figure 4b. After day seven, spheroids did not continue growing to form big spheroids (cancer spheroid), as shown in Figure 4c. Single-cell or small spheroids grown for under seven days showed low relative expression of p-EGFR. Relative expression was calculated by dividing the area of p-EGFR staining (green) by the area of nucleus staining (blue). At all times before the seven days, p-EGFR expression was weak, and the green fluorescence was faint, as shown in Figure 4a. However, when the spheroids were fully grown over 100 μm in diameter, all cancer spheroids in the alginate spot showed overexpression of p-EGFR (Figure 4a). Figure 4d shows that the relative expression levels of p-EGFR were maintained above 90% after seven days. To confirm whether the drugs inhibit p-EGFR, thus killing the cancer cells in the spheroid model, cancer spheroids overexpressing p-EGFR should be exposed to the drugs after seven days. Hence, we decided on drug treatment for seven days in the spheroid model.

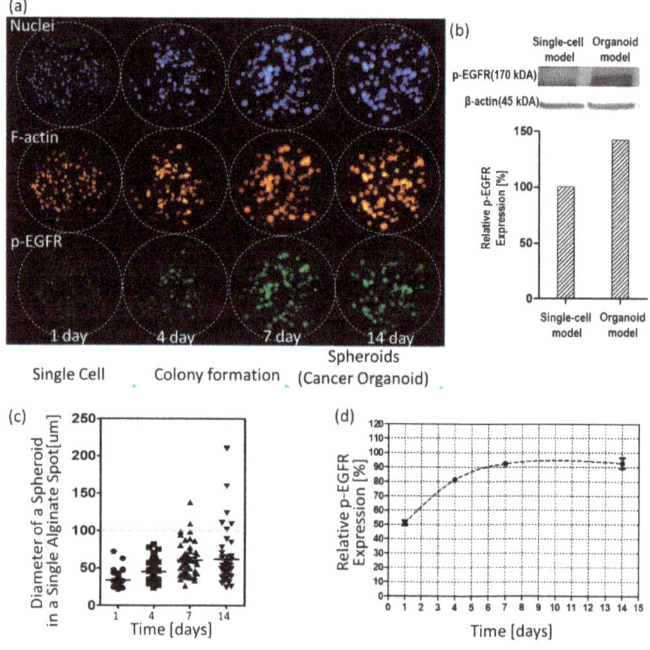

Figure 4. Changes in the nucleus, filamentous actin (F-actin), and p-EGFR in cells and spheroids over time. (**a**) Three color images of the same alginate spots. (**b**) Western blot of p-EGFR expression in single-cell and spheroid models. (**c**) Calculated diameters of spheroids in a single alginate spot over time. (**d**) Relative expression levels of p-EGFR in the spheroid model over time.

3.2. Drug Selection of Targeting p-EGFR Based on the Spheroid Model

Inhibition of p-EGFR occurred early on after administering the drug treatment; immunostaining of the spheroids was conducted at 6, 24, 48, and 144 h after the treatment. Number of cells were effect to whole p-EGFR in an alginate spot. Cell death or cell growth inhibition due to drug may reduce p-EGFR expression compared to that in the control (no drug), even if the drug did not inhibit p-EGFR. Thus, the nucleus of the spheroids should be considered for calculating the relative expression of p-EGFR. After screening 70 drugs (Table 1), cabozantinib was selected as a representative tyrosine phosphorylation inhibitor. Figure 5 shows that cabozantinib inhibited p-EGFR expression. Figure 5a and 5b show that a 36 by 12 spheroid array was formed on the micropillar and that the same micropillar chip was stained with three colors. If the normalized expression of p-EGFR with the drug treatment was significantly different from that in the control (P-value <0.05) and the cytoskeleton was present at >50%, the drugs were classified as p-EGFR inhibitors; thus, cabozantinib was considered an efficient p-EGFR inhibitor in the 3D spheroid model (Figure 5c). If the cytoskeleton was less than 50% after staining (6, 24, 48, and 144 h), normalized p-EGFR was low because the cell did not maintain its structure, and p-EGFR binding to the cell membrane was also disrupted. The cell viability was also measured by staining the spheroid with Calcein AM (Green) (Figure 5d) and used for selecting drugs with p-EGFR expression.

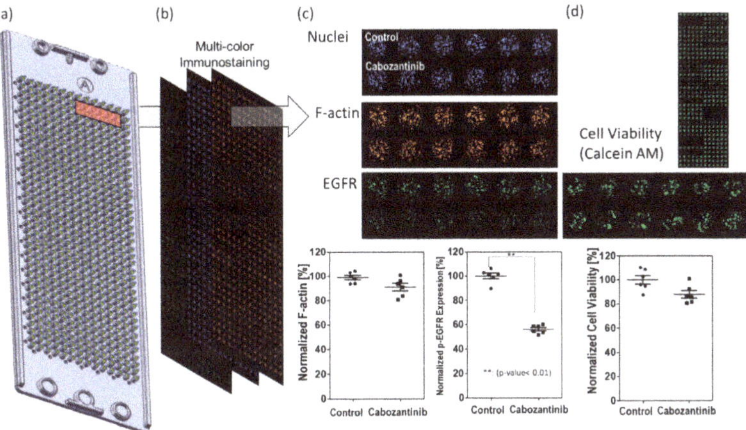

Figure 5. Measurement of p-EGFR inhibition. (**a**) Micropillar chips containing spheroids on the micropillar. (**b**) Three color images were taken of the same micropillar chip. (**c**) Normalized F-actin for detecting the cytoskeleton and normalized expression of p-EGFR for detecting p-EGFR expression. If drug-induced inhibition of normalized p-EGFR is significantly different from that in the control (P-value < 0.05) but not different from that in samples with normalized F-actin, the drugs were classified as p-EGFR inhibitors. The graph shows effective inhibition of p-EGFR on treatment with cabozantinib. (**d**) Cell viability by staining with Calcein AM (Green).

To confirm that the drug inhibited p-EGFR, p-EGFR expression level was observed at 6, 24, 48, and 144 h after drug treatment. Cell viability was also measured six days after drug treatment to determine whether inhibition of p-EGFR caused cell death. Thus, inhibition of p-EGFR and cell viability were essential for identifying the best p-EGFR-targeting drugs. Figure 6 shows the minimum p-EGFR level and cell viability at day seven. The spheroid model showed sensitivity to two drugs (AEE788 (#2), Afatinib (#3)) from among 10 drugs (#2~#11) targeting EGFR out of seventy drugs, whereas the single cell model showed sensitivity to five drugs, as shown in Figure 6a. In the spheroid model, eight drugs targeting EGFR did not evoke a response, which means that the spheroid models show higher resistance to drugs than the single-cell model. Among the 70 drugs, non-EGFR targeting drugs were also screened in the chips. Figure 6b and 6c show the minimum p-EGFR expression and sevenday cell viability of seventy drugs using the singe-cell and spheroid models, respectively. Two non-EGFR targeting drugs (XL147 (#14) and Dovitinib (#39)) showed p-EGFR inhibitor activity and induced cell death in the A549 cancer spheroids as well as the single cells. Figure 6d shows the response change for thirteen drugs in the spheroid model. Thirteen drugs inhibited more than 50% of p-EGFR and induced more than 50% cell death in single-cell models. Cabozantinib (#30), dasatinib (#41), and foretinib (#31) showed similar p-EGFR inhibition, but the spheroids showed high resistance to these drugs. AEE788 (#2), afatinib (#3), regorafenib (#28), bosutinib (#40) did not highly inhibit p-EGFR (phospho Y1092); however, they killed cells in the spheroid model. Tivozanib (#27), AUY922 (#49), dabrafenib (#50), ruxolitinib (#58), and vemurafenib (#60) showed high resistance in the spheroid model. Sixteen drugs inhibited less than 50% p-EGFR expression and induced more than 50% cell death in the single-cell model, but did not induce cell death in the spheroid model (Figure 6e). Although the inhibition of p-EGFR expression was similar, spheroids showed high resistance (high viability) to dacomitinib (#6), Gefitinib (#7), CO-1686 (#11), AZD5363 (#20), Pazopanib (#24), Vandetanib (#29), LY2835219 (#36), AZD4547 (#37), and trametinib (#44) as shown in Figure 6e. Bortezomib (#45) AZD6244 (#43) and

AZD2014 (#16) decreased p-EGFR inhibition but still induced cell death. Gefitinib (#7), CO-1686 (#11), and dacomitinib (#6) did not induce cell death and did not inhibit p-EGFR in the spheroid model. Interestingly, in the spheroid model, XL147 (#14), BEZ235 (#19) inhibited the expression of p-EGFR more than in the single-cell model. Overall, the number of drugs targeting p-EGFR was two (Dovitinib and XL147) according to the spheroid model, which is less than the twenty-nine drugs identified from the single-cell model. Thus, cancer spheroid models can be used to narrow down options and identify highly effective target drugs.

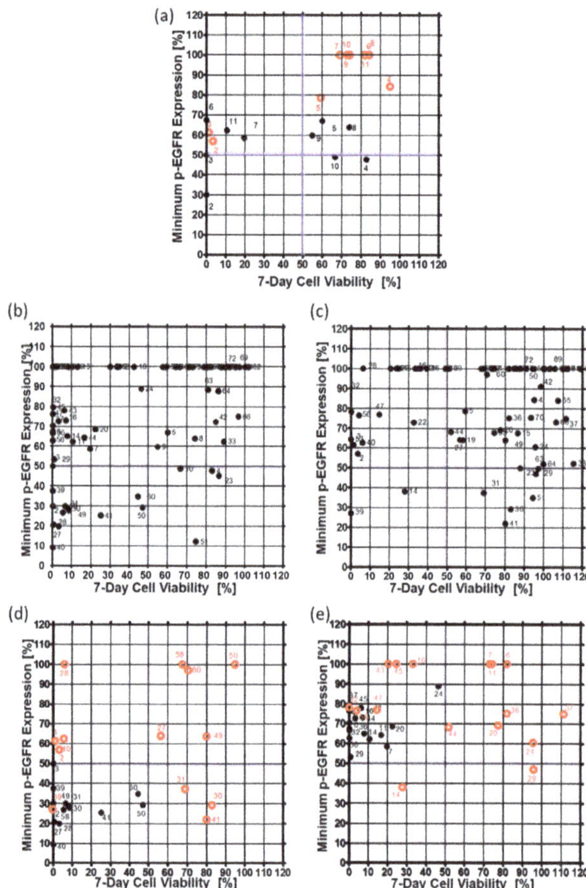

Figure 6. Minimum expression levels of p-EGFR and cell viability on day seven in the single-cell model and the spheroid model. The minimum expression levels of p-EGFR were selected from among the values at 6, 24, 48, and 144 h after drug treatment. (**a**) Expression of p-EGFR and cell viability in the single-cell model. (**b**) Expression of p-EGFR and cell viability in the spheroid model. (**c**) Changes in the expression levels of p-EGFR and cell viability after treatment with thirteen drugs in the single-cell and the spheroid models. Thirteen drugs are under the 50% minimum expression level of p-EGFR and 50% cell viability in the single-cell model. (**d**) Changes in expression of p-EGFR and cell viability due to treatment with sixteen drugs at single-cell and spheroid models. Sixteen drugs were under the 90% minimum expression level of p-EGFR and 50% cell viability in the single-cell model. (**e**) Changes in expression of p-EGFR and cell viability due to treatment with sixteen drugs at single-cell and spheroid models. Sixteen drugs were under the 90% minimum expression level of p-EGFR and 50% cell viability in the single-cell model.

4. Conclusions

Our cancer spheroid array chip was developed using a micropillar and microwell structure designed to evaluate drug efficacy. To form a high-density cancer spheroid array, the encapsulated A549 cells in alginate were grown for over seven days. Cancer spheroids attached to the micropillar were moved to fresh media or staining reagents by placing them in new microwells filled with new reagents. After forming spheroids with diameter greater than 100 µm in a 12 by 36 pillar array chip (25 mm by 75 mm), we confirmed that the A549 cell line showed overexpression of p-EGFR in cancer spheroids. Cancer spheroids were treated with seventy drugs (six replicates) for evaluating drug efficacy. In the single-cell model, eleven drugs were identified as p-EGFR-targeting drugs, but in the cancer spheroid model, only two drugs were identified as highly efficient p-EGFR-targeting drugs. XL147 and Dovitinib showed p-EGFR inhibition and induced the death of A549 cancer spheroids as well as single cells. After comparing the drug response of single-cell and cancer spheroid models, it was shown that the spheroid model could narrow down the list of drug candidates by identifying high-efficacy p-EGFR-targeting drugs.

Cancer spheroid array chips, by enabling the identification of effective drugs from a huge library, could be useful tools for drug discovery. When compared to the 2D cell culture system, cancer spheroids showed higher resistance to the compounds. Besides, immunostaining of cancer spheroid array chips can serve in determining the mechanism of action of drugs in in vivo-like environments. Because animal and clinical trials for many drug candidates are expensive, having a method to pinpoint the best drug candidate can significantly reduce drug development costs and time.

Author Contributions: J.W.C. performed the cell viability assay and data curation and revised the manuscript. S.-Y.L. performed experiments and data curation. D.W.L. supervised the experiments and wrote the paper.

Funding: This research was supported by the National Research Foundation of Korea (NRF). The grant was founded by the Korean government (MSIT: No. 2018R1C1B5045068). This research was also supported by a grant given by the Korea Health Technology R&D Project through the Korea Health Industry Development Institute (KHIDI), funded by the Ministry of Health & Welfare, Republic of Korea (grant number: HI17C-2412-010017).

Conflicts of Interest: The authors declare no conflict of interest.

References

1. Li, G.N.; Livi, L.L.; Gourd, C.M.; Deweerd, E.S.; Hoffman-Kim, D. Genomic and Morphological Changes of Neuroblastoma Cells in Response to Three-Dimensional Matrices. *Tissue Eng.* **2007**, *13*, 1035–1047. [CrossRef] [PubMed]
2. Lv, D.; Hu, Z.; Lu, L.; Lu, H.; Xu, X. Three-Dimensional Cell Culture, A Powerful Tool in Tumor Research and Drug Discovery. *Oncol. Lett.* **2017**, *14*, 6999–7010. [CrossRef] [PubMed]
3. Bokhari, M.; Carnachan, R.J.; Cameron, N.R.; Przyborski, S.A. Culture of HepG2 Liver Cells on Three Dimensional Polystyrene Scaffolds Enhances Cell Structure and Function during Toxicological Challenge. *J. Anat.* **2007**, *211*, 567–576. [CrossRef] [PubMed]
4. Elkayam, T.; Amitay-Shaprut, S.; Dvir-Ginzberg, M.; Harel, T.; Cohen, S. Enhancing the Drug Metabolism Activities of C3a—A Human Hepatocyte Cell Line—By Tissue Engineering within Alginate Scaffolds. *Tissue Eng.* **2006**, *12*, 1357–1368. [CrossRef] [PubMed]
5. Gurski, L.A.; Jha, A.K.; Zhang, C.; Jia, X.; Farach-Carson, M.C. Hyaluronic Acid-Based Hydrogels as 3D Matrices for in Vitro Evaluation of Chemotherapeutic Drugs Using Poorly Adherent Prostate Cancer Cells. *Biomaterials* **2009**, *30*, 6076–6085. [CrossRef] [PubMed]
6. Schyschka, L.; Sánchez, J.J.M.; Wang, Z.; Burkhardt, B.; Müller-Vieira, U.; Zeilinger, K.; Bachmann, A.; Nadalin, S.; Damm, G.; Nussler, A.K. Hepatic 3D Cultures but Not 2D Cultures Preserve Specific Transporter Activity for Acetaminophen-Induced Hepatotoxicity. *Arch. Toxicol.* **2013**, *87*, 1581–1593. [CrossRef] [PubMed]
7. Es, E.; Aboulkheyr, H.; Montazeri, L.; Aref, A.R.; Vosough, M.; Baharvand, H. Personalized Cancer Medicine, an Organoid Approach. *Trends Biotechnol.* **2018**, *36*, 358–371.

8. Jabs, J.; Zickgraf, F.M.; Park, J.; Wagner, S.; Jiang, X.; Jechow, K.; Kleinheinz, K.; Toprak, U.H.; Schneider, M.A.; Meister, M. Screening Drug Effects in Patient-Derived Cancer Cells Links Organoid Responses to Genome Alterations. *Mol. Syst. Biol.* **2017**, *13*, 955. [CrossRef]

9. van de Wetering, M.; Francies, H.E.; Francis, J.M.; Bounova, G.; Iorio, F.; Pronk, A.; van Houdt, W.; van Gorp, J.; Taylor-Weiner, A.; Kester, L. Prospective Derivation of a Living Organoid Biobank of Colorectal Cancer Patients. *Cell* **2015**, *161*, 933–945. [CrossRef]

10. Dekkers, J.F.; Berkers, G.; Kruisselbrink, E.; Vonk, A.; de Jonge, H.R.; Janssens, H.M.; Bronsveld, I.; van de Graaf, E.A.; Nieuwenhuis, E.E.S.; Houwen, R.H.J. Characterizing Responses to CFTR-Modulating Drugs Using Rectal Organoids Derived from Subjects with Cystic Fibrosis. *Sci. Transl. Med.* **2016**, *8*, 344ra84. [CrossRef]

11. Dekkers, J.F.; Wiegerinck, C.L.; de Jonge, H.R.; Bronsveld, I.; Janssens, H.M.; de Groot, K.M.; Brandsma, A.M.; de Jong, N.W.M.; Bijvelds, M.J.C.; Scholte, B.J. A Functional Cftr Assay Using Primary Cystic Fibrosis Intestinal Organoids. *Nat. Med.* **2013**, *19*, 939. [CrossRef] [PubMed]

12. Freedman, B.S.; Brooks, C.R.; Lam, A.Q.; Fu, H.; Morizane, R.; Agrawal, V.; Saad, A.F.; Li, M.K.; Hughes, M.R.; Werff, R.V. Modelling Kidney Disease with Crispr-Mutant Kidney Organoids Derived from Human Pluripotent Epiblast Spheroids. *Nat. Commun.* **2015**, *6*, 8715. [CrossRef] [PubMed]

13. Xinaris, C.; Brizi, V.; Remuzzi, G. Organoid Models and Applications in Biomedical Research. *Nephron* **2015**, *130*, 191–199. [CrossRef] [PubMed]

14. Boj, S.F.; Hwang, C.I.; Baker, L.A.; Chio, I.I.C.; Engle, D.D.; Corbo, V.; Jager, M.; Ponz-Sarvise, M.; Tiriac, H.; Spector, M.S. Organoid Models of Human and Mouse Ductal Pancreatic Cancer. *Cell* **2015**, *160*, 324–338. [CrossRef] [PubMed]

15. Drost, J.; Clevers, H. Organoids in Cancer Research. *Nat. Rev. Cancer* **2018**, *18*, 407. [CrossRef] [PubMed]

16. Baker, L.A.; Tiriac, H.; Clevers, H.; Tuveson, D.A. Modeling Pancreatic Cancer with Organoids. *Trends Cancer* **2016**, *2*, 176–190. [CrossRef] [PubMed]

17. King, S.M.; Quartuccio, S.; Hilliard, T.S.; Inoue, K.; Burdette, J.E. Alginate Hydrogels for Three-Dimensional Organ Culture of Ovaries and Oviducts. *J. Vis. Exp.* **2011**, *52*, e2804. [CrossRef] [PubMed]

18. Weeber, F.; Ooft, S.N.; Dijkstra, K.K.; Voest, E.E. Tumor Organoids as a Pre-Clinical Cancer Model for Drug Discovery. *Cell Chem. Biol.* **2017**, *24*, 1092–1100. [CrossRef] [PubMed]

19. Astashkina, A.; Grainger, D.W. Critical Analysis of 3-D Organoid in Vitro Cell Culture Models for High-Throughput Drug Candidate Toxicity Assessments. *Adv. Drug Deliv. Rev.* **2014**, *69*, 1–18. [CrossRef]

20. Verissimo, C.S.; Overmeer, R.M.; Ponsioen, B.; Drost, J.; Mertens, S.; Verlaan-Klink, I.; van Gerwen, B.; van der Ven, M.; van de Wetering, M.; Egan, D.A. Targeting Mutant Ras in Patient-Derived Colorectal Cancer Organoids by Combinatorial Drug Screening. *Elife* **2016**, *5*, e18489. [CrossRef]

21. Du, G.; Fang, Q.; den Toonder, J.M.J. Microfluidics for Cell-Based High Throughput Screening Platforms—A Review. *Anal. Chim. Acta* **2016**, *903*, 36–50. [CrossRef] [PubMed]

22. Kang, G.; Lee, J.H.; Lee, C.S.; Nam, Y. Agarose Microwell Based Neuronal Micro-Circuit Arrays on Microelectrode Arrays for High Throughput Drug Testing. *Lab Chip* **2009**, *9*, 3236–3242. [CrossRef]

23. Tung, Y.C.; Hsiao, A.Y.; Allen, S.G.; Torisawa, Y.; Ho, M.; Takayama, S. High-Throughput 3D Spheroid Culture and Drug Testing Using a 384 Hanging Drop Array. *Analyst* **2011**, *136*, 473–478. [CrossRef] [PubMed]

24. Kang, J.; Lee, D.W.; Hwang, H.J.; Yeon, S.; Lee, M.; Kuh, H. Mini-Pillar Array for Hydrogel-Supported 3D Culture and High-Content Histologic Analysis of Human Tumor Spheroids. *Lab Chip* **2016**, *16*, 2265–2276. [CrossRef] [PubMed]

25. Lee, L.; Woo, D.; Choi, Y.; Seo, Y.J.; Lee, M.; Jeon, S.Y.; Ku, B.; Kim, S.; Yi, S.H.; Nam, D. High-Throughput Screening (HTS) of Anticancer Drug Efficacy on a Micropillar/Microwell Chip Platform. *Anal. Chem.* **2013**, *86*, 535–542. [CrossRef] [PubMed]

26. Lee, S.Y.; Doh, I.; Nam, D.; Lee, D.W. 3D Cell-Based High-Content Screening (HCS) Using a Micropillar and Microwell Chip Platform. *Anal. Chem.* **2018**, *90*, 8354–8361. [CrossRef] [PubMed]

27. Doh, I.; Kwon, Y.; Ku, B.; Lee, D.W. Drug Efficacy Comparison of 3D Forming and Preforming Sphere Models with a Micropillar and Microwell Chip Platform. *SLAS Discov. Adv. Life Sci. RD* **2019**, *24*, 476–483. [CrossRef] [PubMed]

28. Lee, L.; Woo, D.; Doh, I.L.; Nam, D. Unified 2D and 3D Cell-Based High-Throughput Screening Platform Using a Micropillar/Microwell Chip. *Sens. Actuators B Chem.* **2016**, *228*, 523–528. [CrossRef]

29. Wang, D.; Wang, B.; Wang, R.; Zhang, Z.; Lin, Y.; Huang, G.; Lin, S.; Jiang, Y.; Wang, W.; Wang, L. High Expression of Egfr Predicts Poor Survival in Patients with Resected T3 Stage Gastric Adenocarcinoma and Promotes Cancer Cell Survival. *Oncol. Lett.* **2017**, *13*, 3003–3013. [CrossRef]
30. Hung, M.S.; Chen, I.; Lung, J.; Lin, P.; Li, Y.; Tsai, Y. Epidermal Growth Factor Receptor Mutation Enhances Expression of Cadherin-5 in Lung Cancer Cells. *PLoS ONE* **2016**, *11*, e0158395. [CrossRef]

 *micromachines*

Article

Liver-on-a-Chip-Magnetic Nanoparticle Bound Synthetic Metalloporphyrin-Catalyzed Biomimetic Oxidation of a Drug in a Magnechip Reactor

Balázs Decsi [1], Réka Krammer [1], Kristóf Hegedűs [2], Ferenc Ender [2,3], Benjámin Gyarmati [4], András Szilágyi [4], Róbert Tőtős [5], Gabriel Katona [5], Csaba Paizs [5], György T. Balogh [6,7], László Poppe [1,5,8,*] and Diána Balogh-Weiser [1,4,*]

[1]   Department of Organic Chemistry and Technology, Budapest University of Technology and Economics, 1111 Budapest, Műegyetem rkp. 3, Hungary; decsi.balazs@mail.bme.hu (B.D.); krareka@gmail.com (R.K.)

[2]   SpinSplit Llc., 1082 Budapest, Leonardo da Vinci u. 43b, Hungary; k.hegedus@spinsplit.com (K.H.); ender@eet.bme.hu (F.E.)

[3]   Department of Electron Devices, Budapest University of Technology and Economics, 1117 Budapest, Magyar tudósok krt. 2, Hungary

[4]   Department of Physical Chemistry and Materials Science, Budapest University of Technology and Economics, 1111 Budapest, Műegyetem rkp. 3, Hungary; bgyarmati@mail.bme.hu (B.G.); aszilagyi@mail.bme.hu (A.S.)

[5]   Biocatalysis and Biotransformation Research Centre, Faculty of Chemistry and Chemical Engineering, Babeș-Bolyai University of Cluj-Napoca, 400028 Cluj-Napoca, Arany Janos street 11, Romania; totos.robert@yahoo.com (R.T.); gabik@chem.ubbcluj.ro (G.K.); paizs@chem.ubbcluj.ro (C.P.)

[6]   Department of Chemical and Environmental Process Engineering, Budapest University of Technology and Economics, 1111 Budapest, Budafoki út 8, Hungary; gytbalogh@mail.bme.hu

[7]   Department of Pharmacodynamics and Biopharmacy, University of Szeged, 6720 Szeged, Eötvös u. 6, Hungary

[8]   SynBiocat Llc., 1172 Budapest, Szilasliget u 3, Hungary

*   Correspondence: poppe@mail.bme.hu (L.P.); dweiser@mail.bme.hu (D.B.-W.); Tel.: +36-1-4633299 (L.P.); +36-1-4631285 (D.B.-W.)

Received: 4 September 2019; Accepted: 27 September 2019; Published: 1 October 2019

Abstract: Biomimetic oxidation of drugs catalyzed by metalloporphyrins can be a novel and promising way for the effective and sustainable synthesis of drug metabolites. The immobilization of 5,10,15,20-tetrakis(2,3,4,5,6-pentafluorophenyl)iron(II) porphyrin (FeTPFP) and 5,10,15,20-tetrakis-(4-sulfonatophenyl)iron(II) porphyrin (FeTSPP) via stable covalent or rapid ionic binding on aminopropyl functionalized magnetic nanoparticles (MNPs-NH$_2$) were developed. These immobilized catalysts could be efficiently applied for the synthesis of new pharmaceutically active derivatives and liver related phase I oxidative major metabolite of an antiarrhythmic drug, amiodarone integrated in a continuous-flow magnetic chip reactor (Magnechip).

Keywords: drug metabolism; biomimetic oxidation; microfluidics; organ-on-a-chip; liver-on-a-chip

---

1. Introduction

Continuous flow chemistry is one of the fastest evolving discipline. Its application is widespread in the field of chemistry. It is used not only in petrol chemistry, but also in the commodity chemical industry. Moreover, the interest of the pharmaceutical industry in continuous flow chemistry is getting intensified, because this methodology offers robust and well controlled strategy to produce active substance and drug form [1]. The expenses and time of both, drug development and manufacturing can be decreased with its application offering the easiest way to implement design space approach of drug manufacturing [2,3]. A flow chemistry set is consisted of several components. In the simplest instance, the solutions of reagents are moved through a reactor by a pump. The reactor could contain solid or immobilized catalyst and the flowed-out reaction mass from the reactor is processed, even in other continuous down-stream steps. Continuous flow catalysis opens up the possibility to perform reactions that cannot be performed in traditional, batch circumstances which involves dangerous reagents, unstable intermediates or has extremely high reaction enthalpy [4,5]. The good controllability is due to the great surface to volume ratio, that enables large heat transfer, moreover the plug like operation ensures perfect mixing of the fluid intakes even in laminar conditions, while there is a small amount of liquid in the reactor at the same time. The productivity of tubular reactors differs from the traditional batch reactors. In a continuous reactor the reaction time can be assimilated with the average contact time of the reactants with the catalyst and is specified by the cross-section area and the length of the reactor, by the volume of the catalyst and by the flow rate of the reactants [6]. A disadvantage is that every reagent must be held in solution, no precipitation is allowed because it could cause blocking in the tubing. Continuous flow systems can be classified by the diameter of the reactor. While tubular reactors with over than 500 μm diameter are known as mesoreactors, microfluidics engages microreactors with less than 500 μm diameter [7]. Since the 1970s and the 1980s several microfluidic appliances were appeared such as microfluidic sensors, pumps, and valves characterized this period. The discipline had been evolved since the defining work of Manz and co-workers at the Fifth International Conference on Solid-State Sensors and Actuators (Transducers '89) [8] where they summarized the possibilities of applications of microfluidics.

It should be noted, that the aim of microfluidics is not to miniaturize the size of the appliances (which can have also benefits), but mostly to reduce the diameter of the flow space, with major impact upon the nature of transport and transfer phenomena. The reason of these observations is the microscopic amount of liquid in the tubing. The "micro" prefix is not meaning the size of the chip or the diameter of the conduits, but the small overall volume that causes the changes in the flow ratio [9].

Microfluidic instruments can be used in a wide range of fields, starting with analytical [10], biological [11], and synthetic usage [12]. Special representations of microfluidic appliances are "Lab-on-a-chip" reactors, which conceptions were established by Burns et al. [13]. Their goal was to create a miniaturized device which contains all the necessary components (pumps, valves, tubing, mixers, sensors, detectors etc.) that is suitable to analyze very small volume (nanoliters) of DNA sample. Hereby they have created an integrated, effective, reliable, inexpensive and compact tool, that can analyze DNA in a short time and can be used widespread not only in medical diagnoses but also in agriculture.

"Organ-on-a-chip" reactors, being a subtype of Lab-on-a-chip family, mimic the (coordinated) operation of the living organism's one (or more) organ(s). Enzyme catalyzed biotransformations can be run with the use of them in a well-controlled way. With the immobilization of enzymes on solid supports, a heterogeneous catalytic system can be obtained which is stable enough to reuse it for several time [14]. Organ-on-a-chip reactors can be also used in drug and in preclinical drug research, which is traditionally expensive and time consuming. During the research of in vitro cell based and in vivo experiments were carried out to characterize structure-effect relationship. However, the prior methodology is not able to reveal interactions between tissues and cells, whilst the analysis of the latter is difficult and rises ethical issues. With the use of organ-on-a-chip reactors one could ensure a physiologically relevant in vitro system mimicking the in vivo metabolism of drug candidates [15].

In the human body the main metabolic pathway of drugs and xenobiotics are enzymatic biotransformations, which are generally started by oxidation catalyzed by the CYP450 isoenzymes related oxidative metabolism. The CYP enzymes are mixed type monooxygenase enzymes, located in the endoplasmic reticulum of cells, and in high concentration in the liver. The in vitro methods that are used in preclinical studies are based mostly on hepatocytes and liver microsomes. The latter method is used the most often to study the metabolism of drug molecules of the CYP enzymes. However, in these experiments a complex biological matrix is formed due to the necessity of several coenzymes and their regeneration. This complexity of the matrix makes the analysis difficult and allows only quantitative analysis [16,17]. Therefore, such metabolism mimicking in vitro methods are in the focus of research that can simulate the metabolism of drug molecules without the necessity of complex biological matrix, and can produce metabolites directly from the parent molecule. Synthetic metalloporphyrins can be applied for these purposes. Application of metalloporphyrins is based on their structural resemblance to the heme prosthetic group being present within the active site of the CYP enzymes [18,19]. However, the major disadvantage of the synthetic metalloporphyrins is their easy degradation (autooxidation) under homogeneous oxidative conditions resulting in a short lifetime. This situation can be improved by immobilizing them on solid support by covalent binding or by secondary interactions like ionic bond [20,21]. In our recent work, we showed that meso-tetra(parasulphonato)iron porphyrin could be immobilized on aminopropyl group-modified silica by ionic bond, and the bonded catalyst could be applied in packed bed reactor to perform biomimetic oxidations under continuous flow conditions providing metabolites of antiarrhythmic drug, amiodarone [22].

Another possibility is to use magnetic nanoparticles (MNPs) as solid support to immobilize metalloporphyrins. MNPs proved to be suitable carrier of several types of catalysts [23–25]. Their application opens up the possibility of creating microfluidic magnetic chip reactors in which MNPs can be trapped by external permanent magnets at predesigned positions and reagents are flowed through the MNP-filled microchambers. It was feasible to bind enzymes to MNPs and use them in magnechip reactors [14,26–28]. Previously, phenylalanine ammonia-lyase (PAL) was immobilized on MNPs coated with an aminopropyl group-modified silica shell. The prepared PAL biocatalyst was used to convert phenylalanine and five other unnatural analogues to their respective arylacrylate derivative in a microfluidic magnechip reactor under continuous flow conditions. The catalytic activity of the PAL biocatalyst did not decrease, even after 14 h of continuous operation [14,26].

Combining the two above-mentioned methods, a new technique can be introduced. With the usage of synthetic metalloporphyrins bound to MNPs, a system can be established that can mimic the liver on a chip (Figure 1). In this study, two metalloporphyrin derivatives were immobilized onto MNPs via either ionic or covalent interactions. The immobilized catalysts were integrated in a microfluidic magnetic chip reactor and the biomimetic oxidation of an antiarrhythmic drug, amiodarone was investigated.

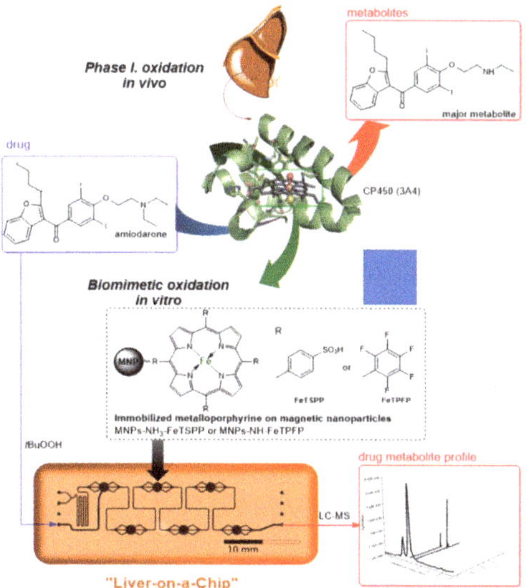

Figure 1. Biomimetic oxidation of amiodarone catalyzed by metalloporphyrins immobilized on magnetic nanoparticles in continuous-flow magnetic chip reactor.

2. Materials and Methods

2.1. Materials

All solvents used in this study were of analytical grade. Methanol (MeOH), trifluoroacetic acid (TFA) and acetic acid were purchased from Merck Ltd. (Budapest, Hungary). Water was obtained from a Millipore (Bedford, MA, USA) Milli-Q water-purification system and used for the preparation of all aqueous solutions. Oxidizing agent *t*-butyl hydroperoxide (*t*-BuOOH was purchased from Sigma-Aldrich (St. Louis, MO, USA). Metalloporphyrines such as 5,10,15,20-tetrakis(2,3,4,5,6-pentafluorophenyl)iron(II) porphyrin (FeTPFP) and 5,10,15,20-tetrakis-(4-sulfonatophenyl)iron(II) porphyrin (FeTSPP) were purchased from Frontier Scientific (Logan, UT, USA). Amiodarone, tris(hydroxymethyl)aminomethane hydrochloride (Tris HCl), $MgCl_2$, glucose-6-phosphate, glucose-6-phosphate dehydrogenase, sodium acetate, potassium chloride (KCl) and NADPH were purchased from Sigma-Aldrich. Human liver microsomes pooled from mixed gender was obtained from Sekisui XenoTech Llc. (Kansas City, KS, USA). Magnetic nanoparticles with aminopropyl functions ($MNPs-NH_2$) were product of SynBiocat Llc. (Budapest, Hungary).

2.2. Methods

2.2.1. HPLC-DAD-MS Analysis

Experiments were carried out on an Agilent 1200 liquid chromatography system coupled with an 6410 QQQ-MS (Agilent Technologies, Palo Alto, CA, USA), equipped with a vacuum degasser, a binary pump, mixer assembly, an auto sampler, a column temperature controller and a diode array detector. Analysis was performed at 45 °C on a Kinetex EVO C_{18} column (50 × 3 mm, 2.6 μm) (Phenomenex), with a mobile phase flow rate of 1.45 mL/min. Composition of eluent A was 0.1% (V/V) trifluoroacetic acid (TFA) in water (pH 1.9), eluent B was a mixture of acetonitrile and water in 95:5 (V/V) with 0.1% (V/V) TFA. A linear gradient of 2–100% B was applied at a range of 0–4.9 min, then 100% B at 4.9–6.0 min. It was followed by a 1.20 min equilibration period prior to the next injection. The injection volume was set at 5 μL and the chromatographic profile was registered at 220 ± 4 nm. The mass spectrometer detector (MSD) operating parameters were as follows: electrospray ionization (ESI) positive ionization, scan ion mode (100–900 m/z), drying gas temperature 350 °C, nitrogen flow rate 11 L/min, nebulizer pressure 40 psi, quadrupole temperature 100 °C, capillary voltage 4000 V, fragmentor voltage 135 V. The results of HPLC-DAD/MS analysis can be found in Supplementary Materials (Supplementary Materials as Figures S1–S21).

2.2.2. Dynamic Light Scattering (DLS) Analysis

Particle size distribution of amino-functionalized MNPs ($MNPs-NH_2$) and FeTPFP or FeTSPP porphyrin immobilized on amino-functionalized MNPs ($MNPs-NH-FeTPFP$ or $MNPs-NH_3-FeTSPP$) was characterized by dynamic light scattering (DLS, Brookhaven BI-200SM Laser Light Scattering Instrument, Holtsville, NY, USA). The samples were sonicated in methanol for 20 min, then analyzed by a laser beam (λ = 488 nm) at 25 °C in three parallel runs.

2.2.3. ζ-Potential Analysis

The zeta potential of amino-functionalized MNPs ($MNPs-NH_2$) and FeTPFP or FeTSPP porphyrin immobilized on amino-functionalized MNPs ($MNPs-NH-FeTPFP$ or $MNPs-NH_3-FeTSPP$) was measured with Zeta Potential Analyzer (Brookhaven) using the Zeta PALS (Phase Analysis Light Scattering) method. Re-dispersed samples were diluted 5-fold in 1 mM KCl aqueous solution. Measurements carried out in a disposable, solvent resistant micro cuvette took 2 min. Zeta potential was calculated from the electrophoretic mobility using Smoluchowski equation.

2.2.4. Covalent Immobilization of FeTPFP on Functionalized Magnetic Nanoparticles (MNPs)

Amino-functionalized magnetic nanoparticles (MNPs-NH$_2$, 10 mg) was sonicated in diglyme (400 µL) for 10 min. Then a solution of 5,10,15,20-tetrakis(2,3,4,5,6-pentafluorophenyl)iron(II) porphyrin (FeTPFP) in diglyme (600 µL, 2.5 mg/mL) was added to the suspension and the mixture was shaken for 72 h at 60 °C. After magnetic separation, the MNPs were washed with isopropanol, distilled water, and methanol and were dried in vacuum cabinet for 4 h.

2.2.5. Ionic Immobilization of FeTSPP on Functionalized MNPs

Amino-functionalized magnetic nanoparticles (MNPs-NH$_2$, 10 mg) were added to methanol:sodium acetate buffer (4:1 V/V, 8 mL, pH = 4.5) and sonicated for 10 min. Then a solution of 5,10,15,20-tetrakis-(4-sulfonatophenyl)iron(II) porphyrin in acetate buffer (4:1 V/V, 8 mL, pH = 4.5 (750 µL, 1.0 mg/mL) was added to the mixture and the suspension was shaken for 5 min at room temperature. After magnetic separation, it was washed with methanol, then dried in vacuum cabinet for 4 h.

2.2.6. Immobilization Yield (Y_I) of MNP-Porphyrines

After the immobilization of FeTPFP or FeTSPP metalloporphyrin, a sample (900 µL) taken directly from the residual binding solvent freed from MNPs was analyzed by a Genesys 2 type ultraviolet-visible (UV-VIS) spectrophotometer (Thermo Fisher Scientific Inc., Waltham, MA, USA) at room temperature. The specific wavelength (λ_{max}) of the corresponding metalloporphyrin was determined (λ_{max} = 407 nm for FeTPFP and λ_{max} = 395 nm for FeTSPP), then calibration curves were also recorded. Immobilization yield (Y_I, %) was calculated from the following:

$$Y_I = \frac{c_{2P}}{c_{1P}} \times 100$$

where c_{1P} is the initial porphyrin concentration, c_{2P} is the residual porphyrin concentration in the binding solution.

2.2.7. Metabolism of Amiodarone (1) by Human Liver Microsomal Reactions

Sodium pyrophosphate (125 µL, 6.38 mg/mL), magnesium chloride (50 µL, 3 mM), glucose-6-phosphate (25 µL, 13 mg/mL), glucose-6-phosphate dehydrogenase (25 µL, 20 IU/mL), Tris-HCl buffer (170 µL, 15.76 mg/mL), human liver microsome (50 µL, final concentration is 1000 µg/mL) and amiodarone solution (in methanol, 5 µL, 0.65 mg/mL) was pipetted in an Eppendorf tube. It was held at 37 °C for 5 min. After that NADPH (50 µL, 3.72 mg/mL) was added to the reaction mixture. It was shaken for 30 min at 37 °C. The reaction was stopped with the addition of methanol (0.5 mL, −20 °C). The microsomes were separated by ultracentrifugation (at 10,000 g, for 5 min). The upper clear phase (0.8 mL) was analyzed by HPLC-DAD-MS method described in Section 2.2.1.

2.2.8. General Method of Homogeneous Biomimetic Batch Reactions of Amiodarone (1)

Amiodarone solution (50 µL, 5.7 mg/mL in methanol:sodium acetate buffer, 4:1 V/V, pH = 4.5, 64 mM), solvent completion (150 µL, methanol:sodium acetate buffer, 4:1 V/V, pH = 4.5), porphyrin solution (50 µL, 0.9 mg/mL in methanol:sodium acetate buffer, 4:1 V/V, pH = 4.5, 64 mM) and oxidizing agent solution (t-BuOOH, 50 µL, 88.2 mM in methanol:sodium acetate buffer, 4:1 V/V, pH = 4.5, 64 mM) was pipetted in an Eppendorf tube. It was shacked for 1 h at room temperature at 400 rpm. The reaction mixture (300 µL) was analyzed by HPLC-DAD-MS method described in Section 2.2.1.

2.2.9. General Method of Heterogeneous Biomimetic Batch Reactions of Amiodarone (**1**)

In Eppendorf tubes porphyrin loaded MNPs (MNPs-NH-FeTPFP or MNPs-NH$_3$-FeTSPP; 2 mg) and amiodarone solution (0.5 mL, 2 mg/mL in methanol:sodium acetate buffer 4:1 V/V, pH = 4.5, 64 mM) was sonicated for 20 min. The concentration was adjusted to 1 mg/mL of amiodarone. The reaction was started with the addition of the oxidizing agent (*t*-BuOOH, 5 molar equivalent). The reaction mixtures were shaken for 1 h at room temperature. After magnetic separation the clear phase (1 mL) was analyzed by HPLC-DAD-MS technique described in Section 2.2.1.

2.2.10. Calculation of the Biocatalytic Parameters

Conversion of the substrate (c, %), biocatalytic activity (U_B, U/g), specific activity (U_P, U·g^{-1}), turnover frequency (TOF, mol$_{prod}$·mol$_{cat}$$^{-1}$·h^{-1}) and space time yield (STY, mg·L^{-1}·h^{-1}) were calculated by using the following equations based on HPLC chromatograms:

$$c\,(\%) = 100 - \frac{n_S}{n_S + n_P} \times 100 \tag{1}$$

where n_S and n_P are the molar amounts of substrate (S) and product(s) (P),

$$U_B(U/g) = \frac{(n_S \times c)}{(t \times m_B)} \tag{2}$$

where U is rate of the substrate conversion (μmol/min), t is reaction time, m_B is the mass of the biocatalyst,

$$U_P(U/g) = \frac{(n_S \times c)}{(t \times m_P)} \tag{3}$$

where m_P is the mass of the porphyrin in the immobilized biocatalyst,

$$TOF(mol/mol \times h) = \frac{n_{products}}{n_{catalyst}} \times \frac{1}{t} \tag{4}$$

$$STY\left(\frac{mg}{l} \times h\right) = \frac{m_p}{V_r \times t} \tag{5}$$

where m_p is the mass of the products in mg, V_r is the volume of the reactor and t is time.

2.2.11. General method for Loading the Microfluidic Magnetic Chip Reactor with Catalyst

Porphyrin-loaded MNPs (6 mg) were added to methanol:sodium acetate buffer (4:1 V/V, pH = 4.5, 64 mM, 2 mL). After sonication, the suspension was loaded in a syringe and was driven through a microfluidic chip to fill up the chambers with nanoparticles following the previously described method, [14,26] outlined here shortly. During the filling of the chip, the MNPs were accumulated in the reaction chambers (diameter 3600 μm, height 110 μm) due to the magnetic field of the permanent magnets located beneath the chambers. Once the last chamber was saturated the process was repeated with the forthcoming upstream chamber until all chambers were filled up.

2.2.12. General Method of Microfluidic Biomimetic Reactions

Biomimetic reactions were carried out by using Magneflow system (SpinSplit LLC, Budapest, Hungary) as follows. Amiodarone (0.5 mg/mL in methanol:sodium acetate buffer, 4:1 V/V, pH = 4.5) and oxidizing agent (5 equiv., 7.35 mM in methanol:sodium acetate buffer, 4:1 V/V, pH = 4.5, 64 mM) solution were driven through the MNP loaded chip reactor 'MagneChip 6/3' (SpinSplit LLC) at four different flow rates (15, 30, 45 and 60 μL/min) following a pre-programmed sequence in SpinStudio (SpinSplit LLC) software. A timeframe of 15 min was provided to ensure the reach of equilibrium before samples were taken. The samples were analyzed by HPLC-DAD-MS technique (Figure 2).

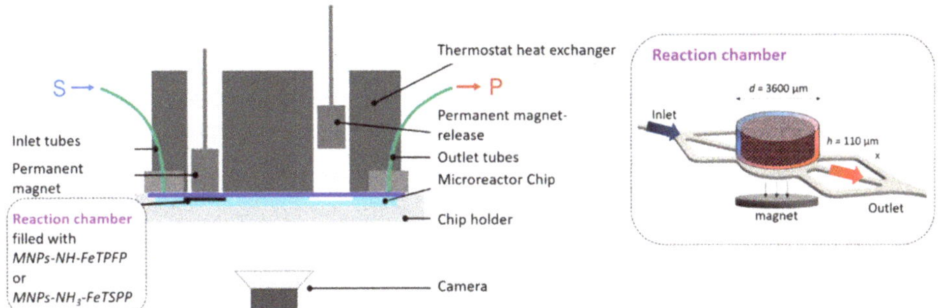

Figure 2. Cross-sectional view of Magneflow chip holder with the microreactor chip. A layer of magnetic nanoparticles (MNP) is formed in the reaction chambers due to the magnetic field applied by moving the permanent magnets toward the microreactor chip. Layer consistency is assessed through the camera image; reactor temperature is maintained by the thermostat heat exchanger.

3. Results and Discussion

3.1. Immobilization of Metaloporphyrin on Functionalized Magnetic Nanoparticles

Two species of synthetic metalloporphyrin (FeTPFP and FeTSPP) were immobilized on aminopropyl-grafted magnetic nanoparticles (MNP-TEOS-NH$_2$) by two different ways. FeTPFP was immobilized by covalent binding based on aromatic nucleophilic substitution between the pentafluorophenyl ring of the porphyrin and amino-group of MNPs. FeTSPP was attached to the MNP surface by ionic interactions between the sulfuric groups of porphyrin and amino-groups of MNPs. To characterize the efficiency of the functionalized MNPs in immobilization of the two type of metalloporphyrins the metalloporphyrin content of the residual binding solution was determined by UV-VIS spectroscopy. Based on the determination of immobilization yields (Y_I), the FeTPFP content of the FeTPFP-MNPs was 0.054 mg/mg catalyst. In case of FeTSPP, the metalloporphyrin content was 0.075 mg/mg immobilized catalyst. These differences can be explained by the rapid and mild binding by ionic forces, in contrast to covalent binding which require more energy.

Colloidal stability of a dispersion consisting nanoparticles is usually a key issue, because the constant stability of the individual particles against aggregation and sedimentation is essential in efficient transport processes, which is essential in heterogenic catalytic reactions. Stability of colloidal systems can be predicted by measuring the zeta potential. The higher absolute value of zeta potential suggests increased stability of the colloidal dispersion and lower tendency towards coagulation or flocculation (-5 mV $< \zeta < 5$ mV brings about rapid coagulation or flocculation) [29]. The zeta potential of all three derivatized MNPs were between -5 and -23 mV, indicating considerable colloidal stability. The largest absolute value (more charged surface) was observed for the amine functionalized particles, while shielding the surface charge either with covalent binding or ionic interactions resulted in lower absolute values of zeta potential. In accordance with the immobilization yields, modification with FeTSPP caused a drop in the zeta potential to an absolute value of around 5 mV. Comparison of the amino-functionalized MNPs (MNPs-NH$_2$) and the two kinds of porphyrin-covered MNPs (MNP-NH-FeTPFP and MNP-NH$_3$-FeTSPP) revealed that the hydrodynamic diameters of porphyrin-modified MNPs were smaller than that of the amino-functionalized MNPs. This means, that even after ultra-sonication the amino-functionalized MNPs tend to aggregate more, because the functional groups on their surface were more ionizable. After immobilization of the metalloporphyrins, most of the ionizable amino functions on the surface were reacted, therefore less ionization could take place (Table 1).

Table 1. Dynamic light scattering (DLS) and ζ-potential data of the modified MNPs.

Type of MNPs	d_P (nm)	ζ-Potential (mV)
MNPs-NH$_2$	429 ± 46	−22.9 ± 0.8
MNPs-NH-FeTPFP	336 ± 25	−16.0 ± 0.8
MNPs-NH$_3$-FeTSPP	296 ± 4	−5.3 ± 1.2

3.2. Metabolism of Amiodarone (**1**)

Amiodarone antiarrhythmic drug was used as model substrate for investigation the biomimetic oxidation catalyzed by metalloporphyrin. In the human body, the CP450-catalyzed oxidation of amiodarone produces *N*-deethyl-amiodarone (**2**) as the major human metabolite, however secondary deethylation resulting primary amine, mono and di-deiodinated, *O*-dealkylated and hydroxylated derivatives (**12**) have been also demonstrated at very low quantity [30–32]. In our recently published study, new derivatives were formed and identified in biomimetic oxidation [22]. The possible structures of metabolites (**2–12**) of amiodarone (**1**) were deduced from HRMS measurements (Figure 3).

The structures of the derivatives (**2–14**) from amiodarone (**1**) indicate, that the oxidative metabolism of amiodarone happened by *N*- and *O*-dealkylations, dehalogenations, hydroxylations or oxidations.

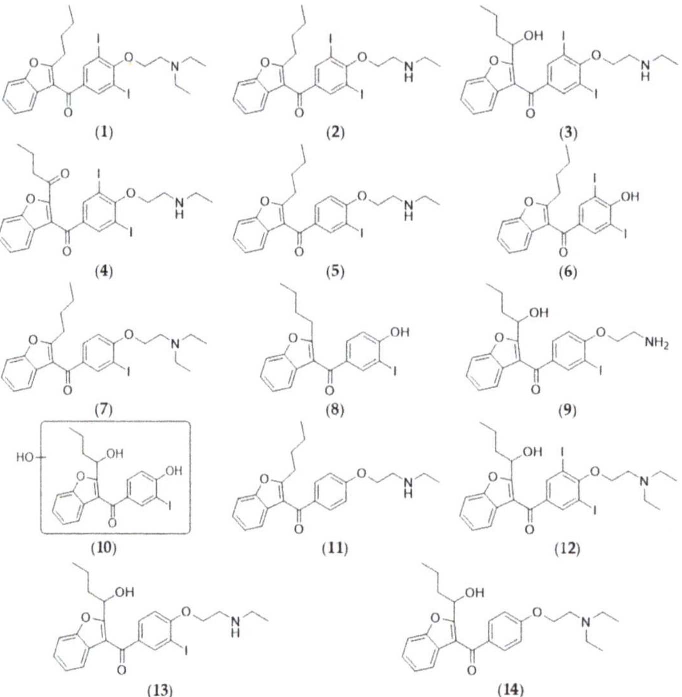

Figure 3. Amiodarone (**1**) and the detected derivatives (the assumed structures were based on their HRMS signal) of amiodarone (**2–14**).

3.3. Biomimetic Oxidation of Amiodarone (**1**) Catalyzed by Metalloporphyrins in Batch Mode

To study the effect of immobilization on the catalytic properties of metalloporphyrines, heterogeneous and homogeneous biomimetic reactions in batch mode were compared under the same conditions (Table 2). Two different synthetic metalloporphyrines (FeTPFP and FeTSPP) were used in our experiments in non-immobilized soluble form (as homogenous reactions) and in immobilized form on MNPs (as heterogeneous reactions). In case of homogenous oxidations, the metabolite profiles of the two metalloporphyrin catalysts were different. With FeTPFP as catalyst, *O*-dealkylation and *N*-deethylation were the main reactions along with the formation of monodehalogenated-amiodarone (**7**) as minor product. With FeTSPP as catalyst, *N*-deethylation, dehalogenation and hydroxylation were the most pronounced and oxo-amiodarone (**4**) was formed from hydroxy-amiodarone (**3**). After immobilization on MNPs the metabolite profile of the metalloporphyrines changed significantly. The same *N*-deethyl-amiodarone (**2**) was formed as main metabolite as in the homogeneous batch reaction, while no further derivatives that formed in the homogeneous reaction (**3**, **4**, and **5**) were generated in the heterogeneous reaction. This means that by immobilization the hydroxylation of monodeethyl-amiodarone become unfavored reaction. In the heterogeneous reaction using immobilized FeTSPP catalyst, novel metabolites were formed involving *O*-dealkylated-amiodaron and further products by subsequent dehalogenations, dealkylations and hydroxylations.

Table 2. Metabolite profile of amiodarone from homogenous (free metalloporphyrin catalyzed: FeTPFP or FeTSPP) and heterogeneous (immobilized metalloporphyrin on MNPs: MNPs-NH-FeTPFP or MNPs-NH$_3$-TSPP) biomimetic oxidations in batch mode.

Metabolite [a]	In Vitro Human Liver Microsomal Investigation	Homogenous Biomimetic Reaction "Free Metalloporphyrin"		Heterogeneous Biomimetic Reaction "Immobilized Metallo-porphyrin on MNPs"	
		FeTPFP	FeTSPP	MNPs-NH-FeTPFP	MNPs-NH$_3$-FeTSPP
amiodarone (**1**)	86.4	8.1	4.3	6.3	53.0
(**2**)	13.3	66.7	66.5	62.6	38.0
(**3**)	-	-	1.0	-	-
(**4**)	-	-	7.6	-	-
(**5**)	-	-	10.5	1.4	1.4
(**6**)	-	24.3	-	25.7	5.9
(**7**)	-	0.9	-	-	-
(**8**)	-	-	-	-	1.3
(**9**)	-	-	-	-	0.4
(**10**)	-	-	-	-	0.1
(**11**)	-	-	-	-	-
(**12**)	0.3	-	-	-	-
(**13**)	-	-	-	-	-
(**14**)	-	-	-	-	-
other [b]	-	-	10.0	4.0	-

[a] the structure and the amount of metabolites were determined by LC-MS measurement, where the ratio of metabolites based relative peak area (%) at $\lambda = 220 \pm 4$ nm on DAD-chromatograms, [b] not identified compounds under limit of detection.

The biocatalytic activity (U_B) and specific biocatalytic activity (U_P) are key parameters of a catalyst: the U_B shows the productivity referring to the mass unit (usually in g) of the total catalyst (non- and immobilized as well) and U_P vale refers to the "active" porphyrin content (usually in g) of the immobilized catalyst. Results in Table 3 shows, that every biomimetic oxidation catalyzed by a metalloporphyrin catalyst provided much higher catalytic activity, than the human liver microsomes-based system. The U_B (and U_P as well) values of the two different metalloporphyrines (FeTPFP and FeTSPP) were comparable in homogenous reaction media, but after immobilization on MNPs significant differences between the U_B (and U_P) values with the two porphyrin could be observed. This result can be explained by the different immobilization yields, binding forces and topological characters of the immobilized catalysts. The increase in specific catalytic activity (U_P) of MNPs-NH-FeTPFP catalyst can be caused by the good dispersion and stabilization of the porphyrins on to MNPs surface.

Table 3. Biocatalytic activity (U_B) and specific biocatalytic activity (U_P) of human liver microsomes (HLM) and metalloporphyrins (FeTPFP or FeTSPP) in their soluble (homogenous catalytic reaction) and MNP immobilized (heterogeneous catalytic reaction) forms in biomimetic oxidation of amiodarone (**1**).

Activity	Homogenous Catalytic Reaction			Heterogenous Catalytic Reaction	
	HLM	FeTPFP	FeTSPP	MNPs-NH-FeTPFP	MNPs-NH$_3$-FeTSPP
U_B (U/g)	0.5	132.7	153.6	12.1	6.1
U_P (U/g)	-	132.7	153.6	223.8	80.8

3.4. Biomimetic Oxidation of Amiodarone (**1**) Catalyzed by MNP-Porphyrines in Continuous-Flow Chip Reactor—Liver-on-a-Chip

Generic PDMS microfluidic chip was used in our experiments, with 6 reaction chamber. After loading the chip, each reaction chamber contained about 250 µg catalyst according to resonance frequency shift measurement based on our previous study [28]. The metabolite profile from chip reactor was different from the batch reactions, because two novel derivatives (**13**, **14**) appeared at significant quantity. The metabolite profile of the two investigated metalloporphyrines (FeTSPP and FeTPFP) were quite similar. The effect of the flow rate could be clearly recognized: The faster the flow rate of amiodarone solution, the less metabolite was formed (Table 4).

Table 4. Metabolite profile of amiodarone from immobilized metalloporphyrin (MNPs-NH-FeTPFP or MNPs-NH$_3$-TSPP) catalyzed biomimetic oxidation in continuous-flow magnechip reactor–"Liver-on-a-Chip".

Metabolites [a]	Flow Rate (µL/min)							
	15		30		45		60	
	FeTPFP	FeTSPP	FeTPFP	FeTSPP	FeTPFP	FeTSPP	FeTPFP	FeTSPP
amiodarone (**1**)	61.0	58.8	75.7	77.0	84.5	84.4	88.4	87.9
(**2**)	0.5	-	0.5	0.3	0.6	0.4	0.8	0.6
(**11**)	10.8	11.5	5.6	5.6	2.7	2.7	1.1	2.2
(**13**)	21.5	23.2	13.4	12.9	8.2	8.5	6.3	6.9
(**14**)	3.9	4.4	3.3	3.0	2.6	2.9	2.3	1.8
other [b]	2.3	2.2	1.5	1.4	1.3	1.2	1.2	0.7

[a] the structure and the amount of metabolites were determined by LC-MS measurement, where the ratio of metabolites based relative peak area (%) at λ = 220 ± 4 nm on DAD-chromatograms, [b] not identified compounds under limit of detection.

To characterize the efficiency of the flow-chip reactor, the turnover frequency (TOF) and the space time yield (STY) values were calculated and compared to batch systems (Figure 4). The turnover values—characterizing the catalyst activity by the number of moles substrate transformed by one mole of catalyst per hour—were in the same range, between 2 and 9, with each of the porphyrin catalyst consistently to the U_P values in Table 3. Notably, the STY—indicating the volumetric productivity of a reactive system—of the microfluidic magnetic chip reactor was remarkably higher than any of the biomimetic oxidation in batch mode.

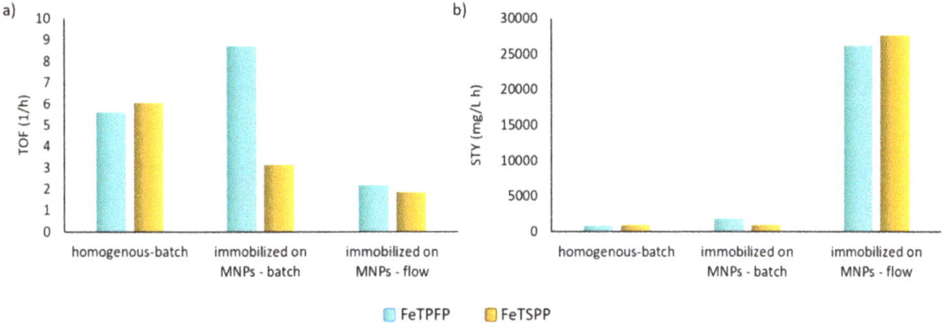

Figure 4. Turnover frequency (TOF) (**a**) and space time yield (STY) (**b**) values of metalloporphyrin (FeTPFP or FeTSPP) catalysts in their soluble (homogenous) or MNP-immobilized (heterogeneous) forms in biomimetic oxidation of amiodarone (**1**) in batch mode or continuous-flow magnetic chip reactor.

4. Conclusions

The investigation of drug metabolites is a key issue during the early stage drug discovery; especially liver related first phase oxidative metabolism catalyzed by CP450 enzymes has a great importance. Traditional animal experiments in vivo or in vitro methods or experiments with liver cells or its microsomes; although they are accepted by FDA, have many limitations, which complicate the effective metabolite production already at analytical scale too. Synthetic metalloporphyrines have high similarity to the active site of CP450, thus they are able to mimic the enzymatic reactions without the complex biological matrix. This study showed that immobilization of metalloporphyrines onto surface-functionalized magnetic nanoparticles can provide biocatalyst with increased activity in the biomimetic oxidation of amiodarone. The type of the metalloporphyrin, the way of immobilization and mode of the process are able to influence the metabolite profile. Integration of these immobilized metalloporphyrines into microfluidic magnetic chip reactors demonstrated that drug metabolites can be produced efficiently in extremely small reactor volumes resulting in excellent volumetric productivity. In addition, not only the in vivo major metabolite can be easily synthetized by the microfluidic magnetic "Liver-on-a-chip" system, but new derivatives can be produced opening up unique novel opportunities for modern drug discovery.

Supplementary Materials: The following are available online at http://www.mdpi.com/2072-666X/10/10/668/s1, Figure S1: Representative HPLC/MS data of human liver microsomal investigation of amiodarone, Figure S2. Representative HPLC/MS data of non-immobilized FeTPFP-catalyzed biomimetic oxidation of amiodarone, Figure S3: Representative HPLC/MS data of non-immobilized FeTSPP-catalyzed biomimetic oxidation of amiodarone, Figure S4: Representative HPLC/MS data of FeTPFP-functionalized magnetic nanoparticles-catalyzed biomimetic oxidation of amiodarone in batch mode, Figure S5: Representative HPLC/MS data of FeTSPP-functionalized magnetic nanoparticles-catalyzed biomimetic oxidation of amiodarone in batch mode, Figure S6: Representative HPLC/MS data of FeTPFP-functionalized magnetic nanoparticles-catalyzed biomimetic oxidation of amiodarone in continuous-flow magnetic chip reactor, Figure S7: Representative HPLC/MS data of FeTSPP-functionalized magnetic nanoparticles-catalyzed biomimetic oxidation of amiodarone in continuous-flow magnetic chip reactor, Figure S8: MS spectrum of amiodarone (**1**), Figure S9: MS spectrum of amiodarone metabolite (**2**), Figure S10: MS spectrum of amiodarone metabolite (**3**), Figure S11: MS spectrum of amiodarone metabolite (**4**), Figure S12: MS spectrum of amiodarone metabolite (**5**), Figure S13: MS spectrum of amiodarone metabolite (**6**), Figure S14: MS spectrum of amiodarone metabolite (**7**), Figure S15: MS spectrum of amiodarone metabolite (**8**), Figure S16: MS spectrum of amiodarone metabolite (**9**), Figure S17: MS spectrum of amiodarone metabolite (**10**), Figure S18: MS spectrum of amiodarone metabolite (**11**), Figure S19: MS spectrum of amiodarone metabolite (**12**), Figure S20: MS spectrum of amiodarone metabolite (**13**), Figure S21: MS spectrum of amiodarone metabolite (**14**).

Author Contributions: Conceptualization: B.D., R.K., K.H., G.T.B., F.E., R.T., L.P. and D.B.-W.; analytical support: B.G., A.S., R.T., G.K, C.P.; resources: G.T.B., C.P. and D.B.-W.; writing—original draft preparation, B.D., K.H., and F.E.; writing—review and editing, G.T.B., L.P. and D.B.-W.; supervision D.B.-W.

Funding: The research reported in this paper has been supported by the National Research, Development and Innovation Fund (TUDFO/51757/2019-ITM, Thematic Excellence Program. Furthermore, we want to thank the financial support from NEMSyB, ID P37_273, Cod MySMIS 103413 funded by the Romanian Ministry for European Funds, through the National Authority for Scientific Research and Innovation (ANCSI) and cofounded by the European Regional Development Fund, Competitiveness Operational Program 2014-2020 (POC). The authors also acknowledge New National Excellence Programme of the Ministry of Innovation and Technology for the financial support, including ÚNKP-19-2-I fellowship of R. Krammer.

Acknowledgments: Authors thank Chemical Department of Gedeon Richter Plc. for the analytical support.

Conflicts of Interest: The authors declare no conflict of interest.

References

1. Baumann, M.; Baxendale, I.R. The synthesis of active pharmaceutical ingredients (APIs) using continuous flow chemistry. *Beilstein J. Org. Chem.* **2015**, *11*, 1194–1219. [CrossRef] [PubMed]

2. May, S.A. Flow Chemistry, Continuous Processing, and Continuous Manufacturing: A Pharmaceutical Perspective. *J. Flow Chem.* **2017**, *7*, 137–145. [CrossRef]

3. Calabrese, G.S.; Pissavini, S. From batch to continuous flow processing in chemicals manufacturing. *AIChE J.* **2011**, *57*, 828–834. [CrossRef]

4. Fortt, R.; Wootton, R.C.R.; de Mello, A.J. Continuous-flow generation of anhydrous diazonium species: Monolithic microfluidic reactors for the chemistry of unstable intermediates. *Org. Process Res. Dev.* **2003**, *7*, 762–768. [CrossRef]

5. Yoshida, J.I.; Takahashi, Y.; Nagaki, A. Flash chemistry: Flow chemistry that cannot be done in batch. *Chem. Commun.* **2013**, *49*, 9896–9904. [CrossRef]

6. Kockmann, N.; Gottsponer, M.; Zimmermann, B.; Roberge, D.M. Enabling continuous-flow chemistry in microstructured devices for pharmaceutical and fine-chemical production. *Chem. Eur. J.* **2008**, *14*, 7470–7477. [CrossRef] [PubMed]

7. Wegner, J.; Ceylan, S.; Kirschning, A. Ten key issues in modern flow chemistry. *Chem. Commun.* **2011**, *47*, 4583–4592. [CrossRef]

8. Manz, A.; Graber, N.; Widmer, H.M. Miniaturized total chemical analysis systems: A novel concept for chemical sensing. *Sens. Actuators B Chem.* **1990**, *1*, 244–248. [CrossRef]

9. Nguyen, N.-T.; Wereley, S.T. *Fundamentals and Applications of Microfluidics*, 2nd ed.; Artech House Books: Norwood, MA, USA, 2006; pp. 1–9, ISBN 9781580539722.

10. Wang, J.; Escarpa, A.; Pumera, M.; Feldman, J. Capillary electrophoresis-electrochemistry microfluidic system for the determination of organic peroxides. *J. Chromatogr. A* **2002**, *952*, 249–254. [CrossRef]

11. Breadmore, M.C.; Wolfe, K.A.; Arcibal, I.G.; Leung, W.K.; Dickson, D.; Giordano, B.C.; Power, M.E.; Ferrance, J.P.; Feldman, S.H.; Norris, P.M.; et al. Microchip-based purification of DNA from biological samples. *Anal. Chem.* **2003**, *75*, 1880–1886. [CrossRef]

12. Greenway, G.M.; Haswell, S.J.; Morgan, D.O.; Skelton, V.; Styring, P. Use of a novel microreactor for high throughput continuous flow organic synthesis. *Sens. Actuators B Chem.* **2000**, *63*, 153–158. [CrossRef]

13. Burns, M.A.; Johnson, B.N.; Brahmasandra, S.N.; Handique, K.; Webster, J.R.; Krishnan, M.; Sammarco, T.S.; Man, P.M.; Jones, D.; Heldsinger, D.; et al. An Integrated Nanoliter DNA Analysis Device. *Science* **1998**, *282*, 484–487. [CrossRef] [PubMed]

14. Ender, F.; Weiser, D.; Nagy, B.; Bencze, C.L.; Paizs, C.; Pálovics, P.; Poppe, L. Microfluidic multiple cell chip reactor filled with enzyme-coated magnetic nanoparticles—An efficient and flexible novel tool for enzyme catalyzed biotransformations. *J. Flow Chem.* **2016**, *6*, 43–52. [CrossRef]

15. Zheng, F.; Fu, F.; Cheng, Y.; Wang, C.; Zhao, Y.; Gu, Z. Organ-on-a-Chip Systems: Microengineering to Biomimic Living Systems. *Small* **2016**, *12*, 2253–2282. [CrossRef] [PubMed]

16. Fasinu, P.; Patrick, J.B.; Rosenkranz, B. Liver-based in vitro technologies for drug biotransformation studies—A review. *Curr. Drug Metab.* **2012**, *13*, 215–224. [CrossRef] [PubMed]

17. Kremers, P. Liver microsomes: A convenient tool for metabolism studies but. In *European Symposium on the Prediction of Drug Metabolism in Man: Progress and Problems*; Boobis, A.R., Kremers, P., Pelkonen, O., Pithan, K., Eds.; Office for Official Publications of the European Communities: Luxembourg, 1999; pp. 38–52.

18. Lohmann, W.; Karst, U. Biomimetic modeling of oxidative drug metabolism: Strategies, advantages and limitations. *Anal. Bioanal. Chem.* **2008**, *391*, 79–96. [CrossRef]

19. Mansuy, D. A brief history of the contribution of metalloporphyrin models to cytochrome P450 chemistry and oxidation catalysis. *C. R. Chim.* **2007**, *10*, 392–413. [CrossRef]

20. Wolak, M.; van Eldik, R. Mechanistic studies on peroxide activation by a water-soluble iron (III)-porphyrin: Implications for O-O bond activation in aqueous and nonaqueous solvents. *Chem. Eur. J.* **2007**, *13*, 4873–4883. [CrossRef]

21. Cooke, P.R.; Lindsay Smith, J.R. Alkene epoxidation catalysed by ligand-bound supported metalloporphyrins. *Tetrahedron Lett.* **1992**, *33*, 2737–2740. [CrossRef]

22. Fődi, T.; Ignácz, G.; Decsi, B.; Béni, Z.; Túrós, G.I.; Kupai, J.; Weiser, D.B.; Greiner, I.; Huszthy, P.; Balogh, G.T. Biomimetic Synthesis of Drug Metabolites in Batch and Continuous-Flow Reactors. *Chem. Eur. J.* **2018**, *24*, 9385–9392. [CrossRef]

23. Liu, R.; Guo, Y.; Odusote, G.; Qu, F.; Priestley, R.D. Core-shell Fe3O4 polydopamine nanoparticles serve multipurpose as drug carrier, catalyst support and carbon adsorbent. *ACS Appl. Mater. Interfaces* **2013**, *5*, 9167–9171. [CrossRef] [PubMed]

24. Lu, Y.; Mei, Y.; Drechsler, M.; Ballauff, M. Thermosensitive core-shell particles as carriers for Ag nanoparticles: Modulating the catalytic activity by a phase transition in networks. *Angew. Chem. Int. Ed.* **2006**, *45*, 813–816. [CrossRef] [PubMed]

25. Vaghari, H.; Jafarizadeh-Malmiri, H.; Mohammadlou, M.; Berenjian, A.; Anarjan, N.; Jafari, N.; Nasiri, S. Application of magnetic nanoparticles in smart enzyme immobilization. *Biotechnol. Lett.* **2016**, *38*, 223–233. [CrossRef] [PubMed]

26. Weiser, D.; Bencze, L.C.; Bánóczi, G.; Ender, F.; Kiss, R.; Kókai, E.; Szilágyi, A.; Vértessy, B.G.; Farkas, Ö.; Paizs, C.; et al. Phenylalanine ammonia-lyase-catalyzed deamination of an acyclic amino acid: Enzyme mechanistic studies aided by a novel microreactor filled with magnetic nanoparticles. *ChemBioChem* **2015**, *16*, 2283–2288. [CrossRef] [PubMed]

27. Stoytcheva, M.; Zlatev, R. *Lab-on-a-Chip Fabrication and Application*; InTech: London, UK, 2016; pp. 157–178, ISBN 978-953-51-2457-3.

28. Ender, F.; Weiser, D.; Vitéz, A.; Sallai, G.; Németh, M.; Poppe, L. In-situ measurement of magnetic nanoparticle quantity in a microfluidic device. *Microsyst. Technol.* **2017**, *23*, 3979–3990. [CrossRef]

29. Kirby, B.J. *Micro-and Nanoscale Fluid Mechanics: Transport in Microfluidic Devices*; Cambridge University Press: Cambridge, MA, USA, 2010.

30. Trivier, J.M.; Libersa, C.; Belloc, C.; Lhermitte, M. Amiodarone N-deethylation in human liver microsomes: Involvement of cytochrome P450 3A enzymes (first report). *Life Sci.* **1993**, *52*, 91–96. [CrossRef]

31. Freedman, M.D.; Somberg, J.C. Pharmacology and pharmacokinetics of amiodarone. *J. Clin. Pharmacol.* **1991**, *31*, 1061–1069. [CrossRef]

32. Berger, Y.; Harris, L. *Amiodarone, Pharmacology–Pharmacokinetics–Toxicology–Clinical Effects*; Médicine Sciences Internationales: Paris, France, 1986; Volume 59, pp. 45–98.

 *micromachines*

Article

Metal and Polymeric Strain Gauges for Si-Based, Monolithically Fabricated Organs-on-Chips

William F. Quirós-Solano [1,2,3,*], Nikolas Gaio [1,3], Cinzia Silvestri [3], Gregory Pandraud [4], Ronald Dekker [1,5] and Pasqualina M. Sarro [1]

1 Department of Microelectronics, Electronic Components, Technology and Materials (ECTM), Delft University of Technology, Mekelweg 4, 2628 CD Delft, The Netherlands
2 School of Electronics Engineering, Instituto Tecnológico de Costa Rica, A.P. 159, 7050 Cartago, Costa Rica
3 BIOND Solutions B.V., Mekelweg 4, 2628 CD Delft, The Netherlands
4 Electrical Sustainable Energy, Photovoltaic Materials and Devices (PVMD), Delft University of Technology, Mekelweg 4, 2628 CD Delft, The Netherlands
5 Philips Research, High Tech Campus, 5656 AE Eindhoven, The Netherlands
* Correspondence: w.f.quirossolano@tudelft.nl; Tel.: +31-(0)-15-278-3753

Received: 24 June 2019; Accepted: 13 August 2019; Published: 15 August 2019

Abstract: Organ-on-chip (OOC) is becoming the alternative tool to conventional in vitro screening. Heart-on-chip devices including microstructures for mechanical and electrical stimulation have been demonstrated to be advantageous to study structural organization and maturation of heart cells. This paper presents the development of metal and polymeric strain gauges for in situ monitoring of mechanical strain in the Cytostretch platform for heart-on-chip application. Specifically, the optimization of the fabrication process of metal titanium (Ti) strain gauges and the investigation on an alternative material to improve the robustness and performance of the devices are presented. The transduction behavior and functionality of the devices are successfully proven using a custom-made set-up. The devices showed resistance changes for the pressure range (0–3 kPa) used to stretch the membranes on which heart cells can be cultured. Relative resistance changes of approximately 0.008% and 1.2% for titanium and polymeric strain gauges are respectively reported for membrane deformations up to 5%. The results demonstrate that both conventional IC metals and polymeric materials can be implemented for sensing mechanical strain using robust microfabricated organ-on-chip devices.

Keywords: organ-on-chip; MEMS; silicon; PDMS; membranes; cell; strain; stress

---

1. Introduction

Organ-on-chip (OOC) aims to become the alternative tool for in vitro screening. Researchers are still facing biological and technological challenges that impede this cutting-edge technology to be adopted as a routine tool in drug development. The limited scalability of current fabrication processes and the lack of self-integrated monitoring are among the technical limitations hindering its adoption [1]. Thus, monolithically microfabricated OOC devices have been developed aiming to overcome such limitations. The so-called *Cytostretch* [2,3], has been developed as a heart on chip with integrated microelectrodes. This platform enables the access of data related to the action potential generated by iPSC-derived cardiomyocytes with also the possibility to precisely stimulate electrically the cell culture.

Other heart-on-chips including mechanical stimulation have also demonstrated their advantage to promote structural organization and maturation of the cells [1–7]. Generally, most devices are based on mechanically flexible materials that enable continuous mechanical stimulation of the different cell cultures [8–12]. Polydimethylsiloxane (PDMS) is typically used as main structural substrate facilitating dynamic stretching, exploiting its intrinsic advantages of biocompatibility, optical transparency and mechanical flexibility [13–15].

Nevertheless, the lack of robustness of most devices is caused by their reliance on bulk optical microscopy and pneumatic transduction. The type and amount of data to be acquired from the cell microenvironment and biological processes is constrained. A complete view of cell behavior, particularly in heart-on-chips, might be obtained if mechanical stress, bioelectrical activities, pH level, and potassium and oxygen concentration could be quantitatively measured and controlled in situ in a spatio-temporal manner.

Unlike other works, the Cytostretch is a modular platform where robust technologies for sensing and actuation can be integrated for such purpose. The possibility to monitor strain on the substrate was preliminary demonstrated with metal strain gauges that allow indirect quantification of the mechanical stress in the flexible membrane. However, previous work did not yet fully exploit the reproducibility, reliability and robustness of the polymer-last approach [2], where complex and electrical microstructures are fabricated prior to PDMS deposition.

This paper presents the development of metal and polymeric strain gauges for in situ monitoring of mechanical strain in the Cytostretch platform. Specifically, the optimization of the fabrication process of metal titanium (Ti) strain gauges and the investigation on an alternative material to improve the robustness and performance of the devices, are presented. Polymeric strain gauges using poly(3,4-ethylenedioxythiophene)(PEDOT:PSS) are investigated to exploit the well-known electronic conduction, high gauge factor, biocompatibility, high transparency and particularly mechanical flexibility of the material [16–21].

2. Strain Gauges on PDMS Membranes

Strain gauges are microstructures commonly used to quantify the strain of a holding substrate subjected to mechanical stress. By monitoring the change in the electrical resistance of these microstructures, the strain experienced by the substrate when stretched can be indirectly obtained. Integrating strain gauges in a microfabricated PDMS-based OOC platform enables the acquisition and quantification of the mechanical strain provided to a cell culture, thus allowing the gathering of more relevant information for biological studies.

2.1. Device Concept and Design

The devices investigated consist of either metal or polymeric strain gauges integrated on a PDMS membrane suspended from a silicon holding frame, a platform based on *Cytostretch* [2,3]. The membrane acts as a stretchable substrate for cell culturing with the strain gauges allowing a continuous electrical monitoring of the strain of the membrane when mechanically stretched. Pneumatic actuation enables the stretching of the membrane by applying a pressure load on its bottom surface. In Figure 1, the architecture of the devices is shown.

The shape, thickness and diameter of the membrane is determined based on the assumption that a semicircular profile of the radial strain is obtained in the membrane when stretched (Equation (1)), thus aiming at providing specific radial strains (0–10%) for pneumatic actuation pressures up to 2 kPa. The circular shape was preferred to exploit the symmetry of the strain distribution along the membrane [22]. The final design comprises a flexible circular 9 μm-thick (t) PDMS membrane of 3 mm in diameter (D). The radial strain of the membrane, ε_r, is given by:

$$\varepsilon_r = \frac{r_{final} - 2r}{2r} = \frac{(r)^2 + d^2}{2rd} arcsin\left(\frac{2rd}{(r)^2 + d^2}\right) - 1 \tag{1}$$

where r and d are the radius and the vertical displacement of the membrane, respectively.

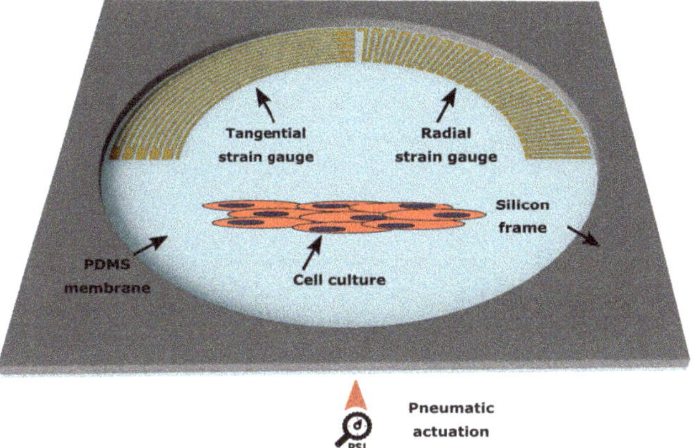

Figure 1. The architecture of the device investigated for stress sensing in a microfabricated PDMS-based OOC platform: Tangential and radial microstructures (strain gauges) on PDMS membranes suspended from a holding silicon frame. The membrane is pneumatically actuated to provide the stretching to a cell culture on its surface.

The position and dimensions of the strain gauges are defined based on the expected distribution of radial and tangential strain on circular membranes. The tangential strain, ε_t, is related to the radial strain according to

$$\varepsilon_t(\varrho) = \varepsilon_r \left(1 - \frac{\varrho^2}{r^2} \right) \tag{2}$$

where ϱ is the distance from the center of the membrane. Consequently, the strain gauges were located close to the edge of the membrane, where the gradient of the tangential strain along the radial direction is high.

Serpentine-like geometries of length (L_r, L_t) were then defined to maximize the microstructures parallel to the expected main strain directions (radial and tangential). Likewise, the width ($W = 20$ μm) was then defined to exploit the expected mechanical behavior and to have final electrical resistances in the range of kΩ, matching with standard resistors necessary for further signal conditioning thickness was defined to comply with the above mentioned electrical criteria and reduced as much as possible to minimize the effect of the strain gauges on the membrane deformation.

2.2. Modelling and Simulation

An initial assessment, trough numerical simulations, was performed to analyze the expected mechanical performance of the envisioned microstructures. The microstructures and the PDMS membrane showed in Figure 1 were modelled and the corresponding solution to the set of equations was obtained using the FEM-based software Comsol Multiphysics®. The mechanical modelling implemented was defined so that the equations can be computationally solved and particularly the boundary conditions match the experimental conditions for the intended operation of the devices and the subsequent electromechanical characterization. As demonstrated in other works, numerical simulation provides better insight on the mechanical behavior of membranes with strain gauges compared to analytical solutions. On thin membranes, the effect of the microstructures on the final deformation of the membrane is better contemplated [22] by numerical solutions.

The modelling of the membrane with strain gauges was based on both linear and non-linear equations of solid mechanics. The non-linearity is relevant for the model as it takes into account the non-elastic behavior of the polymeric membrane [23]. Therefore, it is included in the model by introducing a stress-strain curve in the material properties, based on data previously reported [23,24]. The strain gauges were modelled assuming a linear isotropic and elastic material. Values reported in the literature were used for the Youngs Modulus (90 GPa, 2 GPa) and Poisson ratio (0.31, 0.35) of titanium (Ti) and poly(3,4-ethylenedioxythiophene) polystyrene sulfonate (PEDOT:PSS), respectively [16–18,25].

The boundary conditions established for the model were determined based on the envisioned use of the device for mechanical stretching of cell cultures. A boundary condition of zero displacement on the substrate surface surrounding a circular membrane was considered, as this is the region where the membrane is clamped to the silicon substrate. A boundary force equivalent to the pneumatic pressure applied at the bottom surface of the membrane was included.

Subsequently, the geometry was meshed with (14 thousand elements) a high quality ($\approx$0.83) meshing, to calculate the displacement and strain fields.

The computation was carried out determining the displacement field of each of the elements of the geometry representing the device. In Figure 2a,b, the strain field of a membrane with radial and tangential metal strain gauges actuated with 2 kPa, is shown. In Figure 2c,d, the expected displacement at the center of the membrane and the average strain for both metal and polymeric strain gauges are reported.

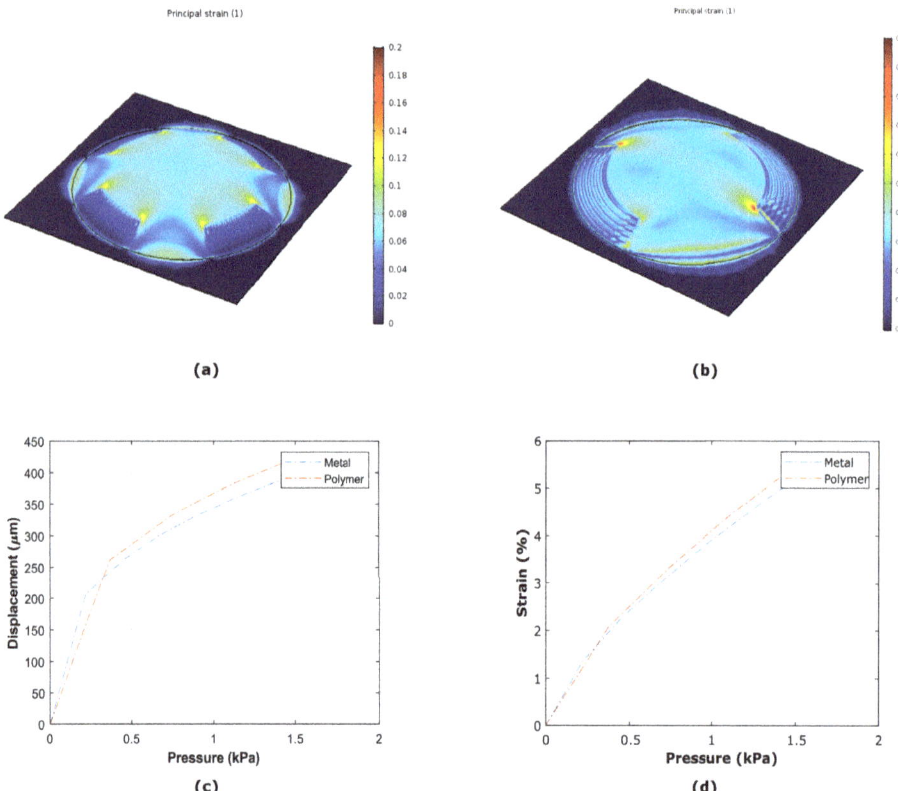

Figure 2. Strain field of the membranes with radial (**a,b**) tangential metal strain gauges for a boundary force corresponding to 2 kPa. (**c**) The curve of the displacement at the center of a membrane with metal (blue line) and polymeric (red line) strain gauges for pressures up to 2 kPa. (**d**) Corresponding average strain of the membrane.

3. Materials and Methods

3.1. Materials

Two materials were investigated, namely a metal, titanium, and a conductive polymer, poly(3,4-ethylenedioxythiophene)(PEDOT:PSS). Titanium was selected given its well-known mechanical behavior, biocompatibility, gauge factor (GF $\simeq$ 0.8) and the possibility to pattern it with conventional lithography and dry etching techniques [26–30].

PEDOT is a polymer derived from ethylene dioxythiophene monomer. The electrical conductivity is caused by the delocalized π-electrons within its chemical structure and the presence of sulfonated polystyrene (PSS). It was adopted due to its known electronic conduction, biocompatibility, high transparency ($\geq$90%) and particularly for its mechanical flexibility (E $\simeq$ 1.2 GPa) [16–18], making it suitable for either sensing or stimulating microstructures [31]. Moreover, previous studies suggest that this polymer can provide a gauge factor in the range of 0.48–17.8 [19–21].

3.2. Fabrication

To fabricate the strain gauges on PDMS membranes, wafer-level fabrication processes were developed based on conventional photolithography and MEMS microfabrication techniques. The nature of the two conductive materials adopted is significantly different, resulting in the development and testing of two different processes to integrate the strain gauges.

3.2.1. Metal Strain Gauges

The fabrication process of the metal strain gauges is schematically depicted in Figure 3. The process starts with the deposition of a 1 μm plasma-enhanced chemical vapor deposition (PECVD) silicon oxide (SiO_2) on the front side of a 100 mm in diameter, 525 μm-thick silicon wafer (Figure 3a). The oxide acts as an etch-stop layer for the Deep Reactive-Ion Etching (DRIE) step used to form the membrane. On the backside of the wafer, a 6 μm PECVD SiO_2 hard-mask layer is deposited and patterned to define the circular shape membranes (Figure 3b).

Then a 300 nm photosensitive polyimide (PI) layer is deposited and patterned. This layer is included to provide electrical isolation as well as protection of the metal lines during the subsequent steps of the process (Figure 3c). Next, a 600 nm-thick Aluminum (Al) and 100 nm-thick titanium (Ti) layer are sputtered on the PI at room temperature. The resistivity of the sputtered materials is 20 μΩ-cm and 106 μΩ-cm, respectively. The Al layer is patterned and selectively etched by using wet etching (Figure 3d). After defining the contacts, the Ti layer is patterned by dry etching (ICP Plasma etcher, Trikon Omega 201) with a 2μm-thick positive resist as a masking layer (Figure 3e).

Then, a layer of PDMS (Dow Corning, Sylgard 184) is spin-coated on the front side. The elastomer and curing agent are mixed in a 10:1 ratio and degassed in a centrifugal vacuum mixing and degassing tool. The polymer is spun in three steps: a first step to spread the material over the silicon wafer at 10 rpm for 10 s; a second step for uniform spreading at 300 rpm for 20 s and a final step at 6000 rpm for 30 s to get the desired 9 μm thickness. The polymer curing is performed for 30 min at 90 °C in a convection oven (Figure 3f). Subsequently, a 100 nm-thick Al layer is sputtered on the PDMS at room temperature. This temperature is set to avoid cracking of the PDMS layer during sputtering due to the high coefficient of thermal expansion of the PDMS (310 μm/m °C). The metal is then patterned by dry etching (ICP Plasma etcher, Trikon Omega 201) to open the areas corresponding to the electrical contacts of the metallic microstructures (Figure 3g). The etching process used was optimized to avoid any issue caused by thermo-mechanical stress. No cracking of the layers is observed when exposing the materials to the thermal gradients during resist deposition and developing steps. Finally, the silicon substrate is etched from the backside by DRIE using a Bosh-based process (Figure 3h). The oxide stop layer is removed by a combination of wet and dry etching (Figure 3i). The Al layer is selectively removed by wet etching in a solution of acetic, phosphoric and hydrofluoric acid (Figure 3j).

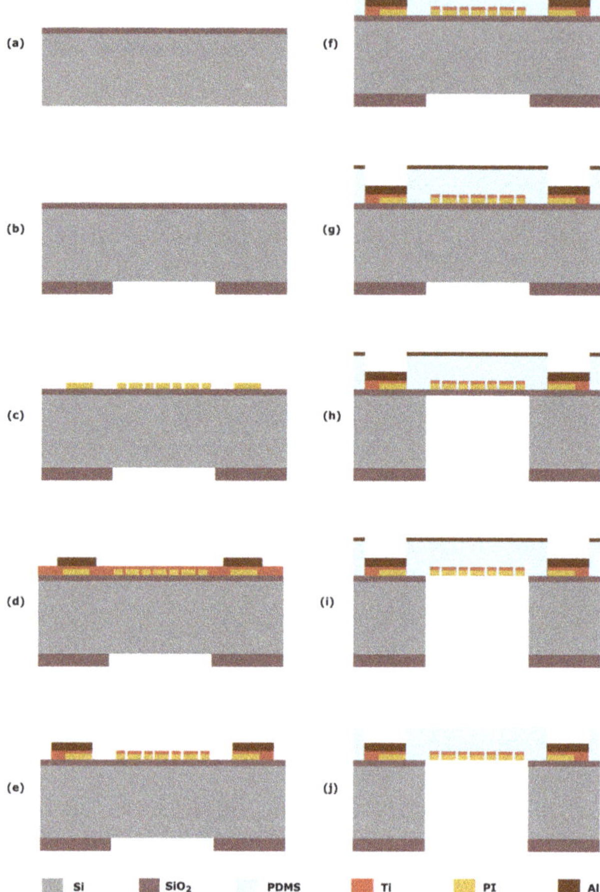

Figure 3. Main steps of the fabrication process for the integration of Ti strain gauges on PDMS membranes. (**a**) Deposition of oxide on the wafer front side. (**b**) Deposition and patterning of the oxide on the wafer back side to define the circular membrane. (**c**) Deposition and patterning of PI layer for electrical and mechanical isolation. (**d**) Deposition of Ti and patterning of electrical contacts (Al). (**e**) Patterning of the metal layer corresponding to the strain gauges (Ti). (**f**) Deposition of PDMS layer. (**g**) Deposition and patterning of the Al masking layer and etching of the PDMS layer to open the electrical contacts. (**h**) Etching of the silicon substrate using a DRIE process. (**i**) Removal of the landing SiO_2. (**j**) Removal of the masking layer (Al) by wet and dry etching.

3.2.2. Polymeric Strain Gauges

Likewise, to fabricate the polymeric (PEDOT:PSS) strain gauges on PDMS membranes, a wafer-level fabrication process was developed based on conventional photolithography and MEMS microfabrication techniques. The steps for the membrane fabrication are similar to the previously described with variations specifically meant to introduce the PEDOT:PSS microstructures.

A 1 µm-thick PECVD SiO_2 is deposited on a 100 mm-Si wafer. Then a layer of 100 nm of Silver (Ag) is deposited by evaporation and subsequently patterned by lift-off to create electrical contacts (Figure 4a). An Al layer is then deposited and patterned to be used as a protective layer during the subsequent polymer etching (Figure 4b). The PEDOT:PSS is deposited by spin coating and cured in an oven at 150 °C for 40 min (Figure 4c). On top of the PEDOT:PSS another Al layer, used as a hard mask during the reactive-ion etching (RIE: O_2, 20 mTorr, 50 W), is sputtered and patterned (Figure 4d,e). The PEDOT-based microstructures are now defined, and metal contacts are exposed. Lastly, the Al masking and protective layer are removed by wet etching using a solution of acetic acid, nitric acid and hydrofluoric acid (PES) (Figure 4f).

Once the conductive polymer is patterned (Figure 4a–f), a PDMS layer is deposited by a two-step spin coating process (Figure 4g). To open the contact pads, the PDMS is etched by RIE, and the SiO_2 as well as the Al are etched as described for the metal strain gauges (Figure 4h–j).

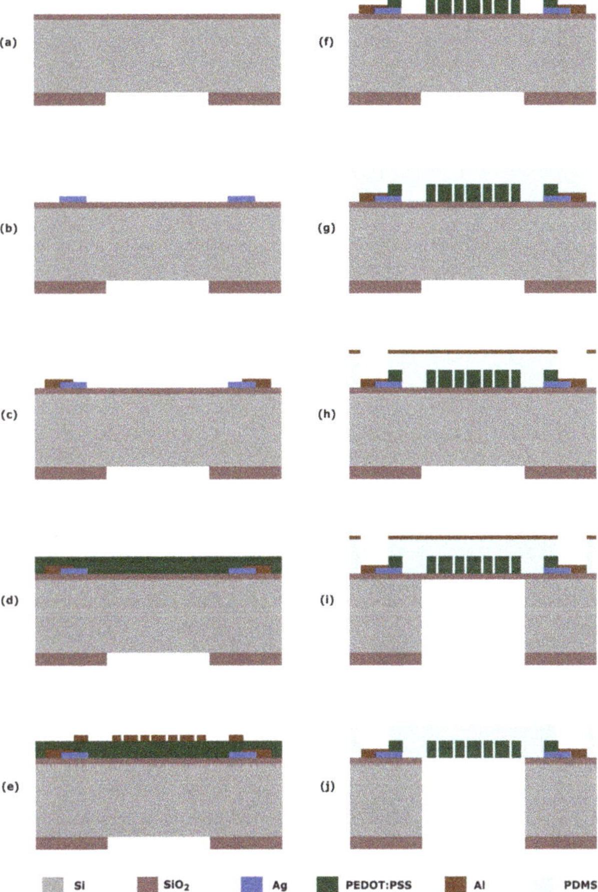

Figure 4. The main steps of the process flow developed for the wafer-scale fabrication of polymeric strain gauges. (**a**) Front and back deposition of the SiO_2 oxide and patterning to define the membranes area. (**b**) Deposition of SiO_2 and Ag on the front side patterning to define the electrical contacts. (**c**) Deposition and patterning of the Al layer to open the electrical contacts to the conductive polymer and to protect the remaining Ag layer for the subsequent etching steps. (**d**) Deposition and curing of the PEDOT:PSS layer. (**e**) Deposition and patterning of the Al masking layer. (**f**) Dry etching of the PEDOT:PSS and removal of Al masking and protective layer from the patterned strain gauges. (**g**) Deposition of the PDMS layer. (**h**) Deposition and patterning of the metallic (Al) masking layer and the PDMS layer to open the electrical contacts. (**i**) Etching of the silicon substrate using a Bosh-based DRIE process. (**j**) Removal of the landing oxide layer and the masking layer (Al) by wet and dry etching.

3.3. Characterization Set-Up

The characterization set-up used to measure the resistance change of the strain gauges subjected to mechanical stress is shown in Figure 5. It consists of three main modules: a 3D-printed holder to interface the silicon chip with a pressure controller, a probe station and external circuitry for data acquisition and signal conditioning.

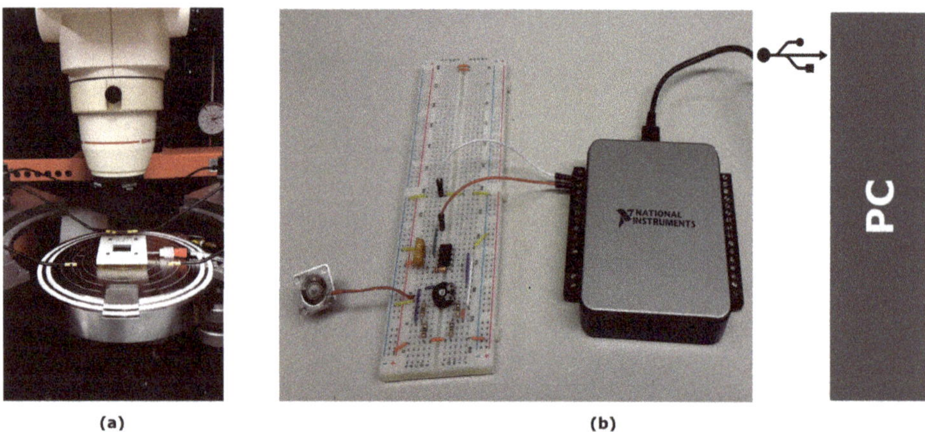

(a) (b)

Figure 5. Measurement set-up developed to characterize the microfabricated strain gauges. (**a**) The electrical signal from the strain gauges is acquired by probing them with a standard probe station. The pneumatic actuation is provided simultaneously through a special coupling holder connected to a pressure source. (**b**) The electrical signal is conditioned and transmitted to a PC for further calculations and data processing.

To couple both the electrical functionality and pneumatic actuation of the devices, a custom-made holder was specifically designed and fabricated by 3D printing. The holder allows for stretching the membrane through the silicon cavity by connecting the chip to a commercial pneumatic pumping system. A commercial Dual AF1 microfluidic pressure and a vacuum pump (Elveflow®, Biotechnology Company, Paris, France) were used to control the pressure. This system makes possible to accurately control the pressure from 0 up to 30 mbar. Moreover, the holder enables electrical connection from the chip to external circuitry as they were both designed such that the Al contact pads on the silicon substrate are easily accessed from the top.

The external circuitry to measure the resistance change is connected to the metallic and polymeric strain gauges contacts through a standard probe station. A Wheatstone bridge is implemented as a first stage for signal conditioning. For this circuit topology, an expression for the resistance of the strain gauges (R_{SG}) as a function of the voltage differences (V, V_{CC}) and the other resistors can be obtained by applying Kirchhoffs voltage law.

The signal (V) is amplified by an operational amplifier (AMP04F, Analog Devices, Norwood, MA, U.S.) with high gain ($G = 100$) and input impedance (in the order of MΩ). The output signal of the amplifier is acquired through an Analog-to-Digital converter (USB-6001, National Instruments, Austin, TX, U.S.), enabling the direct acquisition of the data in a personal computer and the corresponding calculation of the resistance change. The signal is further processed and filtered using a digital low pass filter with the cut-off frequency fixed at 10 Hz. The cut-off frequency is set as low as possible to reduce the high-frequency noise and keep the measurement bandwidth within the expected range of typical biological processes e.g., heart rate: 1–4 Hz.

4. Results

4.1. Microfabrication

4.1.1. Metal Strain Gauges

In Figure 6 a full wafer containing several membranes with strain gauges is shown, demonstrating the wafer-scale capability of the developed process. A close-up of the released membranes (red dash line) and the titanium strain gauges can be also observed. In the background, several fibers are noticed corresponding to the supporting substrate, showing the transparency of the PDMS membrane.

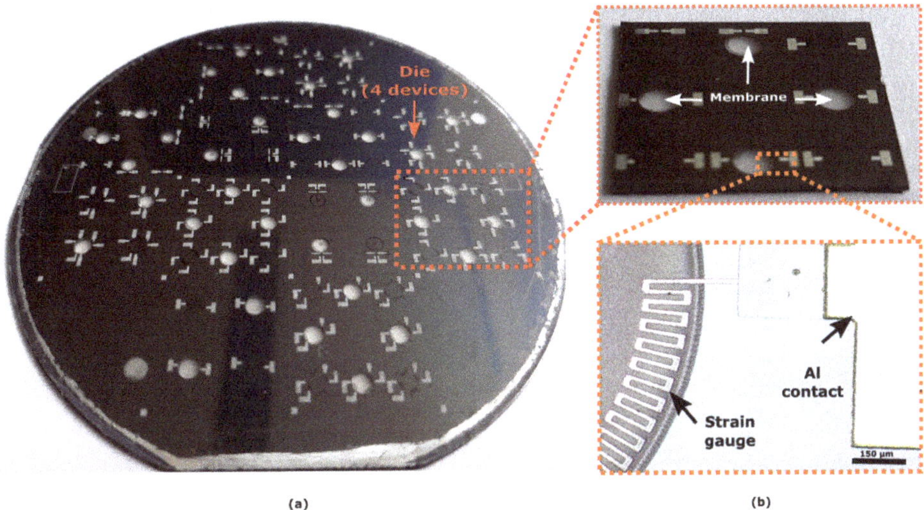

Figure 6. (**a**) A completed wafer containing 36 membranes equipped with strain gauges. (**b**) An optical image of a die containing four devices (**top**) and a close-up of the released membranes with Ti gauges (**bottom**). Scale bars: 150 µm (**bottom**), 3 mm (**top**).

4.1.2. Polymeric Strain Gauges

Several membranes with polymeric strain gauges were realized on a single wafer. In Figure 7 optical images show a close up of the polymeric strain gauges integrated on the membranes. Both radial (Figure 7a) and tangential (Figure 7b) geometries were realized, so to investigate the electromechanical response of the polymeric strain gauges.

Although PEDOT:PSS has already been applied in related devices for neuron cell study, either rigid materials were used as the supporting substrate or fabrication methods lack of compatibility with high scale manufacturing schemes [31], contrary to the advantageous processes here proposed.

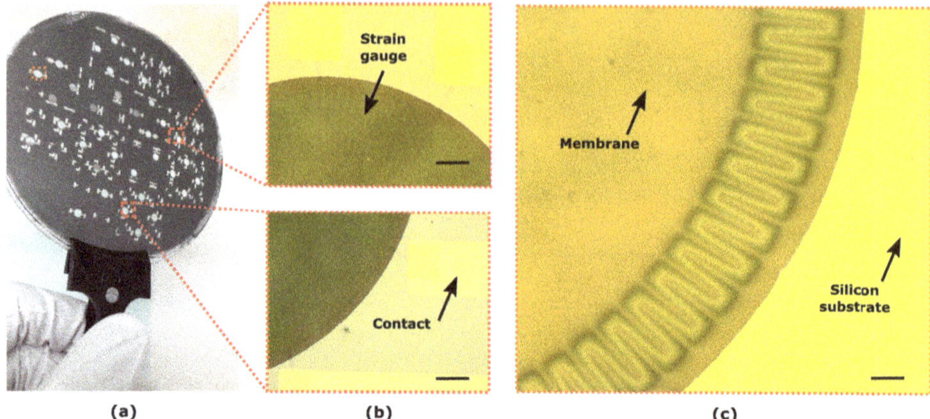

Figure 7. (**a**) A completed wafer containing 36 membranes equipped with polymeric strain gauges. (**b**) Optical images of radial (**top**) and tangential (**bottom**) polymeric strain gauges embedded in 10 μm-thick PDMS. (**c**) A zoom-in perspective illustrating the polymeric strain gauges integrated into the PDMS membranes. Scale bar: 100 μm.

4.2. Electromechanical Characterization

The characterization of the devices was carried out by continuously monitoring the electrical resistance at various stationary pressures. In particular, radial and tangential strain gauges, close to the end PDMS membrane edge, were characterized. The response of the devices was investigated for both Ti and PEDOT:PSS strain gauges.

4.2.1. Electrical Resistance

The resistance change in stationary regime for both tangential and radial strain gauges are reported in Figure 8. The pressure increased from 0 to 3 kPa in steps of approximately 350 Pa, a value slightly higher than the minimum stable change in pressure achieved with the commercial pumping system employed.

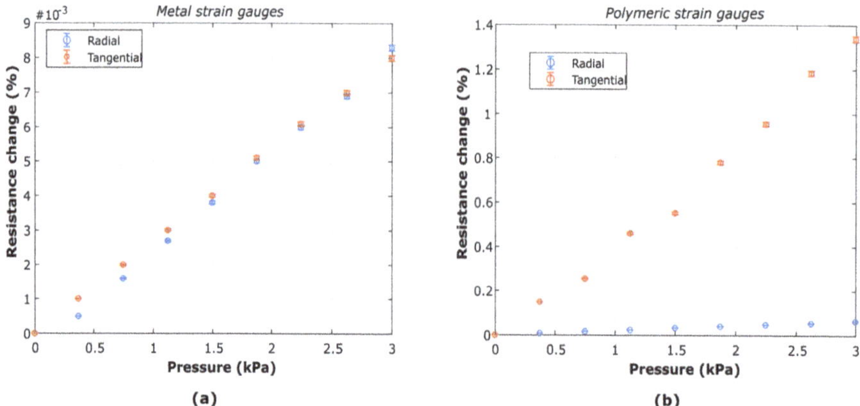

Figure 8. Stationary measurements of resistance change for radial and tangential strain gauges made of (**a**) titanium (Ti) and (**b**) PEDOT:PSS. Error bar: 5%.

An incremental variation of the resistance was observed. For titanium strain gauges, a relative resistance change up to ≈0.008 % over the tested pressure range was measured, for both radial and tangential geometries (Figure 8a).

For PEDOT:PSS strain gauges, a relative resistance change up to ≈1.4% was measured for the tangential geometries, a much higher value than the observed for its metal counterpart. In the case of the radial strain gauge, the resistance change was found to be ≈0.08% (Figure 8b). The measurements were performed under controlled humidity conditions with a relative humidity of ≈48%. Having a controlled humidity is important as it has been shown that both electrical and mechanical properties of PEDOT:PSS vary with humidity [16]. All results were thus obtained under the same environmental conditions. For both geometries and materials, measurements at higher pressure were not possible as the pumping system reached the maximum flow capacity at 3 kPa.

4.2.2. Membrane Displacement and Strain

The displacement at the center of the membrane with polymeric and metal strain gauges was measured for the same pressure ranges. The mechanical stimulus can thus be translated as strain at the surface of the material. For the microfabricated device, the strain can be approximated based on the maximum displacement of the center of the suspended polymer layer [2,3]. In Figure 9a, the displacement for different pressures is shown.

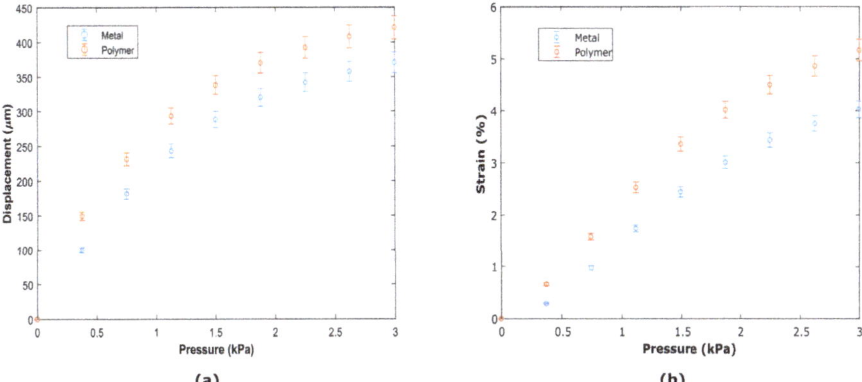

Figure 9. (**a**) Displacement in the center of the membranes with metal and polymeric strain gauges measured optically for different pressures set through the pumping system (1–3 kPa). (**b**) Estimation of the radial strain of the membranes with strain gauges based on the displacement measurements. Error bar: 6%.

The strain on the membrane was also indirectly obtained assuming a semicircular profile and a continuous distribution along the radius. In Figure 9b, the calculated radial strain for the same pressures range used to investigate the resistance change, is shown.

4.2.3. Calibration Curves

The data acquired by the electrical and mechanical characterizations (Figures 8 and 9 have been used to calibrate the measured strain gauges, thus achieving the resistance change of the devices as a function of the total membrane strain. In Figure 10 the calibration curves are shown for both geometries and materials investigated.

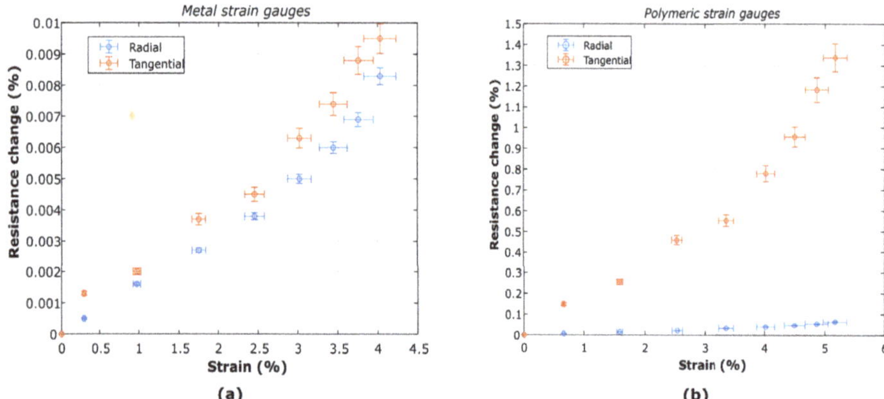

Figure 10. Resistance change as function of the strain on the membrane for radial and tangential strain gauges made of (**a**) titanium (Ti) and (**b**) PEDOT:PSS.

Based on the experimental data, given a resistance change it is possible to establish an approximate transfer function for the actual strain on the membrane. The results of the Ti strain gauges suggest an estimated sensitivity of 4.5 m$\Omega \cdot \mu$m^{-1} and 11.6 m$\Omega \cdot \mu$m^{-1} for radial and tangential geometries, respectively. The tangential geometry showed higher sensitivity, confirming what suggested by theoretical and numerical analysis reported in other works [22]. For polymeric strain gauges, sensitivity of 0.571 $\Omega \cdot \mu$m^{-1} and 62 $\Omega \cdot \mu$m^{-1} for radial and tangential geometries, were obtained, respectively.

5. Discussion

The fabrication processes presented in this paper enabled the development of both metallic and polymeric strain gauges as a potential transduction mechanism for in situ monitoring of strain on PDMS membranes for OOC applications. The electrical resistance of the metallic strain gauges at the end of the fabrication was stable and did not show to be affected by the process. A slight deviation of approximately 2% compared to the designed values was measured. Both radial and tangential geometries (Figure 6b) did not suffer any mechanical disruption after the releasing of the membrane, a critical step to realize the final devices. This demonstrated the robustness and reliability of the fabrication process developed.

The experimental data of resistance change for both metallic radial and tangential devices are within the same order of magnitude, as can be observed in Figure 8. The results show a linear behavior up to 3 kPa, the maximum pressure measurable and supplied by the pneumatic system. However, the data on the displacement shows a non-linear tendency for the same pressure range (Figure 9), causing the non-linearities observed in the calibration curve for strains above 3% (Figure 10). This is due to the saturation of the displacement at the center of the membrane, which might be explained by the difference of a few orders of magnitude in stiffness between the membrane and the metal. Despite these non-linearities, a first linear approximation can be made to establish a transfer function for the actual strain on the membrane given a certain resistance change. Thus, the estimated sensitivity for radial and tangential geometries for strains below 3% is 4.5 m$\Omega \cdot \mu$m^{-1} and 4.4 m$\Omega \cdot \mu$m^{-1}, respectively,

For polymeric strain gauges, the process was more challenging given the nature of the material. The main challenge encountered was to enable and optimize the electrical contact. As it was not possible to establish ohmic contact using readily available metals (Al, Ti, TiN), the process needed to be adapted and optimized. To do so, it was redesigned to minimize long exposure of the materials to water as PEDOT:PPS is highly hygroscopic and degraded easily in contact with water. Given its characteristic reduction potential ($E^0 = +0.8$ V) and the accessibility to Ag deposition and patterning techniques, this material was used to create the electrical contacts. A low reactive metal is adopted due to the high acidity of the sulfonate functional group (PSS), which easily oxidizes most metals used in IC fabrication processes, unavoidably increasing the contact resistance [32]. The addition of such material increased the number of steps and complexity of the process as the material is not standard in the facilities available. Thus, Al was introduced as masking and protection layer so that the Ag contacts were never open when performing the lithography and dry etching steps, safeguarding the processing tools from possible exposure to Ag or unexpected back-sputtering.

Another addressed aspect was the temperature and baking time of the polymer. The effect of these parameters on the electrical contact was investigated. By increasing the baking temperature to 150 °C and the baking time to 30 min, the contact resistance decreased by 20% compared to initial experiments. This indicates the importance of complete removal of water from the polymer to enhance electrical contact. It is worth mentioning that the measurements and fabrication were always carried out under controlled humidity conditions with relative humidity around 48%. Having a controlled humidity was important to keep the consistency of the experiments, as it has been shown that both electrical and mechanical properties of PEDOT:PSS might vary with humidity [22].

Regarding the electrochemical characterization of the polymeric strain gauges, the results also showed a measurable change in resistance of the microstructures when stretching the membrane. In particular, a higher value of relative resistance change (up to 1.4%) was observed compared to metal strain gauges. The tangential structures showed a higher value than the radial ones. This might be correlated with the nature of PEDOT:PSS and the deposition technique used. The polymer has a structure of fiber-like chains with de-localized $\pi-$electrons that enable the conductivity. This suggests that the deposition technique might be inducing a radial arrangement of the fibers as consequence of the spinning, leading to higher resistance changes when stretched. Significant degradation of the devices was observed with stretching and time, suggesting the need for optimizing the patterning of PEDOT:PSS microstructures and to properly characterize the dependence of mechanical properties with environmental conditions. Alternative depositions, such as electro-deposition and encapsulation of the material (e.g., Parylene, Polyimide), can be explored to address these issues and to optimize the polymer deposition, as reported in several works in implantable applications [33–36].

However, it is possible to establish a first linear approximation for the transfer function of polymeric strain gauges. In this case, the sensitivity of $0.571 \ \Omega \cdot \mu m^{-1}$ and $62 \ \Omega \cdot \mu m^{-1}$ for radial and tangential geometries, are obtained respectively for strains below 3%. This indicates a much higher sensitivity for polymeric strain gauges.

In the characterizations performed for both the polymeric and metal strain gauges, more than 50 samples per data point were collected for the mechanical and electrical measurements, with corresponding standard deviations in the 4–5% range, which suggest a high reproducibility and reliability of the devices.

6. Conclusions

So far, the majority of OOCs have relied entirely on bulk optical techniques (immunofluorescence end-point detection, microscope cell imaging) to acquire and analyze the information of cell microenvironments. A complete view of cell responses can be obtained if mechanical strain and other relevant cues could be quantitatively measured in situ and in a spatio-temporal manner. This paper reports wafer-scale microfabrication processes that enable the integration of metal and polymeric strain gauges in monolithically fabricated organs on chips.

The results indicate that both conventional IC metals (Ti) and polymeric materials (PEDOT:PSS) can be used for sensing mechanical strain on flexible substrates for organ-on-chip applications. The transduction behavior and the functionality of the devices were proven. A custom-made set-up allowed to show a resistance change of the devices for different pressures applied to the membranes, demonstrating the functionality of the proposed device. Relative resistance changes of approximately 0.008% and 1.2% for titanium and polymeric strain gauges have been observed, respectively for pressures up to 3 kPa applied to stretch the membranes. Correspondingly, the displacement measurements showed that for such resistance changes a strain of up to $\approx$5 % is induced on the membrane. The results indicate a much higher sensitivity for polymeric strain gauges and a high level of reproducibility.

Author Contributions: Conceptualization, all authors; methodology, W.F.Q.-S.; software, W.F.Q.-S.; validation, W.F.Q.-S.; formal analysis, all authors; investigation, W.F.Q.-S.; data curation, W.F.Q.-S.; Writing—Original draft preparation, all authors; Writing—Review and editing, all authors; visualization, W.F.Q.-S., N.G. and C.S.; supervision, G.P., R.D. and P.M.S.; project administration, P.M.S.; funding acquisition, R.D. and P.M.S.

Funding: William F. Quirós-Solano is partially financed by Instituto Tecnológico de Costa Rica. Nikolas Gaio is financed by Electronic Components and Systems for European Leadership (ECSEL) "InForMed" (No. 2014-2-662155).

Acknowledgments: The authors gratefully acknowledge the technical support and advice of the staff at the TUD-Else Kooi Lab.

Conflicts of Interest: C.S., N.G. and W.F.Q.-S. are founders of the Startup Company BIOND Solutions B.V. (BI/OND), a spin-off from Delft University of Technology. G.P., R.D. and P.M.S. declare no potential conflict of interests.

References

1. van de Stolpe, A. Workshop meeting report Organs-on-Chips: human disease models. *Lab Chip* **2013**, *13*, 3449. [CrossRef] [PubMed]

2. Khoshfetrat Pakazad, S. A novel stretchable micro-electrode array (SMEA) design for directional stretching of cells. *J. Micromech. Microeng.* **2014**, *24*, 34003. [CrossRef]

3. Gaio, N. Cytostretch, an Organ-on-Chip Platform. *Micromachines* **2016**, *7*, 120. [CrossRef] [PubMed]

4. Besser, R.R. Engineered Microenvironments for Maturation of Stem Cell Derived Cardiac Myocytes. *Theranostics* **2018**, *8*, 124–140. [CrossRef] [PubMed]

5. Matsuda, T. N-cadherin-mediated cell adhesion determines the plasticity for cell alignment in response to mechanical stretch in cultured cardiomyocytes. *Biochem. Biophys. Res. Commun.* **2005**, *326*, 228–232. [CrossRef] [PubMed]

6. Matsuda, T. Mechanical control of cell biology. Effects of cyclic mechanical stretch on cardiomyocyte cellular organization. *Prog. Biophys. Mol. Biol.* **2014**, *115*, 93–102.

7. Banerjee, I. Cyclic stretch of embryonic cardiomyocytes increases proliferation, growth, and expression while repressing Tgf-β signaling. *J. Mol. Cell Cardiol.* **2015**, *79*, 133–144. [CrossRef] [PubMed]

8. Taylor, R.E. Sacrificial layer technique for axial force post assay of immature cardiomyocytes. *Biomed. Microdevices* **2013**, *15*, 171–181. [CrossRef] [PubMed]

9. Boudou, T. A Microfabricated Platform to Measure and Manipulate the Mechanics of Engineered Cardiac Microtissues. *Tissue Eng. Part A* **2012**, *18*, 910–919. [CrossRef]

10. Gaitas, A. A device for rapid and quantitative measurement of cardiac myocyte contractility. *Rev. Sci. Instrum.* **2015**, *86*, 34302. [CrossRef]

11. You, J. Cardiomyocyte sensor responsive to changes in physical and chemical environments. *J. Biomech.* **2014**, *47*, 400–409. [CrossRef] [PubMed]

12. Yuan, B. A general approach for patterning multiple types of cells using holey PDMS membranes and microfluidic channels. *Adv. Funct. Mater.* **2010**, *20*, 3715–3720. [CrossRef]

13. Mata, A. Characterization of polydimethylsiloxane (PDMS) properties for biomedical micro/nanosystems. *Biomed. Microdevices* **2005**, *7*, 281–293. [CrossRef] [PubMed]

14. Hongbin, Y. Novel polydimethylsiloxane (PDMS) based microchannel fabrication method for lab-on-a-chip application. *Sens. Actuators B Chem.* **2009**, *137*, 754–761. [CrossRef]

15. Berthier, E. Engineers are from PDMS-land, Biologists are from Polystyrenia. *Lab Chip* **2012**, *12*, 1224–1237. [CrossRef] [PubMed]

16. Lang, U. Mechanical characterization of PEDOT:PSS thin films. *Synth. Met.* **2009**, *159*, 473–479. [CrossRef]

17. Li, Y. Mechanical characterization of PEDOT:PSS thin films. *Trans. Mater. Res. Soc. Jpn.* **2012**, *37*, 303–306. [CrossRef]

18. Greco, F. Ultra-thin conductive free-standing PEDOT/PSS nanofilms. *Soft Matter.* **2011**, *7*, 10642–10650. [CrossRef]

19. Latessa, G. Piezoresistive behaviour of flexible PEDOT:PSS based sensors. *Sens. Actuators B Chem.* **2009**, *139*, 304–309. [CrossRef]

20. Trifigny, N. PEDOT:PSS-based piezo-resistive sensors applied toreinforcement glass fibres for in situ measurement during the compositematerial weaving process. *Sensors* **2013**, *13*, 10749–10756. [CrossRef]

21. Lang, U. Piezoresistive properties of PEDOT:PSS. *Microelectron. Eng.* **2009**, *86*, 330–334. [CrossRef]

22. Schomburg, W.K. The design of metal strain gauges on diaphragms. *J. Micromech. Microeng.* **2004**, *14*, 1101. [CrossRef]

23. Kim, T.K. Microelectronic Engineering Measurement of nonlinear mechanical properties of PDMS elastomer. *Microelectron. Eng.* **2011**, *88*, 1982–1985. [CrossRef]

24. Johnston, I.D. Mechanical characterization of bulk Sylgard 184 for microfluidics and microengineering. *J. Micromech. Microeng.* **2014**, *24*, 35017. [CrossRef]

25. Tsuchiya, T. Young's modulus, fracture strain, and tensile strength of sputtered titanium thin films. *Thin Solid Films* **2005**, *484*, 245–250. [CrossRef]

26. Heckmann, U. New materials for sputtered strain gauges. *Procedia Chem.* **2009**, *1*, 64–67. [CrossRef]

27. Ratner, B.D. A perspective on titanium biocompatibility. In *Titanium in Medicine: Material Science, Surface Science, Engineering, Biological Responses and Medical Applications*; Springer: Berlin/Heidelberg, Germany, 2001; pp. 1–12.

28. Schmid, P. Gauge factor of titanium/platinum thin films up to 350 °C. *J. Proc. Eng.* **2014**, *87*, 172–175. [CrossRef]

29. Dam Madsen, N. Titanium nitride as a strain gauge material. *J. Microelectromech. Syst.* **2016**, *25*, 683. [CrossRef]

30. Millard, R. Implementation of a titanium strain gauge pressure transducer for CTD applications. *Deep Sea Res. Part 1 Oceanogr. Res. Pap.* **1993**, *40*, 1009–1021. [CrossRef]

31. Cho, C.K. Mechanical flexibility of transparent PEDOT:PSS electrodes prepared by gravure printing for flexible organic solar cells. *Sol. Energy Mater. Sol. Cells* **2011**, *95*, 3269–3275. [CrossRef]

32. Voroshazi, E. Influence of cathode oxidation via the hole extraction layer in polymer:fullerene solar cells. *Org. Electron.* **2011**, *12*, 736–744. [CrossRef]

33. Kluba, M. Wafer-Scale Integration for Semi-Flexible Neural Implant Miniaturization. *Proceedings* **2018**, *2*, 941. [CrossRef]

34. Nahvi, M.S. Design, fabrication, and test of flexible thin-film microelectrode arrays for neural interfaces. In Proceedings of the 2017 IEEE 30th Canadian Conference on Electrical and Computer Engineering (CCECE), Windsor, ON, Canada, 30 April–3 May 2017; pp. 1–4.

35. Wang, K. Towards circuit integration on fully flexible parylene substrates. In Proceedings of the 2009 Annual International Conference of the IEEE Engineering in Medicine and Biology Society, Minneapolis, MN, USA, 3–6 September 2009; pp. 5866–5869.

36. Mimoun, B. Flex-to-Rigid (F2R): A Generic Platform for the Fabrication and Assembly of Flexible Sensors for Minimally Invasive Instruments. *IEEE Sens. J.* **2013**, *13*, 3873–3882. [CrossRef]

 *micromachines*

Article

Permeability of Epithelial/Endothelial Barriers in Transwells and Microfluidic Bilayer Devices

Timothy S. Frost [1,*], Linan Jiang [2], Ronald M. Lynch [1,3] and Yitshak Zohar [1,2]

[1] Department of Biomedical Engineering, University of Arizona, Tucson, AZ 85721, USA
[2] Department of Aerospace and Mechanical Engineering, University of Arizona, Tucson, AZ 85721, USA
[3] Department of Physiology, University of Arizona, Tucson, AZ 85721, USA
* Correspondence: tfrost@email.arizona.edu

Received: 25 June 2019; Accepted: 11 August 2019; Published: 13 August 2019

Abstract: Lung-on-a-chip (LoC) models hold the potential to rapidly change the landscape for pulmonary drug screening and therapy, giving patients more advanced and less invasive treatment options. Understanding the drug absorption in these microphysiological systems, modeling the lung-blood barrier is essential for increasing the role of the organ-on-a-chip technology in drug development. In this work, epithelial/endothelial barrier tissue interfaces were established in microfluidic bilayer devices and transwells, with porous membranes, for permeability characterization. The effect of shear stress on the molecular transport was assessed using known paracellular and transcellular biomarkers. The permeability of porous membranes without cells, in both models, is inversely proportional to the molecular size due to its diffusivity. Paracellular transport, between epithelial/endothelial cell junctions, of large molecules such as transferrin, as well as transcellular transport, through cell lacking required active transporters, of molecules such as dextrans, is negligible. When subjected to shear stress, paracellular transport of intermediate-size molecules such as dextran was enhanced in microfluidic devices when compared to transwells. Similarly, shear stress enhances paracellular transport of small molecules such as Lucifer yellow, but its effect on transcellular transport is not clear. The results highlight the important role that LoC can play in drug absorption studies to accelerate pulmonary drug development.

Keywords: barrier permeability; epithelial–endothelial interface; paracellular/transcellular transport

---

1. Introduction

Drug discovery is becoming slower and more expensive over time, a trend referred to as Eroom's law (Moore's law spelled backward), despite major progress in technologies such as high-throughput screening and computational drug design [1]. Following this trend, the number of new drugs produced per billion US \$ has halved every decade since 1950 [1,2]. The increasing difficulty and cost of developing a new compound can be traced back to current methods of new compound screening. In vitro cell cultures and animal models are the major means used to select promising drugs for clinical trials. Experience has shown that these models are poor predictors for compound success in human clinical trials. In vitro cell models in particular have had a limited impact, since they are overly simple failing to recreate the complex tissue microenvironment in vivo, especially the interactions between multiple cell types [3,4]. Animal models, while presenting in vivo tissue-level complexity, fail to provide the proper tissue microenvironment because animals do not possess the same anatomy or physiology as humans; consequently, these models have less than 8% successful translation to therapies in some cancer trials [5,6]. These serious shortcomings underline the critical need to develop new in vitro biomimetic systems that better represent the in vivo human physiological conditions in effort of hastening medical innovation [7].

Many pharmaceutical companies are turning to microfluidic devices as a means to streamline the pre-clinical phases of new compound screening and development [8,9]. Among this new generation of microfluidic devices are microphysiological systems (MPSs), often called organs-on-chips, which have begun to add additional insights as well as increased physiological accuracy to the screening of new compounds [10]. MPSs are in vitro models that capture important aspects of in vivo organ function through the use of specialized culture microenvironments [11]. These microfluidic systems, incorporating small plastic devices, are designed to model certain human organs to trigger more accurate cellular responses [12,13]. Indeed, MPS technology has begun to accelerate medical research across several fields. MPSs are used for modeling of diseases, such as cancer, which is emerging as a prominent application driving development of complex systems with higher-order tissue functions [11]. In drug discovery, these models are utilized to perform high-throughput assays to assess drug viability, optimizing clinical trials, and potentially reducing R&D costs to develop new compounds [14].

MPSs can be used to evaluate the entire life cycle of a compound inside a human body by culturing different human cell types in different chambers on the chip [15,16], and are often used to provide a unique window into the behavior of a tissue that cannot be observed in detail in vivo [17–19]. MPSs are particularly useful in exploring the dynamic permeability across barrier-tissue interfaces such as blood–brain, vascular-endothelial, and lung-blood barrier. Tissue barrier permeability is essential in the selective transport across the interface, is maintained through tight cell–cell junctions, and is controlled by growth factors, cytokines, and other stress related molecules [20]. Understanding the drug life cycle in the body begins with the absorption of a drug into the bloodstream. The intricate process of cellular transport is highly tailored to each organ and is dependent on the cell types being used as well as the expression of any proteins involved in the transport of a drug into the blood stream. Pulmonary drug delivery has recently become an attractive route for medical therapy as it is both minimally invasive and able to quickly interact with a large blood volume [21]. In order to study pulmonary absorption, static transwell inserts have been used as a standard assay for many years [22]. However, these static conditions lack the fluid-flow induced shear stress to mimic blood or air flow in human body. Media flow is an essential characteristic of lung-on-a-chip (LoC) devices allowing the introduction of mechanical forces in an effort to provide a more physiologically relevant model system [23]. Transport of bio-species ranging from small molecules such as water and ions, to large proteins, and even whole cells is significantly more complex in systems with media flow. On top of diffusion, convection is added as another mass transport mechanism. Furthermore, the shear stress exerted on the cellular layer may modify the tissue barrier properties through alteration of inter-cellular junctions and cellular interactions with the extracellular matrix [24–26]. With the growing interest in microfluidic pulmonary models to study drug absorption, a thorough evaluation of the lung–blood barrier is needed. Recently, some permeability measurements were reported in a small airway-on-a-chip model [23], and vascular-endothelial barrier permeability was characterized using a biomimetic microfluidic blood vessel model [27]. To date, little has been done to analyze the transport of larger molecules within an organ-on-a-chip device. In this work we utilize microfluidic membrane bilayer devices, typically used for LoC applications, to form epithelial/endothelial cell barriers. The permeability of various molecules, via different transport pathways, is characterized with and without epithelial/endothelial monolayers or bilayers, and the results are compared with corresponding data collected in traditional transwell inserts. In the current work, both epithelial and endothelial cell layers are submerged in liquid media although, for human lung models, an air-liquid interface is an important feature affecting molecular diffusion in the epithelial side of the interface and subsequently the permeability.

2. Methods

Two experimental setups have been utilized in this work: (i) commercially available transwell inserts under static condition to eliminate the effect of shear stress, and (ii) in-house fabricated microfluidic devices under fluid flow accounting for flow-induced shear stress. Epithelial and endothelial cell layers were cultured in the transwells inserts and microfluidic devices to characterize the permeability of various bio-species based on their concentration measurements. The experimental error analysis associated with these measurements has been reported elsewhere [28].

2.1. Transwell Insert Models

Either epithelial or endothelial cells were cultured in the top compartments (often called apical or luminal) of commercially-available 12-well transwells (Corning, Corning, NY, USA) on polyester membranes with 0.4 µm diameter pores and 0.5% porosity. Once the cultured cell monolayers had reached confluency, media in the top compartments were replaced with 500 µL solutions of cell media mixed with various fluorescently-labeled molecules at pre-determined initial concentrations. After filling the bottom compartments (often called basolateral or abluminal) with 1500 µL fresh cell media, the transwells were placed in an incubator for two hours. Then, liquid samples of 100 µL were removed from both compartments, 5 from the top and 15 from the bottom, and the corresponding fluorescent intensities of all samples were measured using a BioTek Synergy 2 Plate Reader (BioTek, Winooski, VT, USA). Molecular concentrations in the collected samples were then estimated based on the fluorescent intensity measurements using pre-established calibration curves. The average value among the samples collected from each compartment was taken as the molecular concentration. The experiments were repeated without cell monolayers in the transwells. Both the initial concentration and incubation time were determined to keep the concentration measurements within the linear range of the calibration curves to avoid signal saturation.

2.2. Microfluidic Devices

Microfluidic devices were fabricated using soft-lithography techniques. Polydimethylsiloxane (PDMS, Sylgard 184 Silicone Elastomer, Dow Corning, Midland, MI, USA) resin was poured over an aluminum mold and allowed to cure for 24 h at 55 °C. The cured PDMS substrate with grooves, each about 35 mm long and cross-section area of 500 µm × 1000 µm, was peeled off the aluminum mold. A polyester track etched membrane, about 20 µm thick with 1% porosity and 0.8 µm diameter pores, was plasma treated and soaked in a solution of 5% (v/v) (3-Aminopropyl)triethoxysilane (APTES, MP Biomedicals, Solon, OH, USA) in water at 55 °C for 1 hour for leakage-free sealing [29]. While the membrane was drying, pairs of microchannel grooves were oxygen-plasma treated for 60 s at 1000 mTorr (Harrick Plasma, Ithaca, NY, USA). Each pair of PDMS microchannels, coated to prevent small molecule absorption [30], was bonded together with a membrane sandwiched between the overlapping channel segments about 20 mm in length. Once firm bonding had been achieved, tubing adapters were placed over the punched holes to serve as inlet/outlet connectors to the external fluid handling system. An image of a fabricated device along with a schematic of a device operation is shown in Figure 1

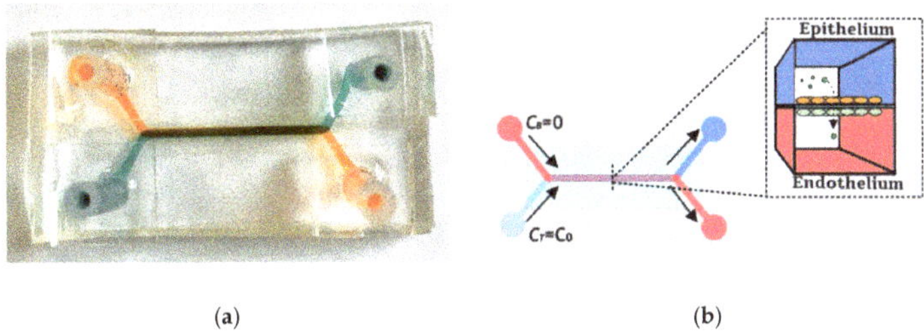

(a)   (b)

Figure 1. A microfluidic bilayer device: (**a**) A photograph of a fabricated and packaged microfluidic bilayer device with blue dye in the top and red dye in the bottom microchannel, and (**b**) a schematic of a microfluidic bilayer device in operation; the top epithelial microchannel is separated from the bottom endothelial microchannel channel by a porous membrane. A syringe pump is used to drive cell medium mixed with fluorescent molecular markers through the top channel and only cell medium through the bottom channel at equal rates, the fluorescent markers are transported across the membrane, from the top to the bottom microchannel and downstream along both microchannels.

2.3. Cell Cultures

Epithelial and endothelial cell lines were chosen to represent human lung–blood barrier interface. The epithelium was composed of carcinoma alveolar epithelial cells (A549; ATCC Manassas, VA, USA), cultured in RPMI media supplemented with 10% fetal bovine serum (FBS, Gibco, Waltham, MA, USA) (v/v) and 1% penicillin streptomycin (Sigma Aldrich, St. Louis, MO, USA). The A549 cell line has characteristic features of pulmonary epithelium, which has a role in the oxidative metabolism of drugs in the lung. This cell line has indeed been extensively utilized in toxicology studies, including its properties on permeable supports, because of its potential target for drug delivery of macro-molecules [31].The endothelium was composed of human umbilical vein cells (HUVEC; ATCC Manassas, VA, USA), cultured in F-12K media (Kaighn's modification, Gibco, Waltham, MA, USA) with 10% FBS, 1% penicillin streptomycin, 50 μg/mL endothelial cell growth supplement (ECGS, Sigma Aldrich, St. Louis, MO, USA) from bovine neural tissue, and 100 μg/mL heparin (Sigma Aldrich, St. Louis, MO, USA). HUVEC cells were selected for this characterization phase of LoC models, since they are most commonly utilized to represent an endothelium based on human derived primary endothelial cells [32]. The cell passage number of the A549 and HUVEC culture was 40–60 and 20–30, respectively. Cell media inside the transwells and microfluidic devices was replenished every 48 h. All cell cultures were maintained at 37 °C, 5% CO_2, 95% air, and 100% relative humidity.

Epithelial and endothelial cells were seeded in either transwells or microfluidic devices at a density of 2×10^6 cells/mL. HUVEC cell attachment and proliferation required membrane surface coating with collagen. A 200 μg/mL type 1 collagen solution (Gibco, Waltham, MA, USA) was prepared in 0.01 M acetic acid and perfused across the membrane surface. After one hour of curing at an ambient temperature, the collagen solution was replaced with 1x HBSS followed by 30 min incubation time in the cell culture environment to adjust the system pH to a biocompatible level. The HUVEC media for cells cultured in transwell or microfluidic devices had an elevated ECGS concentration (150 μg/mL) to promote a fully confluent monolayer [33].

2.4. Tracer Molecules

Molecules can be transported through confluent cell monolayers via two major routes paracellular and transcellular as illustrated in Figure 2. A confluent monolayer refers to cells in culture, which are in contact forming a cohesive sheet of adhering neighboring cells. It is thought that contact inhibition of proliferation is a characteristic of a confluent monolayer, where cells stop proliferating upon contact formation [34]. Adherens junctions (AJs) and tight junctions (TJs) comprise two modes of cell–cell adhesion that provide different functions. While AJs initiate cell–cell contacts and mediate the maintenance of the contacts, TJs regulate the paracellular pathway for the movement of ions and solutes in-between cells [35]. In contrast, transcellular transport involves the passage of solutes through the cells crossing both the apical and basolateral membranes. The transcellular transport includes passive diffusion through ion channels, active carrier mediated transportation such as organic anion/cation transporters, and transcytosis of macromolecules such as transferrin and insulin [36]. Various fluorescently labeled molecules were used to study both molecular pathways in the transwells and in the microfluidic devices.

Fluorescently labeled tracer molecules commonly used in permeability studies were chosen as biomarkers for paracellular and transcellular transport. Lucifer yellow (0.44 kDa; Invitrogen, Carlsbad, CA, USA) has previously been used as a small molecule paracellular transport marker [31,37]. However, it should also be noted that a growing body of literature suggests that transcellular passage of Lucifer yellow may also occur via organic anion transporters [38]. Dextrans, non-digestible sugars, were selected to represent typical agents that are transported via the paracellular route through cell tight junctions [27,39]. Expression by epithelial or endothelial cells of membrane receptors mediating transcytosis of dextran has not been reported thus far; however, pinocytosis of dextran was observed [40], in which molecules are captured in vesicles on one side of the cell, drawn across the cell, and ejected on the other side. Since pinocytosis is a very slow process, transcellular transport of dextran is negligible compared to its paracellular transport [41]. To explore potential molecular-size effect, 4-dextran (4 kDa) and 70-dextran (70 kDa) (Invitrogen, Carlsbad, CA, USA) were chosen to represent the intermediate and large-size agents, respectively. FITC-transferrin (78 kDa; Rockland Antibodies, Limerick, PA, USA) was used to represent a typical protein that is transported via the transcellular route, transcytosis, where epithelial A549 and endothelial HUVEC cells uptake the molecules by binding transferrin to its membrane receptor expressed by both cell lines [31,39,42,43]. FITC-transferrin and 70-dextran molecules are about the same in size and, therefore, may share similar paracellular transport level.

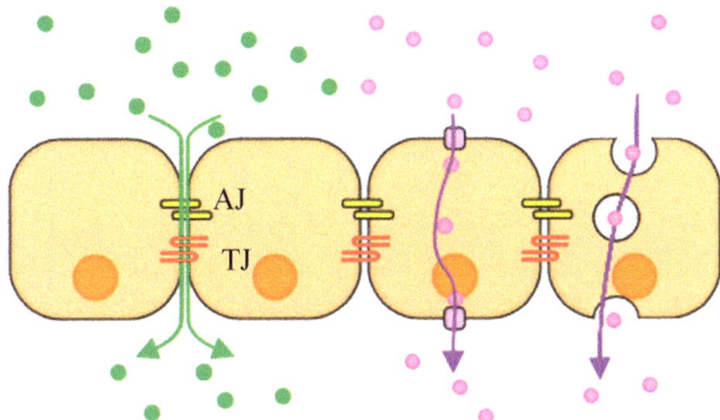

Figure 2. A schematic illustrating two cellular transport mechanisms: Paracellular (green) occurs as molecules diffuse through adherens (AJ) and tight junctions (TJ) between cells because of concentration gradient, and transcellular (purple) takes place when cells uptake and release molecules through their bodies via transport proteins or as vesicles.

2.5. Concentration Measurements in Microfluidic Devices

Epithelial or endothelial cell monolayers were cultured with their specific medium, described in Section 2.3, on the separation membranes in the top channels. Co-cultures of cell bilayers were constructed by growing epithelial and endothelial cells on opposite surfaces of the separation membranes. HUVEC cell suspensions were first seeded with endothelial medium in the top channels on collagen-coated surface membranes. Once the HUVEC cell monolayers reached full confluency, typically within 1–2 days, the bottom channels were perfused with suspensions of A549 cells in the endothelial medium. The tubing adapters were temporarily plugged, and the microfluidic devices were flipped upside down. The inverted devices were suspended over a small water bath for 24 h allowing the establishment of epithelial and endothelial confluent monolayers. The devices were then flipped back to an upright position and the tubing adapters were un-plugged. Live cell staining was performed using CellTracker fluorescent probes (Invitrogen, Carlsbad, CA, USA), within the microfluidic devices, and cells were immunostained for known intercellular adhesion biomarkers to confirm confluency of the cell layers.

In all cases, with or without cells, the top and bottom channels were filled with cell media solution mixed with various tracer molecules at pre-determined initial concentrations and cell media with zero molecular concentration, respectively. Both microchannels were then connected to a PHD Ultra syringe pump (Harvard Apparatus, Holliston, MA, USA) to drive the media in each channel at the same constant flow rate to minimize trans-membrane pressure differences. A549 cell medium was flown through both channels only when epithelial monolayers were tested. In all other conditions, including cell bilayers, HUVEC cell medium was flown through both channels. The temporal fluorescent intensities because of decreasing concentrations at the top and increasing concentrations at the bottom microchannel outlets, with and without cells, were recorded at five minutes intervals. The channel outlets were exposed to a light source for a short time, less than 5 s, to limit photo bleaching, and calibration curves were used to relate the measured fluorescent intensities to molecular concentrations. For steady-state experiments, similar to the transient experiments, microfluidic devices with and without confluent cell monolayers or bilayers were connected to the syringe pump under several constant flow rates. Following the development of steady-state molecular concentration distributions, 500 µL effluent samples from both microchannels were collected and the corresponding fluorescent-light intensities were measured using the BioTek Synergy 2 well plate reader. The sample collection time was 25, 10, 5, and 2.5 h for 20, 50, 100, and 200 µL/h flow rate, respectively, with the microfluidic devices placed in while the syringe pump kept outside of an incubator. As in the transwell experiments, the same calibration curves were used to relate the light intensities to molecular concentrations. The initial inlet concentrations in the microfluidic temporal and steady experiments were the same as those applied in the transwell experiments. It was important to determine the time to reach steady state from the temporal experiments to ensure collection of effluent samples under steady-state conditions. The tested flow rate range was 20–200 µL/h with a corresponding wall shear stress range of 0.0012–0.012 dyne/cm^2.

3. Results and Discussion

A549 and HUVEC cell monolayers as well as A549/HUVEC bilayers were successfully cultured in microfluidic devices as shown by the bright-field images in Figure 3a–c. Since it is difficult to distinguish between different cell types in bright-field images of bilayers cultured on opposite surfaces of a membrane, live cell staining was performed within a microfluidic device to confirm successful establishment of an A549/HUVEC cell co-culture.

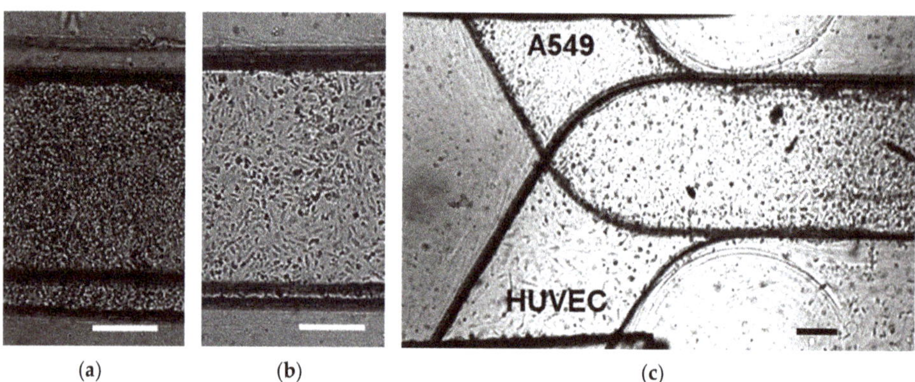

(a)   (b)   (c)

Figure 3. Microscopic bright-field images of: (**a**) confluent A549 epithelial cell monolayer, (**b**) confluent human umbilical vein cells (HUVEC) endothelial cell monolayer, and (**c**) confluent A549/HUVEC cell bilayer cultured in microfluidic devices; scale bars are 250 µm.

Cell layer confluency is typically verified via staining particular proteins involved in cell–cell adhesion. Previous work has suggested that A549 form tighter barriers with the presence of intercellular junction proteins [44,45]. Tight junctions comprise transmembrane proteins (occludin, claudins, JAM), cytoplasmic attachment proteins (ZO-1, ZO-2, ZO-3), and cytoskeleton protein F-actin [46]. The ZO-1 cytoplasmic protein was selected for staining as a biomarker for tight junctions in the A549 cells. The abundance of the red signal in Figure 4a, corresponding to the stained ZO-1 proteins (Proteintech, Rosemont, IL, USA), is evidence for the formation of tight junctions and confluency of the A549 cell monolayer. Adherens junctions in endothelial cells are mainly composed of cadherins and catenins [47]. The vascular endothelial cadherin (VE-cadherin, Cell Signaling Technology, Beverly, MA, USA), a calcium-dependent cell–cell adhesion transmembrane glycoprotein, was selected for staining as a biomarker for adherens junctions in the HUVEC cells. The green signal in Figure 4b, corresponding to the stained VE-cadherins, demonstrates the formation of Adherens junctions and confluency of the HUVEC cell monolayer.

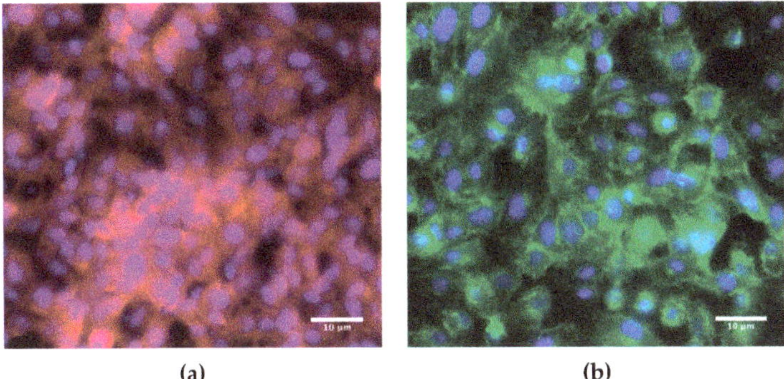

(a)   (b)

Figure 4. Fluorescence microscope images of: (**a**) A549 epithelial cells immunostained for ZO-1 cytoplasmic proteins (red) as a biomarker for tight junctions, and (**b**) HUVEC endothelial cells immunostained for VE-cadherin transmembrane proteins (green) as a biomarker for adherens junctions with standard Hoechst staining (blue) of both A549 and HUVEC cell nuclei. The images were taken from slide-mounted membranes, which were removed from the microfluidic devices following cell fixing and staining.

3.1. Molecular Transport in Transwells

Diffusion in transwells is a process continuing until an equilibrium state is reached, at which the concentration at the bottom equals the concentration at the top compartment with zero gradient. The parameter used for assessing molecular transport is the relative temporal concentration at the bottom compartment, $C(t)$, defined as:

$$\frac{C(t)}{C_0} = \frac{C_B(t)}{C_T(t) + 3 \cdot C_B(t)} \tag{1}$$

where $C_B(t)$ and $C_T(t)$ are the average molecular concentrations measured at the bottom and top compartments, respectively, with initial conditions $C_B(t=0) = 0$ and $C_T(t=0) = C_0$. Initial concentrations for Lucifer yellow, 4-dextran, 70-dextran, and transferrin molecules were $C_0 = 100, 60, 1$, and 1 μM, respectively. Since the bottom compartment volume is three times larger than the top compartment volume, the equilibrium concentration is expected to be a quarter of the top initial concentration, i.e. $C_B(t \rightarrow \infty) = C_T(t \rightarrow \infty) = C_0/4$, independent of molecule or cell type. Therefore, to elucidate the effect of different cell monolayers on the transport of various molecules, experiments were terminated after two hours to ensure that the concentration at the bottom compartment, while high enough to allow reliable measurements, is still smaller than the equilibrium level.

The measured relative concentrations of three paracellular-transport markers, Lucifer yellow, 4-dextran, and 70-dextran, and one transcellular-transport marker, transferrin, are compared in Figure 5 with and without confluent cell monolayers. All relative concentrations measured at the bottom compartments are smaller than the equilibrium concentration, $C_{t2} = C(t = 2\ h) < 0.25C_0$, indicating that the diffusion process is still ongoing after the two-hours incubation time. Without cells, black columns, the relative concentration decreases with increasing molecular size. Based on the Stokes–Einstein equation, molecular diffusivity depends on its size as follows [48,49]:

$$D = \frac{k_B \cdot T}{6\pi\mu r} \qquad \text{with} \qquad r = \left(\frac{3M}{4\pi\rho N}\right)^{1/3} \tag{2}$$

where k_B is the Boltzmann constant, T is the temperature, μ and ρ are the fluid viscosity and density, respectively; r is a radius calculated based on the molecular weight, M, and the Avogadro number, N. Following the convection-diffusion equation:

$$\frac{\partial c}{\partial t} + \mathbf{u}\cdot\nabla c \;=\; D\nabla^2 c \tag{3}$$

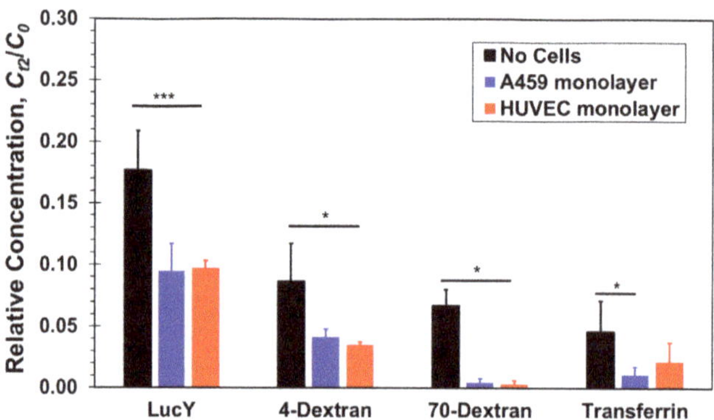

Figure 5. Relative concentrations of four molecular markers, Lucifer yellow, 4-dextran, 70-dextran, and transferrin, measured at the bottom compartments of transwells with and without confluent A549 epithelial or HUVEC endothelial cell monolayer after a 2-h incubation time. Significance determined by Student's *t*-test; *$P < 0.05$; ***$P < 0.001$; $n > 3$.

The spatiotemporal concentration is directly proportional to the molecular diffusion coefficient, D, where $\mathbf{u}$ is the velocity vector. Lucifer yellow is the smallest molecule in this study with the highest diffusion coefficient, $5.0 \times 10^{-10}\,\mathrm{m}^2/\mathrm{s}$, while the diffusion coefficients of the large 70-dextran and transferrin molecules are much smaller, $5.6 \times 10^{-11}\,\mathrm{m}^2/\mathrm{s}$ and $5.85 \times 10^{-11}\,\mathrm{m}^2/\mathrm{s}$, respectively, and the diffusion coefficient of 4-dextran is intermediate about $1.35 \times 10^{-10}\,\mathrm{m}^2/\mathrm{s}$. Thus, in the absence of cells, the measured relative concentration decreases because of decreasing diffusivity with increasing molecular size.

In the presence of cell monolayers, either epithelial A549 or endothelial HUVEC, the measured relative concentrations, C_{t2}/C_0, are smaller than those measured in the absence of cells (Figure 5). Lucifer yellow and both dextran molecules have been widely considered as paracellular markers [27,31]. Hence, their mass transfer rate is generally not aided by active membrane transport proteins; it is rather a passive diffusion through the cellular junctions and membrane pores due to local concentration gradients. The total mass transferred from the top to the bottom compartment, and the resulting bottom concentration, is the integral over time of the product of molecular flux and flux area. Therefore, since the cell monolayers present additional resistance connected in series with the membrane resistance to molecular transport through the combined barrier, the molecular flux decreases in the presence of cells and with it the molecular concentration. Furthermore, in the case of the large 70-dextran, the relative concentration with cells is at the background level suggesting that the confluent monolayers essentially blocked molecular transport between the two compartments. For the smaller molecules, the concentrations measured in the presence of cells were found to be about half of the measured concentrations without cells but clearly above background levels. Since neither A549 nor HUVEC cells are known to uptake these molecules, they penetrated either one of the confluent monolayers via the paracellular route. The junctions of the confluent A549 and HUVEC cell monolayers appear to be tight enough to block the transfer of large proteins but leaky to small proteins amounting to a reduction of the separation-membrane porosity. HUVEC cell layers are known to be leaky with low paracellular resistance [50], and the leaky junctions are associated with cell proliferation or turnover (mitosis) and cell death (apoptosis) [18,51]. A549 cell layers have also been reported to be much leakier than other conventional epithelial cell lines [52]. Indeed, the leakiness of A549 cell monolayers limits their utility in examining the transport of low molecular weight drugs [31]. Therefore, it is not surprising that for these three paracellular markers, no clear difference has been observed between paracellular transport through a confluent endothelial HUVEC and an epithelial A549 monolayer.

The size and diffusivity of FITC-transferrin and large 70-dextran are about the same. Therefore, in the absence of an alternative transport mechanism, paracellular transport of both molecules through a confluent cell monolayer should be similar. Indeed, the transferrin concentration measurements with cells are smaller than the measured concentration without cells. However, while the concentration level of 70-dextran with cells is nearly zero, the measured concentrations of transferrin are statistically significant above background level. Assuming that a confluent monolayer of either A549 or HUVEC cells renders paracellular transport of transferrin negligible, similar to 70-dextran, the elevated transferrin concentration measurements can be attributed to the transcellular transport not available for dextran molecules. Hence, it seems that transferrin receptors expressed by both A549 and HUVEC cells actively facilitate transcellular transport of transferrin molecules not through the tight junctions of the confluent monolayers. The transferrin relative concentration with HUVEC cells is about double that with A549 cells, suggesting either higher transport rate or higher receptor density in HUVEC cells.

3.2. Molecular Convection–Diffusion in Microfluidic Devices

The convection–diffusion experiments in the microfluidic device start with initial conditions of zero molecular concentration at the bottom channel inlet and a uniform concentration at the top channel inlet, C_0. Molecular transport from the top to the bottom channel starts after imposing the same constant flow rate through both channels. The time-dependent molecular concentrations are estimated from the average concentration measurements, $c(x,y,z;t)$, which were recorded at the top and bottom channel outlets yielding:

$$C_T(t) = \left[\int_{-W/2}^{W/2} \int_{h}^{H+h} c(x = L/2, y, z; t) dy dz \right] / (W \cdot H) \tag{4}$$

and

$$C_B(t) = \left[\int_{-W/2}^{W/2} \int_{-H-h}^{-h} c(x = L/2, y, z; t) dy dz \right] / (W \cdot H) \tag{5}$$

where H, W, and L are the channel height, width, and overlapping length, respectively, and h is half the membrane thickness. The origin of the coordinate system is chosen to be at the center of the device such that the fluid domain is confined to the top: $-L/2 \leq x \leq L/2$, $h \leq y \leq H+h$, $-W/2 \leq z \leq W/2$, and bottom channel: $-L/2 \leq x \leq L/2$, $-H-h \leq y \leq -h$, $-W/2 \leq z \leq W/2$. Because of the convection effect, a steady-state concentration distribution is developed before complete mixing of the two fluid flows such that $C_\infty = C_B(t \rightarrow \infty) < C_0/2$. Initial concentrations in microfluidic devices were identical to those in the transwell experiments, i.e., $C_0 = 100, 60, 1,$ and 1 µM for Lucifer yellow, 4-dextran, 70-dextran, and transferrin molecules, respectively.

3.2.1. Transient Response Characterization

The temporal measurements of Lucifer yellow concentrations at the bottom channel outlets of the microfluidic devices with and without cells, under a flow rate of 20 µL/h, are summarized in Figure 6. The measured concentrations are normalized by the steady-state concentration, $C_\infty = 0.2C_0$, and a characteristic time scale of $\tau = 25$ min. In all cases, the system had reached a steady state within one hour. More importantly, the collapse of the data indicates that the bottom-channel concentration increase with time, $C_B(t)$, in the absence of cells is very much the same as in the presence of cells, either an A549 or HUVEC monolayer or an A549/HUVEC bilayer. Thus, within experimental error, each of the three cell-culture combinations has negligible effect on the transport rate of Lucifer yellow molecules from the top to the bottom channel. In contrast, in the transwell static experiments, the concentration at the bottom compartment with cells was found to be about half of that with no cells, and is discussed next.

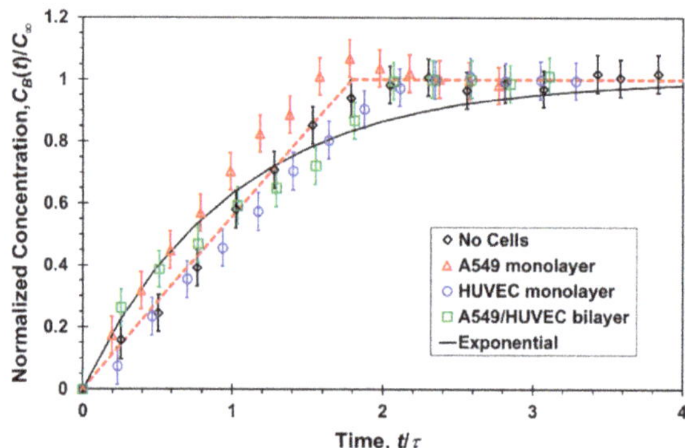

Figure 6. Normalized Lucifer yellow concentration measured at the bottom channel outlet of a microfluidic device as a function of time, under a 20 µL/h flow rate in each channel, with and without confluent cell mono/bilayers; the experimental data (symbols) are fitted with a linear (red dash line) and exponential function (black solid line).

3.2.2. Steady-state Molecular Transport

The flow rate, Q, is the most dominant parameter controlling the molecular transport rate between the two channels as it can vary from a very low rate resulting in complete mixing of the two streams, i.e., $C_\infty \to C_0/2$ as $Q \to 0$, to a flow rate high enough to render cross-membrane diffusion negligible such that $C_\infty \to 0$ as $Q \to \infty$. Therefore, the average molecular concentration at the bottom channel outlet was measured under different flow rates once steady state concentration distributions had been established within the microfluidic devices. The relative concentrations of the four molecules, C_∞/C_0, in the absence and presence of cell cultures are detailed in Figure 7 as a function of the flow rate. In general, in all cases with and without cells, the concentration is roughly inversely proportional to the flow rate decreasing with increasing flow rate. Under high flow rate, $Q > 100$ µL/h, the measured concentrations are so low such that it is very difficult to elucidate the effect of the cell cultures on the ensuing molecular transport because of the large experimental error. Phase contrast imaging revealed no changes in the confluence or morphology of monolayers cultured under the tested flow rate range, $Q = 30$–200 µL/h.

The steady concentrations measured at the bottom channel outlets under a 20 µL/h flow rate, however, are high enough to facilitate a meaningful discussion and are depicted in Figure 8. In microfluidic devices, similar to the results in the transwells, the concentrations of the 4-dextran molecules in the presence of cells are smaller than the concentration without cells; namely, in both cases, leaky junctions of the confluent cell layers failed to block paracellular transport of intermediate-size molecules but rather presented an added resistance akin to decreasing membrane porosity. Also, the concentrations of the 70-dextran in the presence of cells are within the background noise indicating that the cellular junctions are tight enough to prevent paracellular transport of large molecules. On the other hand, in contrast with the transwells findings, the Lucifer yellow relative concentrations with and without cells are about the same within experimental error. Since Lucifer yellow has been utilized as a paracellular marker, its uptake by A549 and HUVEC cells is assumed to be negligible. Thus, if indeed its transcellular transport is insignificant, it seems that the shear stress enhances paracellular transport of Lucifer yellow to such a level practically diminishing the resistance of the cell layers to small-molecule transport.

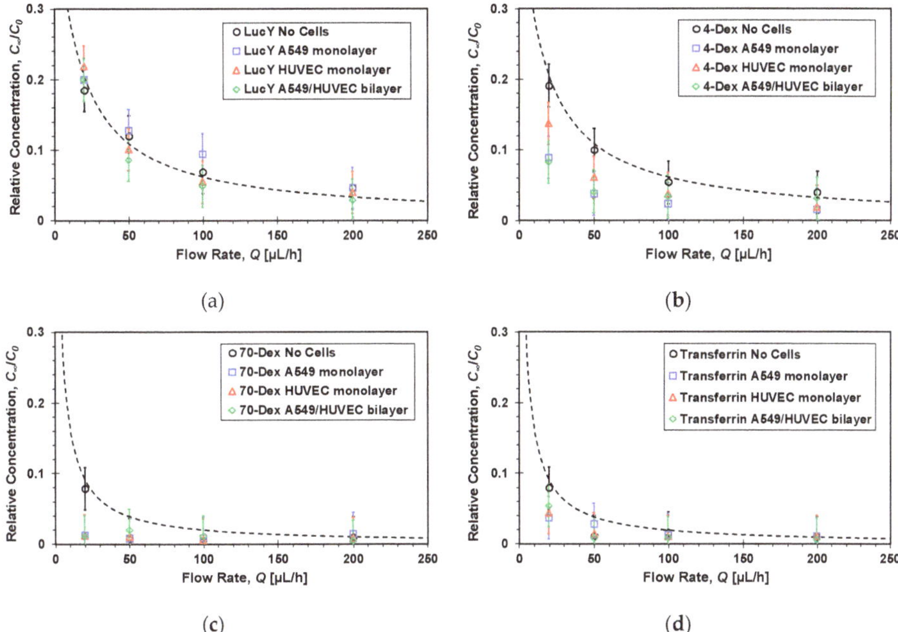

Figure 7. Flow rate effect on the steady-state relative concentrations of: (**a**) Lucifer yellow, (**b**) 4-dextran, (**c**) 70-dextran, and (**d**) transferrin markers measured at the bottom channel outlets of microfluidic devices, because of molecular transport from top to bottom channel, with and without confluent cell mono/bilayers; dash lines are functions inversely proportional to the flow rate fitted to the no cell experimental data with $R^2 = 0.9$.

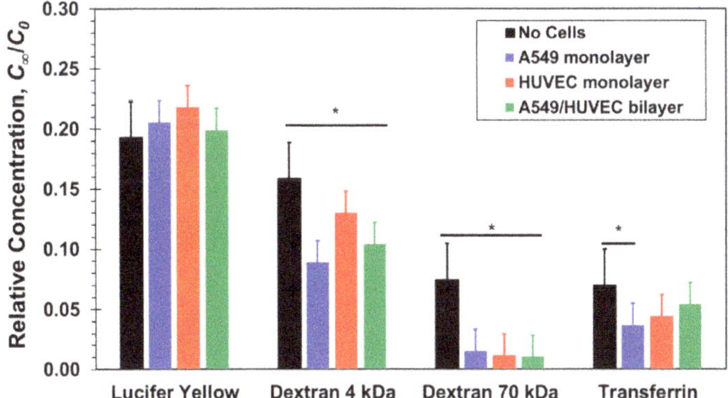

Figure 8. Steady-state relative concentrations of four molecular markers, Lucifer yellow, 4-dextran, 70-dextran, and transferrin, measured at the bottom channel outlets of microfluidic devices with and without confluent cell mono/bilayers under a 20 µL/h flow rate in each channel. Significance determined by Student's *t*-test; *$P < 0.05$; $n > 3$.

The concentration of transferrin is significantly larger than that of 70-dextran. Both molecules are similar in size and, since applied shear stress had no effect on transport of large 70-dextran molecules through cellular junctions, paracellular transport of transferrin is also expected to be negligible. Therefore, the measured transferrin concentration can be attributed to molecular transport via the transcellular route, i.e., transcytosis, as was observed in transwells. It is difficult to discern the contributions of the different cell layers to the transferrin transport because of the relatively large experimental error. However, in comparison with the transwells results, the small difference between transferrin concentrations with and without cells suggest that the shear stress in microfluidic devices can perhaps enhance transferrin transcellular transport by the active protein transporters.

3.3. Barrier Permeability Characterization

Barrier-tissue interfaces are semi-permeable allowing selective transportation of molecules such as water, ions, and nutrients across the interface. Such a permeability barrier is maintained through cell–cell junctions and is controlled by growth factors, cytokines, and other stress-related molecules. Barrier permeability is a dynamic process influenced by many factors including exposure to inflammatory agents and gene products. Permeability is a measure of the average molecular flux across a barrier interface area. In microfluidic devices, under steady-state conditions, the apparent permeability is given by:

$$P_{am} = (C_\infty/C_0)(Q/A) \tag{6}$$

where A is the barrier interface area. However, in transwells, experiments are terminated prior to reaching an equilibrium of uniform concentration in top and bottom compartment; therefore, the permeability is calculated as follows:

$$P_{at} = (C_{t2}/C_0)(V/A)(1/t) \tag{7}$$

C_{t2} is the measured concentration at the bottom compartment with a volume of $V = 1500$ µL following diffusion time of $t = 2$ h. The permeability values estimated based on Equations (6) and (7) in the range of 10^{-6}–10^{-5} cm/s are in close agreement with previously published data [12,23,27,31]. The molecular size effect on the permeability, mainly because of paracellular transport, is demonstrated in Figure 9. The estimated permeability in transwells after two hours of diffusion (Figure 9a) and in microfluidic devices at steady-state under a 20 µL/h flow rate (Figure 9b) are plotted as a function of the molecular weight. The separation membrane permeability—without cells—is inversely proportional to the molecular size because of decreasing diffusivity with increasing size. The barrier permeability—with cell monolayers cultured on the separation membrane—decreases roughly logarithmically with increasing molecular size within experimental error. In static transwells, with no shear stress, the permeabilities of all three paracellular markers with a confluent A549 or HUVEC cell monolayer are markedly smaller than without cells; no clear difference between the epithelial- and the endothelial-monolayer permeability can be discerned. The trends in microfluidic devices with shear stress are similar to those in transwells, except for the Lucifer yellow permeability in microfluidic devices which is about the same with and without the cell layers. The permeability of the 70-dextran is close to zero in both systems. Since its transcellular transport is negligible, its paracellular transport through the cell TJs is also negligible because of its large size. Transcellular transport of 4-dextran is assumed to be negligible, similar to 70-dextran, independent of dextran molecular size. Therefore, because of its smaller size, the 4-dextran permeability in both transwells and microfluidic devices is likely due to paracellular transport through leaky TJs.

The effect of shear stress on permeability can be delineated from comparing the transwells with the microfluidic results. However, the microfluidic permeability was estimated under steady molecular flux independent of time, while the molecular flux in transwells is time-dependent decreasing gradually from its maximum at the start of the diffusion process to zero at equilibrium when the concentrations at the top and bottom compartments are the same. Therefore, a direct comparison between the permeability results obtained in transwells and microfluidic devices is not appropriate. Instead, it is more appropriate to compare normalized permeability results for each system. The ratio between the permeability with and without cell layers in transwells and microfluidic devices, P_C/P_N, is shown in

Figure 10 as a function of molecular size. The trends in both systems are similar as the permeability ratio decreases with increasing molecular size indicating that the paracellular transport is adversely affected as the size of the molecules increases. However, the microfluidics slope is significantly higher than the transwells slope highlighting the additional effect of shear stress. The results suggest that the paracellular-transport dependence on molecular size can be regulated to a certain degree by applying shear stress. Indeed, it is known that TJs form regulated, selectively permeable barriers between two distinct compartments. TJs do not just represent static structural elements but they are dynamically regulated to control paracellular solute and ion transport in diverse physiologic states [53].

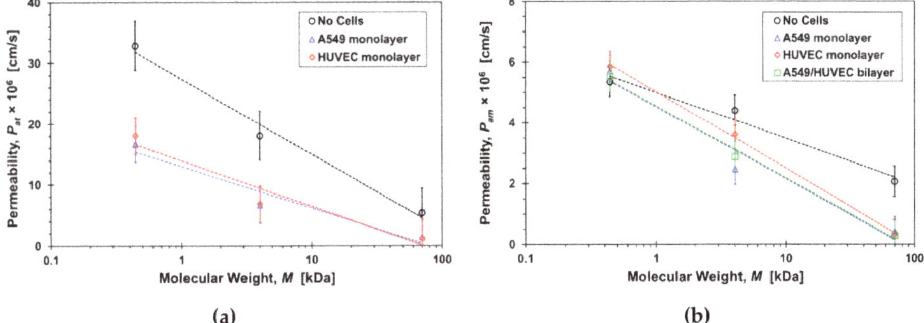

(a)   (b)

Figure 9. Molecular size effect on the permeability of paracellular markers, Lucifer yellow, 4-dextran, and 70-dextran, with and without cells in: (**a**) transwells after 2-hrs incubation time, P_{at}, and (**b**) microfluidic devices under a 20μL/h flow rate, P_{am}. Dash lines are logarithmic functions fitted to the corresponding data sets with $R^2 > 0.9$.

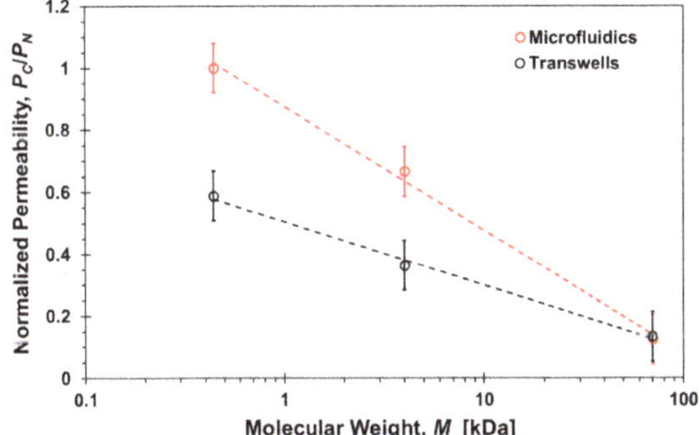

Figure 10. Molecular size effect on the ratio between the permeability with and without cells, P_C/P_N, of paracellular markers, Lucifer yellow, 4-dextran, and 70-dextran, in transwells after 2-h incubation time and microfluidic devices under a 20 μL/h flow rate. Dash lines are logarithmic functions fitted to the corresponding data sets with $R^2 > 0.9$.

While there is a general consensus that shear stress affects paracellular transport, the actual resulting effect is not clear. Shear stress was observed to enhance barrier function in human airway epithelial cells reducing paracellular permeability [54], and the permeability of bovine aortic endothelial monolayers was reported to be enhanced by application of shear [55]. Elsewhere it was suggested that the effect depends on the shear stress level as high shear stress suppresses mitosis and apoptosis while low shear stress supports both processes [51]. Thus, cellular turnover rates and apoptosis rates, and by association the prevalence of leaky junctions and permeability, will be greater in low shear stress regions. In a recent study, permeability was found to increase at the onset of flow and slowly plateaus to a baseline value [27]. Similarly, here the paracellular permeability of smaller molecules is enhanced as the permeability ratio of 4-dextran almost doubled, increasing from about 1/3 in static transwells to about 2/3 in microfluidic devices, due to the applied shear stress.

The permeability results of Lucifer yellow in microfluidic devices are somewhat enigmatic. In transwells, its permeability with either A549 or HUVEC cell monolayer is about half of its permeability without cells, which can be attributed to the effect of paracellular transport similar to the 4-dextran. However, the permeability of Lucifer yellow in microfluidic devices is about the same with and without cell mono or bilayers. While Lucifer yellow has been widely used as a paracellular transport marker, previous work suggested that it could be transported via a transcellular route as well. Cells with organic anion transporters (OATs) may uptake Lucifer yellow since it is anionic in solution. The presence of OATs on A549 epithelial and HUVEC endothelial cells has been reported along with the suggestion that some transcellular transport of Lucifer yellow may occur [56,57]. To determine whether OATs play a significant role in Lucifer yellow molecular transport, a common OAT blocker (probenecid, Santa Cruz Biotechnology, Dallas, TX USA) was used to hinder any OAT activity through competitive inhibition [58]. Confluent A549 monolayers cultured in a transwell top compartment and a microfluidic device top channel were exposed to media solutions mixed with 1 μM Lucifer yellow and 1 μM probenecid. Lucifer yellow concentrations were then measured at the transwell bottom compartment, after two-hour incubation time, and at the device bottom channel outlet, after reaching steady state, similar to previous experiments.

As shown in Figure 11, within experimental error, the measured relative concentrations in both transwells and microfluidic devices are about the same with and without probenecid treatment in presence of a A549 cell monolayer. This may be consistent with more recent reports that OATs are not expressed by HUVEC cells [59,60]. Nevertheless, the contribution of OATs to Lucifer yellow transcellular transport is negligible with or without shear stress. Paracellular seems to be the dominant transport route resulting in reduced permeability through the membrane with confluent A549 monolayer in transwells, $P_C/P_N < 1$. In microfluidic devices, on the other hand, the Lucifer yellow permeability ratio is about one, $P_C/P_N \cong 1$. Thus, the shear stress could only enhance the paracellular transport as observed for 4-dextran, with vanishing effect of the cell monolayer, or augment the enhanced paracellular transport by activating a transcellular transport route other than OATs.

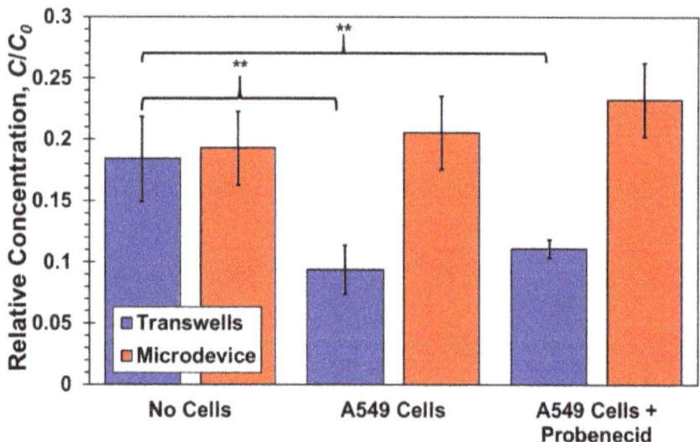

Figure 11. Lucifer yellow relative concentrations with and without a confluent A549 cell monolayer, with and without exposure to probenecid, measured at the bottom compartments of transwells, after a 2-h incubation time, and at the bottom channel outlets of microfluidic devices, under a 20 μL/h flow rate in each channel. Significance determined by Student's *t*-test; *$P < 0.05$; **$P < 0.01$; $n > 3$.

Finally, the permeability ratio for transferrin also increases from $P_C/P_N = 0.35$ in transwells (Figure 5) to about 0.55 in microfluidic devices (Figure 8). Here again, the role of shear stress in transcellular transport is not very clear. Physiologically laminar shear stress was observed to downregulate the expression of transferrin receptor reducing transcytosis [61], but elevated steady shear stress has also been shown to increase intracellular uptake enhancing endocytosis [62]. Fluid shear stress is known to stimulate both apical and basolateral expression and trafficking of protein transporters [63], which may account for the higher transferrin permeability ratio in microfluidic devices with shear stress.

4. Conclusions

Epithelial A549 and endothelial HUVEC cell were successfully co-cultured in a microfluidic membrane bilayer device to model the lung–blood barrier interface. The microfluidic devices enabled characterization of the epithelial/endothelial barrier permeability using measured concentrations resulting because of paracellular and transcellular molecular transport. The bilayer interface permeability for four different molecules was compared with the permeability in microfluidic devices, at steady state, and transwells, after two-hours incubation time, without and with either A549 or HUVEC cell monolayers. The microfluidic transient response reveals that a steady-state molecular distribution is achieved as a balance between convection and diffusion in less than an hour. The molecular flux through the porous membrane in microfluidic devices and transwells, without cells, is inversely proportional to the molecular size because of decreasing diffusivity with increasing size. Paracellular transport of 70-dextran is negligible because of its large size and, in the absence of an active dextran transporter, its slow pinocytosis can also be neglected. As a result, the measured 70-dextran concentrations in microfluidic devices and transwells were within the background noise. Transcellular transport of 4-dextran is similar to that of 70-dextran; however, because of its smaller size, the paracellular transport via tight junctions is not negligible. The permeability of 4-dextran with cells in microfluidic devices and transwells reduces to about two-thirds and one-thirds, respectively, in comparison to the permeability with no cells. The higher ratio in microfluidic devices can be attributed to the flow-induced shear stress known to enhance the leaky tight junctions. Paracellular transport of transferrin is similar to that of 70-dextran; however, because of the expression of transferrin membrane receptors by A549 and HUVEC cells, its transcytosis is significant. Similar to 4-dextran, transferrin permeability ratio in microfluidic devices is higher than in transwells indicating that the transferrin transport pathway may be more active when exposed to shear stress. Lucifer yellow permeability through confluent A549 cell monolayers in transwells and microfluidic devices with and without probenecid treatment, known to block organic anion transporters, is about the same. This suggests that transcellular transport of Lucifer yellow is negligible and, as a small molecule, paracellular is its dominant transport mechanism. Lucifer yellow permeability in transwells with cells is reduced to about half of the permeability without cells. However, in microfluidic devices with shear stress, the permeability with and without cells is about the same.

In summary, the ability to precisely monitor transport properties of bio-species in microfluidic devices with multiple layers of cells, in a more physiological microenvironment, has been demonstrated. Using this technology, we can evaluate the shear stress effect on permeability that cannot be observed in standard static transwell inserts.

Author Contributions: Conceptualization T.F., L.J., Y.Z.; Data Curation T.F., Y.Z.; Formal Analysis T.F., Y.Z.; Funding Acquisition T.F., Y.Z.; Methodology T.F., L.J., Y.Z.; Resources T.F., L.J.; Supervision L.J., R.L., Y.Z.; Validation T.F., Y.Z.; Visualization T.F., Y.Z.; Writing T.F., L.J., R.L., Y.Z.

Funding: This work was supported by the Arizona Biomedical Research Commission through grant ABRC ADHS14-082983, the UA GPSC Research and Project grant, and by the UA/NASA Space Grant Graduate Fellowship.

References

1. Scannell, J.W.; Blanckley, A.; Boldon, H.; Warrington, B. Scannell 2012 diagnosing the decline in pharmaceutical R&D efficiency. *Nat. Rev. Drug Discov.* **2012**, *11*, 191–200. [PubMed]
2. Van Norman, G.A. Overcoming the Declining Trends in Innovation and Investment in Cardiovascular Therapeutics. Beyond FROOM's Law. *JACC Basic Transl. Sci.* **2017**, *2*, 613–625. [CrossRef] [PubMed]

3. Huh, D.; Hamilton, G.A.; Ingber, D.E. From 3D cell culture to organs-on-chips. *Trends Cell Biol.* **2011**, *21*, 745–754. [CrossRef] [PubMed]

4. Brandon, E.F.A.; Raap, C.D.; Meijerman, I.; Beijnen, J.H.; Schellens, J.H.M. An update on in vitro test methods in human hepatic drug biotransformation research: Pros and cons. *Toxicol. Appl. Pharmacol.* **2003**, *189*, 233–246. [CrossRef]

5. Mak, I.W.; Evaniew, N.; Ghert, M. Lost in translation: animal models and clinical trials in cancer treatment. *Am. J. Transl. Res.* **2014**, *6*, 114–118. [PubMed]

6. Sena, E.; Bart van der Worp, H.; Howells, D.; Macleod, M. How can we improve the pre-clinical development of drugs for stroke? *Trends Neurosci.* **2007**, *30*, 433–439. [CrossRef] [PubMed]

7. Chuang-stein, C.; D'Agostino, R. Statistics in Drug Development. In *Pharmaceutical Statistics Using SAS: A Practical Guide*; Dmitrienko, A., Chuang-Stein, C., Eds.; SAS: Cary, NC, USA, 2007; pp. 1–6.

8. Maguire, T.J.; Novik, E.; Chao, P.; Barminko, J.; Nahmias, Y.; Yarmush, M.L.; Cheng, K.-C. Design and Application of Microfluidic Systems for In Vitro Pharmacokinetic Evaluation of Drug Candidates. *Curr. Drug Metab.* **2009**, *10*, 1192–1199. [CrossRef]

9. Muthaiyan Shanmugam, M.; Subhra Santra, T. Microfluidic Devices in Advanced Caenorhabditis elegans Research. *Molecules* **2016**, *21*, 1006. [CrossRef]

10. Bhatia, S.N.; Ingber, D.E. Microfluidic organs-on-chips. *Nat. Biotechnol.* **2014**, *32*, 760–772. [CrossRef]

11. Edington, C.D.; Chen, W.L.K.; Geishecker, E.; Kassis, T.; Soenksen, L.R.; Bhushan, B.M.; Freake, D.; Kirschner, J.; Maass, C.; Tsamandouras, N.; et al. Interconnected Microphysiological Systems for Quantitative Biology and Pharmacology Studies. *Sci. Rep.* **2018**, *8*, 4530. [CrossRef]

12. Huh, D.; Matthews, B.D.; Mammoto, A.; Montoya-Zavala, M.; Hsin, H.Y.; Ingber, D.E. Reconstituting organ-level lung functions on a chip. *Science* **2010**, *328*, 1662–1668. [CrossRef]

13. Huh, D.; Leslie, D.C.; Matthews, B.D.; Fraser, J.P.; Jurek, S.; Hamilton, G.A.; Thorneloe, K.S.; McAlexander, M.A.; Ingber, D.E. A human disease model of drug toxicity-induced pulmonary edema in a lung-on-a-chip microdevice. *Sci. Transl. Med.* **2012**, *4*, 159ra147. [CrossRef]

14. Marx, U.; Walles, H.; Hoffmann, S.; Lindner, G.; Horland, R.; Sonntag, F.; Klotzbach, U.; Sakharov, D.; Tonevitsky, A.; Lauster, R. 'Human-on-a-chip' Developments: A Translational Cutting- edge Alternative to Systemic Safety Assessment and Efficiency Evaluation of Substances in Laboratory Animals and Man? *Lab. Anim.* **2012**, *40*, 235–257. [CrossRef] [PubMed]

15. Zhao, Y.; Kankala, R.; Wang, S.-B.; Chen, A.-Z. Multi-Organs-on-Chips: Towards Long-Term Biomedical Investigations. *Molecules* **2019**, *24*, 675. [CrossRef] [PubMed]

16. Visone, R.; Gilardi, M.; Marsano, A.; Rasponi, M.; Bersini, S.; Moretti, M. Cardiac Meets Skeletal: What's New in Microfluidic Models for Muscle Tissue Engineering. *Molecules* **2016**, *21*, 1128. [CrossRef]

17. Benam, K.H.; Dauth, S.; Hassell, B.; Herland, A.; Jain, A.; Jang, K.-J.; Karalis, K.; Kim, H.J.; MacQueen, L.; Mahmoodian, R.; et al. Engineered In Vitro Disease Models. *Annu. Rev. Pathol. Mech. Dis.* **2015**, *10*, 195–262. [CrossRef]

18. Higuita-Castro, N.; Nelson, M.T.; Shukla, V.; Agudelo-Garcia, P.A.; Zhang, W.; Duarte-Sanmiguel, S.M.; Englert, J.A.; Lannutti, J.J.; Hansford, D.J.; Ghadiali, S.N. Using a Novel Microfabricated Model of the Alveolar-Capillary Barrier to Investigate the Effect of Matrix Structure on Atelectrauma. *Sci. Rep.* **2017**, *7*, 11623. [CrossRef]

19. Kimura, H.; Sakai, Y.; Fujii, T. Organ/body-on-a-chip based on microfluidic technology for drug discovery. *Drug Metab. Pharmacokinet.* **2018**, *33*, 43–48. [CrossRef]

20. Lal, B.K.; Varma, S.; Pappas, P.J.; Hobson, R.W.; Durán, W.N. VEGF Increases Permeability of the Endothelial Cell Monolayer by Activation of PKB/akt, Endothelial Nitric-Oxide Synthase, and MAP Kinase Pathways. *Microvasc. Res.* **2001**, *62*, 252–262. [CrossRef]

21. Tiwari, G.; Tiwari, R.; Sriwastawa, B.; Bhati, L.; Pandey, S.; Pandey, P.; Bannerjee, S.K. Drug delivery systems: An updated review. *Int. J. Pharm. Investig.* **2012**, *2*, 2–11. [CrossRef]

22. Ghaffarian, R.; Muro, S. Models and Methods to Evaluate Transport of Drug Delivery Systems Across Cellular Barriers. *J. Vis. Exp.* **2013**, 50638. [CrossRef]

23. Benam, K.H.; Villenave, R.; Lucchesi, C.; Varone, A.; Hubeau, C.; Lee, H.-H.; Alves, S.E.; Salmon, M.; Ferrante, T.C.; Weaver, J.C.; et al. Small airway-on-a-chip enables analysis of human lung inflammation and drug responses in vitro. *Nat. Methods* **2016**, *13*, 151–157. [CrossRef]

24. Shyy, J.Y.-J.; Chien, S. Role of Integrins in Endothelial Mechanosensing of Shear Stress. *Circ. Res.* **2002**, *91*, 769–775. [CrossRef]

25. Ingber, D.E. Mechanical Signaling and the Cellular Response to Extracellular Matrix in Angiogenesis and Cardiovascular Physiology. *Circ. Res.* **2002**, *91*, 877–887. [CrossRef]

26. Mehta, D.; Malik, A.B. Signaling Mechanisms Regulating Endothelial Permeability. *Physiol. Rev.* **2006**, *86*, 279–367. [CrossRef]

27. Thomas, A.; Wang, S.; Sohrabi, S.; Orr, C.; He, R.; Shi, W.; Liu, Y. Characterization of vascular permeability using a biomimetic microfluidic blood vessel model. *Biomicrofluidics* **2017**, *11*, 024102. [CrossRef]

28. Frost, T.S.; Estrada, V.; Jiang, L.; Zohar, Y. Convection-diffusion molecular transport in a microfluidic bilayer device with a porous membrane. *Submitt. Microfluid. Nanofluidics* 2019.

29. Song, K.-Y.; Zhang, H.; Zhang, W.-J.; Teixeira, A. Enhancement of the surface free energy of PDMS for reversible and leakage-free bonding of PDMS–PS microfluidic cell-culture systems. *Microfluid. Nanofluidics* **2018**, *22*, 135. [CrossRef]

30. van Meer, B.J.; de Vries, H.; Firth, K.S.A.; van Weerd, J.; Tertoolen, L.G.J.; Karperien, H.B.J.; Jonkheijm, P.; Denning, C.; IJzerman, A.P.; Mummery, C.L. Small molecule absorption by PDMS in the context of drug response bioassays. *Biochem. Biophys. Res. Commun.* **2017**, *482*, 323–328. [CrossRef]

31. Foster, K.A.; Oster, C.G.; Mayer, M.M.; Avery, M.L.; Audus, K.L. Characterization of the A549 Cell Line as a Type II Pulmonary Epithelial Cell Model for Drug Metabolism. *Exp. Cell Res.* **1998**, *243*, 359–366. [CrossRef]

32. Man, S.; Ubogu, E.E.; Williams, K.A.; Tucky, B.; Callahan, M.K.; Ransohoff, R.M. Human Brain Microvascular Endothelial Cells and Umbilical Vein Endothelial Cells Differentially Facilitate Leukocyte Recruitment and Utilize Chemokines for T Cell Migration. *Clin. Dev. Immunol.* **2008**, *2008*, 1–8. [CrossRef]

33. Maciag, T.; Hoover, G.; Weinstein, R. High and Low Molecular Weight Forms of Endothelial Cell Growth Factor. *J. Biol. Chem.* **1982**, *257*, 5333–5336.

34. McClatchey, A.I.; Yap, A.S. Contact inhibition (of proliferation) redux. *Curr. Opin. Cell Biol.* **2012**, *24*, 685–694. [CrossRef]

35. Hartsock, A.; Nelson, W.J. Adherens and tight junctions: Structure, function and connections to the actin cytoskeleton. *Biochim. Biophys. Acta Biomembr.* **2008**, *1778*, 660–669. [CrossRef]

36. Considine, R.V. Plasma Membrane, Membrane Transport, and Resting Membrane Potential. In *Medical Physiology: Principles for Clinical Medicine*; Williams & Wilkins: Philadelphia, PA, USA, 2013; ISBN 1-4511-1039-1.

37. Nemoto, R.; Yamamoto, S.; Ogawa, T.; Naude, R.; Muramoto, K. Effect of Chum Salmon Egg Lectin on Tight Junctions in Caco-2 Cell Monolayers. *Molecules* **2015**, *20*, 8094–8106. [CrossRef]

38. Masereeuw, R.; Moons, M.M.; Toomey, B.H.; Russel, F.G.M.; Miller, D.S. Active Lucifer Yellow Secretion in Renal Proximal Tubule: Evidence for Organic Anion Transport System Crossover. *J. Pharmacol. Exp. Ther.* **1999**, *289*, 8.

39. Azizi, P.M.; Zyla, R.E.; Guan, S.; Wang, C.; Liu, J.; Bolz, S. S.; Hcit, B.; Klip, A.; Lee, W.L. Clathrin-dependent entry and vesicle-mediated exocytosis define insulin transcytosis across microvascular endothelial cells. *Mol. Biol. Cell* **2015**, *26*, 740–750. [CrossRef]

40. Matsukawa, Y.; Lee, V.H.L.; Crandall, E.D.; Kim, K.-J. Size Dependent Dextran Transport across Rat Alveolar Epithelial Cell Monolayers. *J. Pharm. Sci.* **1997**, *86*, 305–309. [CrossRef]

41. Michael Danielsen, E.; Hansen, G.H. Small molecule pinocytosis and clathrin-dependent endocytosis at the intestinal brush border: Two separate pathways into the enterocyte. *Biochim. Biophys. Acta Biomembr.* **2016**, *1858*, 233–243. [CrossRef]

42. Kobayashi, S.; Kondo, S.; Kazuhiko, J. Permeability of Peptides and Proteins in Human Cultured Alveolar A549 Cell Monolayer. *Pharm. Res.* **1995**, *12*, 1115–1119. [CrossRef]

43. Fotticchia, I.; Guarnieri, D.; Fotticchia, T.; Falanga, A.P.; Vecchione, R.; Giancola, C.; Netti, P.A. Energetics of ligand-receptor binding affinity on endothelial cells: An in vitro model. *Colloids Surf. B Biointerfaces* **2016**, *144*, 250–256. [CrossRef]

44. Li, Y.; Gao, A.; Yu, L. Monitoring of TGF-β 1-Induced Human Lung Adenocarcinoma A549 Cells Epithelial-Mesenchymal Transformation Process by Measuring Cell Adhesion Force with a Microfluidic Device. *Appl. Biochem. Biotechnol.* **2016**, *178*, 114–125. [CrossRef]

45. Nalayanda, D.D.; Puleo, C.; Fulton, W.B.; Sharpe, L.M.; Wang, T.-H.; Abdullah, F. An open-access microfluidic model for lung-specific functional studies at an air-liquid interface. *Biomed. Microdevices* **2009**, *11*, 1081–1089. [CrossRef]

46. Feng, S.; Zou, L.; Wang, H.; He, R.; Liu, K.; Zhu, H. RhoA/ROCK-2 Pathway Inhibition and Tight Junction Protein Upregulation by Catalpol Suppresses Lipopolysaccaride-Induced Disruption of Blood-Brain Barrier Permeability. *Molecules* **2018**, *23*, 2371. [CrossRef]

47. Harris, E.S.; Nelson, W.J. VE-cadherin: at the front, center, and sides of endothelial cell organization and function. *Curr. Opin. Cell Biol.* **2010**, *22*, 651–658. [CrossRef]

48. Erickson, H.P. Size and Shape of Protein Molecules at the Nanometer Level Determined by Sedimentation, Gel Filtration, and Electron Microscopy. *Biol. Proced. Online* **2009**, *11*, 32–51. [CrossRef]

49. Ravetto, A.; Hoefer, I.E.; den Toonder, J.M.J.; Bouten, C.V.C. A membrane-based microfluidic device for mechano-chemical cell manipulation. *Biomed. Microdevices* **2016**, *18*, 31. [CrossRef]

50. Bischoff, I.; Hornburger, M.C.; Mayer, B.A.; Beyerle, A.; Wegener, J.; Fürst, R. Pitfalls in assessing microvascular endothelial barrier function: impedance-based devices versus the classic macromolecular tracer assay. *Sci. Rep.* **2016**, *6*, 23671. [CrossRef]

51. Tarbell, J.M. Shear stress and the endothelial transport barrier. *Cardiovasc. Res.* **2010**, *87*, 320–330. [CrossRef]

52. Fujita, T.; Kawahara, I.; Yamamoto, M.; Yamamoto, A.; Muranishi, S. Comparison of the Permeability of Macromolecular Drugs Across Cultured Intestinal and Alveolar Epithelial Cell Monolayers. *J. Pharm. Pharmacol.* **1995**, *1*, 231–234.

53. Bruewer, M.; Nusrat, A. Regulation of Paracellular Transport across Tight Junctions by the Actin Cytoskeleton. In *Tight Junctions*; Gonzalez-Mariscal, L., Ed.; Springer US: Boston, MA, USA, 2006; pp. 135–145. ISBN 978-0-387-36673-9.

54. Sidhaye, V.K.; Schweitzer, K.S.; Caterina, M.J.; Shimoda, L.; King, L.S. Shear stress regulates aquaporin-5 and airway epithelial barrier function. *Proc. Natl. Acad. Sci. USA* **2008**, *105*, 3345–3350. [CrossRef]

55. Kang, H.; Cancel, L.M.; Tarbell, J.M. Effect of shear stress on water and LDL transport through cultured endothelial cell monolayers. *Atherosclerosis* **2014**, *233*, 682–690. [CrossRef]

56. Seki, S.; Kobayashi, M.; Itagaki, S.; Hirano, T.; Iseki, K. Contribution of organic anion transporting polypeptide OATP2B1 to amiodarone accumulation in lung epithelial cells. *Biochim. Biophys. Acta Biomembr.* **2009**, *1788*, 911–917. [CrossRef]

57. Ito, S.; Osaka, M.; Higuchi, Y.; Nishijima, F.; Ishii, H.; Yoshida, M. Indoxyl Sulfate Induces Leukocyte-Endothelial Interactions through Up-regulation of E-selectin. *J. Biol. Chem.* **2010**, *285*, 38869–38875. [CrossRef]

58. Cole, L.; Coleman, J.; Kearns, A.; Morgan, G.; Hawes, C. The organic anion transport inhibitor, probenecid, inhibits the transport of Lucifer Yellow at the plasma membrane and the tonoplast in suspension- cultured plant cells. *J. Cell Sci.* **1991**, *99*, 545–555.

59. Mishima, M.; Hamada, T.; Maharani, N.; Ikeda, N.; Onohara, T.; Notsu, T.; Ninomiya, H.; Miyazaki, S.; Mizuta, E.; Sugihara, S.; et al. Effects of Uric Acid on the NO Production of HUVECs and its Restoration by Urate Lowering Agents. *Drug Res.* **2016**, *66*, 270–274. [CrossRef]

60. Tumova, S.; Kerimi, A.; Williamson, G. Long term treatment with quercetin in contrast to the sulfate and glucuronide conjugates affects HIF1α stability and Nrf2 signaling in endothelial cells and leads to changes in glucose metabolism. *Free Radic. Biol. Med.* **2019**, S0891584919302576. [CrossRef]

61. Wasserman, S.M.; Mehraban, F.; Komuves, L.G.; Yang, R.-B.; Tomlinson, J.E.; Zhang, Y.; Spriggs, F.; Topper, J.N. Gene expression profile of human endothelial cells exposed to sustained fluid shear stress. *Physiol. Genom.* **2002**, *12*, 13–23. [CrossRef]

62. Kudo, S.; Ikezawa, K.; Ikeda, M.; Oka, K.; Tanishita, K. Albumin concentration profile inside cultured endothelial cells exposed to shear stress. *Am. Soc. Mech. Eng. Bioeng. Div.* **1997**, *35*, 547–548.

63. Duan, Y.; Weinstein, A.M.; Weinbaum, S.; Wang, T. Shear stress-induced changes of membrane transporter localization and expression in mouse proximal tubule cells. *Proc. Natl. Acad. Sci. USA* **2010**, *107*, 21860–21865. [CrossRef]

Brief Report

Prototyping a Versatile Two-Layer Multi-Channel Microfluidic Device for Direct-Contact Cell-Vessel Co-Culture

Li-Jiun Chen [1], Bibek Raut [1], Nobuhiro Nagai [2], Toshiaki Abe [2] and Hirokazu Kaji [1,3,*]

[1] Department of Finemechanics, Graduate School of Engineering, Tohoku University, 6-6-01 Aramaki, Aoba-ku, Sendai 980-8579, Japan

[2] Division of Clinical Cell Therapy, United Centers for Advanced Research and Translational Medicine (ART), Tohoku University Graduate School of Medicine, 2-1 Seiryo, Aoba-ku, Sendai 980-8575, Japan

[3] Department of Biomedical Engineering, Graduate School of Biomedical Engineering, Tohoku University, 6-6-01 Aramaki, Aoba-ku, Sendai 980-8579, Japan

* Correspondence: kaji@biomems.mech.tohoku.ac.jp; Tel.: +81-22-795-4249

Received: 20 December 2019; Accepted: 8 January 2020; Published: 10 January 2020

Abstract: Microfluidic devices are gaining increasing popularity due to their wide applications in various research areas. Herein, we propose a two-layer multi-channel microfluidic device allowing for direct-contact cell-vessel co-culture. Using the device, we built a co-culture model of the outer blood-retina barrier (oBRB), mimicking the in vivo retinal pigment epithelial cells-Bruch membrane-fenestrated choroids. To demonstrate the versatility of the design, we further modified the device by inserting platinum electrodes for trans-epithelial electrical resistance (TEER) measurement, demonstrating the feasibility of on-chip assessment of the epithelial barrier integrity. Our proposed design allows for direct-contact co-culture of cell–cell or cell–vessel, modifiable for real-time evaluation of the state of the epithelial monolayers.

Keywords: microfluidics; microfabrication; organ-on-a-chip; trans-epithelial electrical resistance; multi-culture

1. Introduction

Organ-on-a-chip, a subset of lab-on-a-chip, is often associated with replicating tissue- or organ-level functions in integrated microfluidic devices to answer fundamental research questions such as cell physiology or to explore disease pathophysiology [1]. Several microfabrication techniques exist, which are used for different research applications, ranging from fabrication of biosensors and disease models to vascularization of engineering tissues for medical studies [2–4]. Narrowing down on medical applications, many in vitro models have been built utilizing bio-microelectromechanical systems (Bio-MEMS) or static transwell system to study different human organs. MEMS and microfluidic on-chip models have been covered and readers are referred to the reviews [1,3–7].

Focusing on microvessels and vasculature development, microfluidic models of microvessels and barrier properties have been extensively studied. Some of the models particularly focus on recapitulating the functionalities, i.e., the restrictive barrier properties or the responsiveness of the in vitro microvessels to exogenous stimulations. Wang et al. performed a tri-culture of pericytes, astrocytes, and endothelial cells inside microfluidic devices with embedded Ag/AgCl wire to characterize the restrictive blood–brain barrier (BBB) [8]. To study neuroinflammation of BBB, Herland et al. performed a tri-culture of astrocytes, pericytes, and endothelial cells in a single microchannel and subjected the model under tumor necrosis factor-alpha (a pro-inflammatory cytokine) [7]. Functional responses of the microvessel under stimulation (endothelial permeability and cytokine release) were analyzed. Using a similar approach, Price et al. tested the effect of mechanical factors on the restrictiveness of the endothelial barriers using a single microvessel [9]. On the other hand, multi-channel microfluidic devices that allowed examinations on the vascular formation and tracking morphology of microvessels over the course of experiments and under different stimulations have been developed and used in various applications [10–14].

Based on the pre-established design of a multichannel microfluidic device, we modified the device based on the Jeon research group [15] and the Kamm research group [16] into top-down, direct-contact cell–cell or cell–vascular co-culturing (Figure 1). Using the device, we built a model of the outer blood–retina barrier (oBRB), mimicking the in vivo retinal pigment epithelial cells-Bruch membrane-fenestrated choroids using human retinal pigmental epithelial cells (ARPE-19) and human umbilical vein endothelial cells (HUVEC). To demonstrate the versatility of the device, we slightly modified the design by embedding four Platinum electrodes (diameter 300 μm) for trans-epithelial electrical resistance (TEER) measurement. Using the modified device, we quantified the TEER of ARPE monolayers and the oBRB model. The device proposed in this study can be used as a multi-culturing cell and/or vessel model for various applications with the potential of connecting individual fluidic devices in a fluidic network, mimicking human physiology as demonstrated in the work of Edington et al. [17].

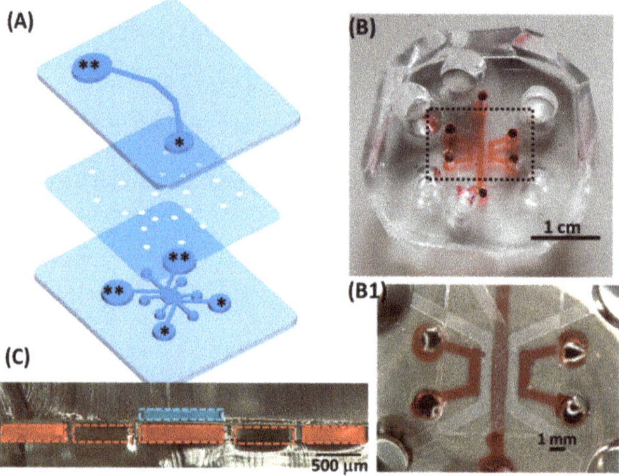

Figure 1. Configuration of the proposed device. (**A**) Exploded view; medium reservoirs are marked with asterisks (** as inlets, * as outlets when loading cells or media). (**B**) Top view of the assembled device, with dye water in some of the lower microchannels. (**B1**) Magnified view of the black dashed box in (**B**), showing the central part of the device overlaying the single top microchannel and multiple lower microchannels. (**C**) Microscopic cross-sectional view of the assembled two-layer device. The dashed blue line indicated the top microchannel; each dashed red box indicated the lower microfluidic channels separated by micropillars. Red rectangles in the lower channel corresponded to where red dye water was loaded. Together, the blue and red rectangles indicate the potential culturing chambers. The detailed dimension is in Figure A1.

2. Materials and Methods

2.1. Cell Culture

GFP expressing human umbilical vein endothelial cells (GFP-HUVECs) were purchased from Angio-Proteomie (cAP-0001GFP, Boston, MA, USA) and routinely subcultured at 4×10^3 cells/cm^2 in 75-cm^2 tissue culture-grade flasks coated with 0.1% gelatin. Human lung fibroblasts (NHLF) was purchased from Lonza (CC-2512, Basel, Switzerland) and subcultured based on the supplier's protocol.

Human retinal pigmented epithelial cells (ARPE-19) were kindly provided by Leonard Hjelmeland (Department of Ophthalmology, Section of Molecular and Cellular Biology, University of California, Davis, CA, USA) on 27 November 2000, and used after receiving approval from the Tohoku University Environmental & Safety Committee (no. 2014MdA-232-5). NHLF and ARPE-19 were cultured in Dulbecco's Modified Eagle Medium (DMEM) supplemented with 10% fetal bovine serum (FBS, S1400-500, Biowest, France) and 1% Antibiotic-Antimycotic (100X, Gibco, 15240062) while GFP-HUVECs were cultured in Endothelial Growth Medium (Lonza CC-3162, Basel, Switzerland). All cultures were maintained at 37 °C in a humidified atmosphere containing 5% CO_2. The passage number used for the experiments for each of the respective cell types was passage 20–30 for ARPE-19, passage 6–9 for NHLF, and passage 4–9 for GFP-HUVECs.

2.2. Fabrication of the Microfluidic Device

The microfluidic device used in the study has an upper microchannel and multiple lower microchannels, separated by a polyethylene terephthalate (PET) porous membrane of 8 μm in diameter (Falcon #353093). The device was made by conventional photolithography: the photomasks for the upper microchannel were designed in LayoutEditor (Justpertor GmbH, Unterhaching, Germany) and patterned by the Heiderlberg DWL66 laser writer. Thereafter, a master mold was made by spin-coating a 200 μm thick layer of SU-8 2100 (Microchem, Newton, MA, USA) on a silicon wafer and patterned by UV-lithography. Microchannels were made by liquid poly(dimethyl siloxane) (PDMS; Sylgard 184, Dow Corning, Midland, MI, USA) mixed at 10:1 ratio (wt/wt) of base polymer to curing agent, poured over the silicon master mold, degassed under vacuum pressure and cured in a 70 °C oven for at least 3 h. The cured PDMS was then cut into slabs and removed from the mold. The creation of the master mold and PDMS soft lithography for the lower microchannel followed the same procedure, except the lower PDMS slabs were made thin (2 mm in thickness) as opposed to the upper PDMS slabs. The thin lower PDMS slab was for microscopic observation while the upper PDMS slab was made thick to hold cell culturing media. The inlets and outlets of PDMS slabs were punched with the respective dimensions (Figure A1A,B), tapped to remove dust and impurities.

The membrane for the device was cut into 7 mm × 10.5 mm (width × length) out of the conventional PET membrane from Falcon transwell inserts (Falcon #353093). The membrane was bonded to the upper PDMS following the protocol developed previously [18]. After baking at 120 °C for an hour, the upper and lower channels treated with oxygen plasma were bonded. After a device was assembled, it was kept in a 120 °C oven for at least an hour. The bonding between the upper and lower PDMS slabs was checked (no leakage of ethanol) before use.

2.3. Design Prototype: Outer Blood–Retina Barrier Model

Prior to seeding cells into microfluidic devices, 10 μg of fibronectin (Wako 063-05591) reconstituted in 1 mL PBS(-) filtered through 0.45 μm millipore was used to coat the upper microchannel (overnight in a 37 °C incubator) and washed with PBS-once before cell seeding. The formation of microvasculature was based on published work from previous studies [13,19]. Briefly, GFP-HUVECs and NHLF were detached from the culturing surfaces and made into suspensions of 7×10^6 cells/mL. Fibrinogen precursor (Sigma-Aldrich F8630, St. Louis, MO, USA) 10 mg reconstituted in 1 mL of PBS(-) and filtered through 0.45 μm millipore was mixed with 3 times volume of GFP-HUVECs (at 7×10^6 cells/mL) thoroughly. Thrombin (Sigma-Aldrich T7513, St. Louis, MO, USA) pre-aliquoted at 50 U/mL (two μL for 0.25 mg fibrinogen used) was mixed thoroughly with the cell–fibrinogen mixture which was immediately loaded into the central lower microchannel (Figure 1C, cross-section). The devices were kept in the 37 °C incubator for 5 min for the fibrin gel to polymerize. Thereafter, NHLF (at 7×10^6 cells/mL) mixed with fibrinogen was loaded into both side channels (Figure 1C) following the same procedure, resulting in GFP-HUVECs and NHLF dispersed in 2.5 mg/mL fibrin (Figure 2B). After confirmation of the gelation, EGM supply was initiated by filling the inlet reservoirs (** in Figure 1A) and aspirating the medium from the outlets (* in Figure 1A) using a tip-cut 1 mL pipette. The microvasculature was cultured for 36 h before ARPE-19 were seeded, forming the proposed oBRB model (Figure 3). In seeding ARPE-19 into the upper channel, the medium in the upper channel was aspirated before cell seeding (at 10^7 cells/mL, 10 µL per device). The devices were left in a 37 °C incubator for 2 h for ARPE-19 to attach, after which 100 µL of DMEM was supplied and aspirated from the other end of the upper channel to wash out unattached ARPE-19. The devices were then returned to the incubator for an additional 2 h before loading the medium. The culture media (DMEM for the upper channel, EGM for lower channels) were refreshed every 48 h.

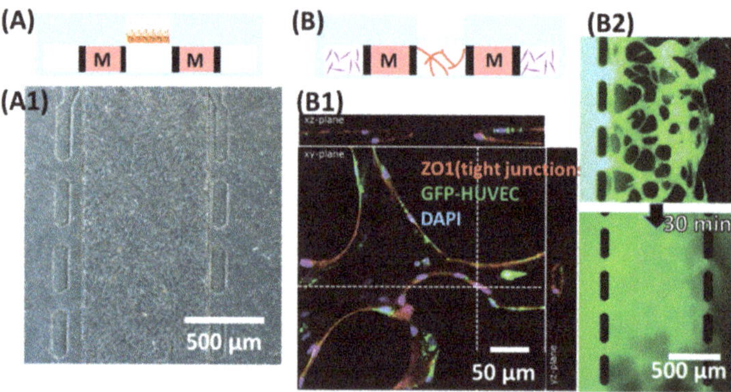

Figure 2. Demonstration of individual culturing chambers. (**A**) Culturing the human retinal pigmental epithelial cell (ARPE-19) monolayer in the upper channel. (**A1**) Microscopic view focusing on the upper channel, showing monolayer of ARPE. (**B**) Formation of microvascular lumens in the lower channels. "M" (medium channels) corresponded to * (lower channel) in Figure 1A. (**B1**) Confocal microscopic image staining for zonula occludens-1 (ZO1) and 4',6-diamidino-2-phenylindole (DAPI). (**B2**) Snapshots taken after loading 70 kDa of FITC-dextran into one side of the lower channel immediately (**top**) and 30 min after (**bottom**).

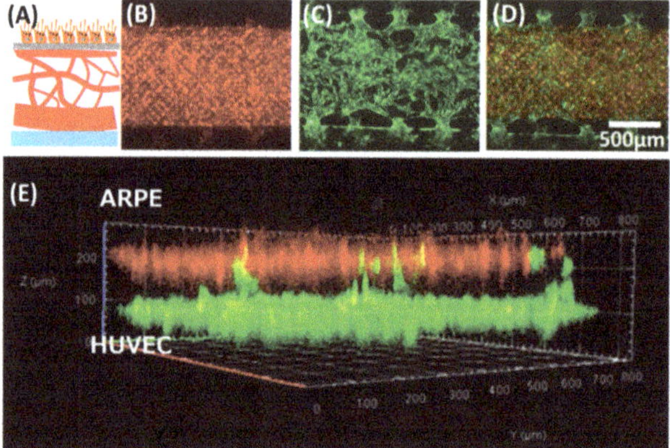

Figure 3. The oBRB model prototype built using the proposed device. (**A**) Schematic showing the anatomical structure of oBRB (retinal pigment epithelial cells–Bruch membrane-choroids). (**B–E**) tri-culture of ARPE–human umbilical vein endothelial cells (HUVEC)-human lung fibroblasts (NHLF). (**B**) ARPE cells stained in red were cultured in the upper channel. (**C**) GFP-HUVEC microvessels were cultured in the lower channel. (**D**) Overlay of (**B**) and (**C**). (**E**) Confocal microscopic view of the cross-section of the co-culture model.

2.4. Modification of the Device for Trans-Epithelial Electrical Resistance (TEER) Measurement

The masks of the upper and lower channels were redesigned with four microgrooves for placement of two electrodes in each of the upper and lower channels (Figure 4A, for dimensions, see Figure A1). The microgrooves for platinum (Pt) wire were designed and made at 200 µm in width, but Pt wire of 300 µm in diameter was used as it fits firmly inside the PDMS microgroove. In the last step of fabrication, four pieces of 3 cm long Pt wires were inserted into one device. The protrusion of wire inside the microchannel is approximately 1 mm (Figure 4D). After insertion, a drop of uncured PDMS was applied to the electrode channel's entrance and cured to minimize the potential displacement of electrodes during handling alligator clippers for measurement (Figure 4E). Each microfluidic device underwent quality checks for the following: (1) porous membrane covered the entire lower microchannel length, except inlets and outlets, (2) the tip of an electrode protruded 1 mm into the microchannel, (3) the upper and the central lower channel were properly aligned. Devices that met the criteria were then kept sterilized until being used for experiments. The individual layer and the assembled device are shown in Figure 4.

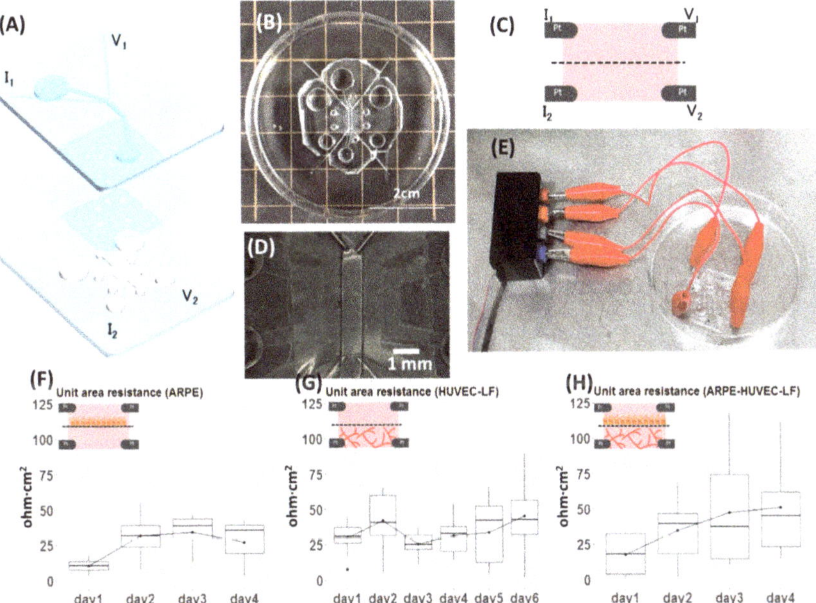

Figure 4. Demonstration of the versatility of the proposed design, with modification for measuring TEER. (**A**) Exploded view of the device. (**B**) Top view of the assembled device. (**C**) Schematic view of the cross-section showing the position of voltage and current electrodes. (**D**) Magnified view of (**B**) focusing on the position of electrodes inside the device. (**E**) Part of the system set-up showing the connection of the device to the electrode adapter, which is connected to the EVOM instrument for TEER measurement. (**F–H**) Results of TEER measurement reported as unit area resistance. The inset for each plot shows what was measured of (**F**) TEER of ARPE monolayer (n = 4). (**G**) TEER of co-culture of HUVEC and NHLF (n = 10). (**H**) TEER of the oBRB model (n = 11).

2.5. TEER Measurement Setup

The Pt wires from the device were connected to four ports of an adaptor (Figure 4E) which was connected to the EVOM (World Precision Instruments, Sarasota, FL, USA) set to Ohm mode. To be consistent in the approach, the Pt wires were connected to current (I) and voltage (V) ports in the order shown in Figure 4A for all the devices. Due to variability in TEER readings between devices, the blank (acellular device filled with culture media: DMEM in the upper channel, EGM in the lower channel) for each sample was recorded before every experiment. Culture media were equilibrated to room temperature (kept 20 min in the clean bench) and readings were recorded in all cases to minimize variations in results.

2.6. Data Collection and Analysis

2.6.1. Staining for Fluorescence Imaging

Microscopic images were captured using a fluorescent microscope (Olympus, Tokyo, Japan) or a confocal microscope (LSM700, Carl Zeiss MicroImaging, Gina, Germany). As part of the assessment of the growth of cells inside a microfluidic device, ARPE-19 cells were stained with Cell TrackerTM Fluorescent Probes red (5 µM, Molecular Probes) according to the manufacturer's staining protocols. ARPE-19 were stained prior to seeding into devices.

2.6.2. TEER Measurements

Samples (devices) used for experiments were individually labeled and the TEER readings and microvasculature morphology specific to each device were recorded. The number of samples is indicated in the corresponding figure captions. Due to the relatively small sample size, we did not test for the statistical significance of the difference in TEER readings between different days of culture. TEER values were extracted and reported as a unit area of resistance (ohm·cm^2) by first subtracting the raw values from the blank and multiplied the resulting value by area in cm^2. The area was calculated as width × length between electrodes (0.08 cm × 0.5 cm). Readings from gel-free, acellular devices were taken as the blank for the ARPE monolayers, while the blank for HUVEC-NHLF and ARPE-HUVEC-NHLF cultures was taken as the readings 30 min after seeding (i.e., cells embedded in fibrin gel) to take the effect of gel contributing TEER into account. The device was individually labeled and the blank and readings specific to each device were used in the calculation.

3. Results and Discussion

3.1. The Device as the Outer Blood–Retina Model

In building the model, we first evaluated the formation of ARPE monolayer in the device (Figure 2A). Separately, the formation of lumens of the HUVEC vasculature was confirmed by immunostaining of tight junction (Figure 2B1) and loading a diameter of 6 µm microbeads in one side of the inlet of the device (Video S1). Images of the microvascular network focused on a z-plane in Figure 2B2 were captured after flowing Fluorescein isothiocyanate (FITC)-dextran 70 kDa at 0.5 mg/mL in one side of the lower channel inlet. Figure 2B shows the schematic cross-sectional view of the multi-channel microfluidic device in which endothelial cells (HUVEC) and the supporting cells (NHLH) in two side channels were cultured (media channels were indicated M in Figure 2). After confirming the ability to culture epithelial cells and the vasculature, we attempted to demonstrate the feasibility of the multi-culture of cells and/or vessels using our proposed device and built a model of the outer blood–retina barrier (oBRB). The design of the microfluidic device was an extension based on previously developed models by Noo Li Jeon and Roger Kamm research groups where models have been used to study microvessels. Here, we further modified it to enable culturing of the ARPE-19 directly on the top of HUVEC microvasculature, mimicking oBRB where in vivo retinal pigment epithelial cells and the juxtaposed choriocapillaris are separated by the Bruch membrane (Figure 3). We adopted a larger pore size here as a demonstration of the model; the diameter of the pore, however, is customizable by using commercial membranes of different diameters. We had previously built the cell–cell oBRB model using ARPE-19 and HUVEC cells [20], characterizing the behavior of cells under pathological angiogenesis. We anticipate the prototype we proposed here to be used to characterize the epithelial monolayer microvasculature such as BRB or BBB, under pathological microenvironments.

3.2. Versatility of the Design: Modification for TEER Measurement

To demonstrate the versatility of our proposed microfluidic device, we modified it to allow for direct TEER measurement. TEER provides a quick, non-invasive, label-free and real-time indication of the state of cells as an alternative to perfusion of fluorescent-labeled molecules such as FITC-dextran in the evaluation of the barrier property. Both evaluation methods have their pros and cons: TEER offers an instant, real-time assessment of the state of cells while permeability assay is a simple yet elegant way that is typically used as an end-point assessment of the barrier integrity of a cell monolayer. In the designed device, two Pt wire in the upper channel corresponded for voltage and current measurements, the same configuration for the lower central channel (Figure 4A,C and Figure A1). The four-electrode measurement system was adopted here to be compatible with the EVOM instrument. The advantage of the four-electrode system is the negligible effect of double-layer capacitance, as the current supply and voltage measuring electrodes are separated [21]. As an alternative, the two-electrode measurement system can also be used here if we wish to take the impedance spectroscopy using a potentiostat. The two-layered microfluidic TEER device and the connection to the EVOM instrument are shown in Figures 4E and A1E.

Many microfluidic TEER models had been proposed; a square culturing chamber was adopted by Helm et al. [22], while some measured the TEER of cells cultured in long microchannels [23,24]. Here, we further modified our device for TEER evaluation of a top-down cell-vessel model. We first measured the TEER of ARPE-19 monolayers and HUVEC-NHLF separately before conducting experiments on the oBRB models. TEER readings of ARPE monolayers (Figure 4F) increased by days after seeding, corresponding to the proliferation of cells inside the device. There was no obvious trend from the co-culture of HUVEC-NHLF (Figure 4G), due to the nature of the fenestrated vessels. We also evaluated the ARPE-19 of the oBRB model by measuring the TEER (Figure 4H), where the readings indicated the proliferation of ARPE-19. TEER values for HUVEC-NHLF culture and the oBRB model were higher than ARPE monolayer culture, perhaps due to the presence of fibrin gel and the influence of microvessels. The readings, however, were subjected to fluctuations arising from changes in temperature, and movement of electrodes during clipping. According to the datasheet of EVOM from World Precision Instruments, there is a 10% variability in TEER reading using the commercial STX2 chopstick electrodes on the transwell TEER measurement due to non-uniform current density flowing across the membrane, arising from variations in the electrode position. Such variation is expected to be even higher in the microfluidic system [25]. In our system, one of the possible causes of variations in reading between devices could be due to variations in the length of protrusion of electrodes inside the microchannel, although we attempted to keep it the same (1 mm) as much as possible. This can be further improved with micro-patterned electrodes. In our device, as the distance between measuring electrodes is 5 mm, we did not anticipate non-uniformity in the current density but suspected the fluctuations in readings as the result of differences in temperature when reading was taken [22].

3.3. Design of the Proposed Microfluidic Device

The feasibility of the proposed device as a platform for top-down multi-culture of cell-vessel was demonstrated by building a model of the human outer blood-retina barrier (oBRB). In addition, the versatility of the device was shown by modifying the device, allowing TEER evaluation of a cell monolayer. Although we only demonstrated the prototype of oBRB here, the device can be modified for specific organ-on-a-chip applications and studying physiological responses after taking allometric scaling and fluidic shear stress into account. Cell-vascular models are widely applicable to the human organ system, ranging from the blood–brain barrier (BBB) and glial interactions to interactions of trophoblasts and the blood vessels in the placenta. We have been developing integrated TEER measurement of the microfluidic system to evaluate the state of monolayer cells in real-time; such a system is applicable to evaluating the response of a monolayer to its environment, such as the BBB, the airway epithelium under shear stress, and mechanisms regulating intestinal epithelium.

The model we proposed here is customizable and adaptable to different organs of origin, with potential extensibility for quantification of physical parameters such as barrier resistance and shear stress. We foresee that our proposed device has potential in disease modeling and drug testing.

4. Conclusions

We proposed a microfluidic device consisting of a single upper microchannel and multi-lower microchannels for direct-contact cell–vessel co-culture. The co-culture model was demonstrated by making the outer blood–retina barrier, made up of retinal pigment epithelial cells, endothelial cells, and the supporting cells. The versatility of the device was demonstrated by modifying the device for the insertion of Pt electrodes for TEER measurement. We measured the TEER of the ARPE monolayer and that of the oBRB model (ARPE–HUVEC–NHLF culture). Although only one aspect of the versatility design was demonstrated in this work, it is however, unlimited but extensible for investigation of the effect of shear stress on the multi-culture model by additionally gluing PDMS ports to the in/outlets of the device, or connect modular devices in an integrated fluidic network for studying physiological responses.

Supplementary Materials: The following are available online at http://www.mdpi.com/2072-666X/11/1/79/s1, Video S1: Perfusion of microbeads demonstrating the formation of lumens.

Author Contributions: Conceptualization, H.K., L.-J.C., and B.R.; methodology, fabrication of the devices, and data analysis, L.-J.C. and B.R.; writing—original draft preparation, L.-J.C., writing—reviewing and editing, B.R., T.A., N.N., and H.K., funding acquisition, H.K. All authors have read and agreed to the published version of the manuscript.

Funding: This work is funded by a Grant-in-Aid for Scientific Research (B) (17H02752) from the Ministry of Education, Culture, Sports, Science and Technology (MEXT), Japan.

Acknowledgments: The authors extend their appreciation to Shun Ito for the initial development of the device and Taisei Amanokura for help in the fabrication of devices).

Conflicts of Interest: The authors declare no conflict of interest.

Appendix A

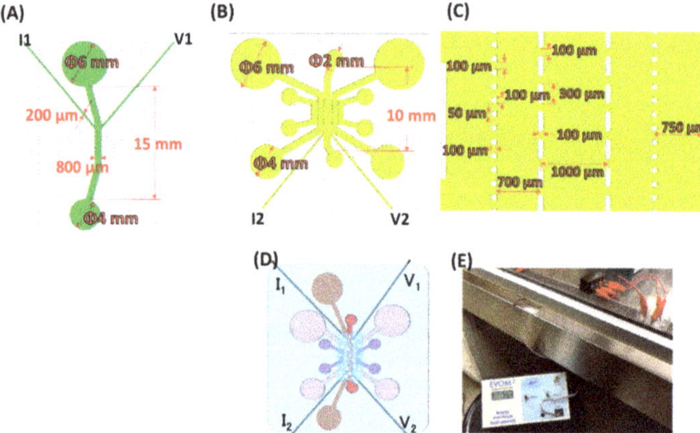

Figure A1. Dimensions of the microchannels for trans-epithelial electrical resistance (TEER) measurement. (**A**) Upper channel. (**B**) Lower channel. (**C**) Magnified view of (**B**), showing dimensions of multi-channels. (**D**) Top view of the assembled TEER device. (**E**) Experimental set-up for TEER measurement showing the connection between the microfluidic device and the EVOM instrument.

References

1. Perestrelo, A.R.; Águas, A.C.P.; Rainer, A.; Forte, G. Microfluidic organ/body-on-a-chip devices at the convergence of biology and microengineering. *Sensors* **2015**, *15*, 31142–31170. [CrossRef] [PubMed]
2. Chen, L.-J.; Kaji, H. Modeling angiogenesis with micro- and nanotechnology. *Lab Chip* **2017**, *17*, 4186–4219. [CrossRef] [PubMed]
3. Weibel, D.B.; DiLuzio, W.R.; Whitesides, G.M. Microfabrication meets microbiology. *Nat. Rev. Microbiol.* **2007**, *5*, 209–218. [CrossRef] [PubMed]
4. Voldman, J.; Grey, M.; Schmidt, M. Microfabrication in biology and medicine. *Annu. Rev. Biomed. Eng.* **1999**, *1*, 401–425. [CrossRef]
5. Hulkower, K.I.; Herber, R.L. Cell migration and invasion assays as tools for drug discovery. *Pharmaceutics* **2011**, *3*, 107–124. [CrossRef]
6. Sackmann, E.K.; Fulton, A.L.; Beebe, D.J. The present and future role of microfluidics in biomedical research. *Nature* **2014**, *507*, 181–189. [CrossRef]
7. Herland, A.; Meer, A.D.; van der FitzGerald, E.A.; Park, T.-E.; Sleeboom, J.J.F.; Ingber, D.E. Distinct contributions of astrocytes and pericytes to neuroinflammation identified in a 3D human blood-brain barrier on a chip. *PLoS ONE* **2016**, *11*, e0150360. [CrossRef]
8. Wang, J.D.; Khafagy, E.-S.; Khanafer, K.; Takayama, S.; ElSayed, M.E.H. Organization of endothelial cells, Pericytes, and astrocytes into a 3D microfluidic in vitro model of the blood-brain barrier. *Mol. Pharm.* **2016**, *13*, 895–906. [CrossRef]
9. Price, G.M.; Wong, K.H.K.; Truslow, J.G.; Leung, A.D.; Acharya, C.; Tien, J. Effect of mechanical factors on the function of engineered human blood microvessels in microfluidic collagen gels. *Biomaterials* **2010**, *31*, 6182–6189. [CrossRef]
10. Jeon, J.S.; Bersini, S.; Gilardi, M.; Dubini, G.; Charest, J.L.; Moretti, M.; Kamm, R.D. Human 3D vascularized organotypic microfluidic assays to study breast cancer cell extravasation. *Proc. Natl. Acad. Sci. USA* **2015**, *112*, 214–219. [CrossRef]
11. Osaki, T.; Sivathanu, V.; Kamm, R.D. Engineered 3D vascular and neuronal networks in a microfluidic platform. *Sci. Rep.* **2018**, *8*, 5168. [CrossRef] [PubMed]
12. Adriani, G.; Ma, D.; Pavesi, A.; Kamm, R.D.; Goh, E.L.K. A 3D neurovascular microfluidic model consisting of neurons, astrocytes and cerebral endothelial cells as a blood–brain barrier. *Lab Chip* **2017**, *17*, 448–459. [CrossRef] [PubMed]
13. Kim, S.; Lee, H.; Chung, M.; Jeon, N.L. Engineering of functional, perfusable 3D microvascular networks on a chip. *Lab Chip* **2013**, *13*, 1489–1500. [CrossRef] [PubMed]
14. Yeon, J.H.; Ryu, H.R.; Chung, M.; Hu, Q.P.; Jeon, N.L. In vitro formation and characterization of a perfusable three-dimensional tubular capillary network in microfluidic devices. *Lab Chip* **2012**, *12*, 2815–2822. [CrossRef] [PubMed]
15. Kim, J.; Chung, M.; Kim, S.; Jo, D.H.; Kim, J.H.; Jeon, N.L. Engineering of a biomimetic pericyte-covered 3d microvascular network. *PLoS ONE* **2015**, *10*, e0133880. [CrossRef] [PubMed]
16. Chen, M.B.; Whisler, J.A.; Fröse, J.; Yu, C.; Shin, Y.; Kamm, R.D. On-chip human microvasculature assay for visualization and quantification of tumor cell extravasation dynamics. *Nat. Protoc.* **2017**, *12*, 865–880. [CrossRef]
17. Edington, C.D.; Chen, W.L.K.; Geishecker, E.; Kassis, T.; Soenksen, L.R.; Bhushan, B.M.; Freake, D.; Kirschner, J.; Maass, C.; Tsamandouras, N.; et al. Interconnected microphysiological systems for quantitative biology and pharmacology studies. *Sci. Rep.* **2018**, *8*, 4530. [CrossRef]
18. Aran, K.; Sasso, L.A.; Kamdar, N.; Zahn, J.D. Irreversible, direct bonding of nanoporous polymer membranes to PDMS or glass microdevices. *Lab Chip* **2010**, *10*, 548–552. [CrossRef]
19. Jeon, J.S.; Bersini, S.; Whisler, J.A.; Chen, M.B.; Dubini, G.; Charest, J.L.; Moretti, M.; Kamm, R.D. Generation of 3D functional microvascular networks with human mesenchymal stem cells in microfluidic systems. *Integr. Biol.* **2014**, *6*, 555–563. [CrossRef]
20. Chen, L.-J.; Ito, S.; Kai, H.; Nagamine, K.; Nagai, N.; Nishizawa, M.; Abe, T.; Kaji, H. Microfluidic co-cultures of retinal pigment epithelial cells and vascular endothelial cells to investigate choroidal angiogenesis. *Sci. Rep.* **2017**, *7*, 3538. [CrossRef]

21. Theile, M.; Wiora, L.; Russ, D.; Reuter, J.; Ishikawa, H.; Schwerk, C.; Schroten, H.; Mogk, S. A simple approach to perform TEER measurements using a self-made volt-amperemeter with programmable output frequency. *JoVE J. Vis. Exp.* **2019**, *5*, e60087. [CrossRef] [PubMed]

22. Van der Helm, M.W.; Odijk, M.; Frimat, J.P.; van der Meer, A.D.; Eijkel, J.C.; van den Berg, A.; Segerink, L.I. Direct quantification of transendothelial electrical resistance in organs-on-chips. *Biosens. Bioelectron.* **2016**, *15*, 924–929. [CrossRef] [PubMed]

23. Henry, O.Y.F.; Villenave, R.; Cronce, M.J.; Leineweber, W.D.; Benz, M.A.; Ingber, D.E. Organs-on-chips with integrated electrodes for trans-epithelial electrical resistance (TEER) measurements of human epithelial barrier function. *Lab Chip* **2017**, *17*, 2264–2271. [CrossRef] [PubMed]

24. Odijk, M.; van der Meer, A.D.; Levner, D.; Kim, H.J.; van der Helm, M.W.; Segerink, L.I.; Frimat, J.-P.; Hamilton, G.A.; Ingberb, D.E.; van den Berga, A. Measuring direct current trans-epithelial electrical resistance in organ-on-a-chip microsystems. *Lab Chip* **2015**, *15*, 745–752. [CrossRef]

25. Douville, N.J.; Tung, Y.-C.; Li, R.; Wang, J.D.; El-Sayed, M.E.H.; Takayama, S. Fabrication of two-layered channel system with embedded electrodes to measure resistance across epithelial and endothelial barriers. *Anal. Chem.* **2010**, *15*, 2505–2511. [CrossRef]

MDPI

St. Alban-Anlage 66

4052 Basel

Switzerland

Tel. +41 61 683 77 34

Fax +41 61 302 89 18

www.mdpi.com

Micromachines Editorial Office

E-mail: micromachines@mdpi.com

www.mdpi.com/journal/micromachines